T0390402

Nanophotonics and Plasmonics

An Integrated View

SERIES IN OPTICS AND OPTOELECTRONICS

Recent titles in the series

Handbook of GaN Semiconductor Materials and Devices
Wengang (Wayne) Bi, Hao-chung (Henry) Kuo, Pei-Cheng Ku, and Bo Shen Piprek (Eds.)

Handbook of Optoelectronic Device Modeling and Simulation: Fundamentals, Materials, Nanostructures, LEDs, and Amplifiers – Volume One
Joachim Piprek (Ed.)

Handbook of Optoelectronic Device Modeling and Simulation: Lasers, Modulators, Photodetectors, Solar Cells, and Numerical Methods – Volume Two
Joachim Piprek (Ed.)

Nanophotonics and Plasmonics: An Integrated View
Dr. Ching Eng (Jason) Png and Dr. Yuriy Akimov

Handbook of Solid-State Lighting and LEDs
Zhe Chuan Feng (Ed.)

Optical Microring Resonators: Theory, Techniques, and Applications
V. Van

Optical Compressive Imaging
Adrian Stern

Singular Optics
Gregory J. Gbur

The Limits of Resolution
Geoffrey de Villiers and E. Roy Pike

Polarized Light and the Mueller Matrix Approach
José J Gil and Razvigor Ossikovski

Light—The Physics of the Photon
Ole Keller

Advanced Biophotonics: Tissue Optical Sectioning
Ruikang K Wang and Valery V Tuchin (Eds.)

Handbook of Silicon Photonics
Laurent Vivien and Lorenzo Pavesi (Eds.)

Microlenses: Properties, Fabrication and Liquid Lenses
Hongrui Jiang and Xuefeng Zeng

Nanophotonics and Plasmonics

An Integrated View

Edited by

Ching Eng Png
Yuriy Akimov

CRC Press is an imprint of the
Taylor & Francis Group, an **informa** business
A TAYLOR & FRANCIS BOOK

CRC Press
Taylor & Francis Group
6000 Broken Sound Parkway NW, Suite 300
Boca Raton, FL 33487-2742

Printed on acid-free paper
Version Date: 20170713

International Standard Book Number-13: 978-1-4987-5867-3 (Hardback)

Visit the Taylor & Francis Web site at
http://www.taylorandfrancis.com

and the CRC Press Web site at
http://www.crcpress.com

Contents

Preface

With recent developments of nanotechnologies, the photonics community found a powerful incentive that eventually led to the formation of two new fields — *nanophotonics* and *plasmonics* — studying nanostructured materials for manipulation and control of light at nanometer scale. Started from exploration of electromagnetic waves supported by nanosized waveguides and then moving toward the use of localized eigenmodes, nanophotonics and plasmonics created a number of opportunities in the applied sciences and even made their impact in the market.

Nanophotonics and plasmonics, being closely related fields, have common physics, targets, and methodologies, but deal with different materials. Nanophotonics is mainly focused on optically transparent materials such as dielectrics and semiconductors, while optically opaque materials such as metals are the choices of plasmonics. Despite strong similarity in physics of these phenomena, the two fields were developed within different communities in slightly different ways. While nanophotonics focused on propagating eigenwaves for integrated optics and optical communication with a little exposure to nanoscale resonators, plasmonics studied both propagating and localized eigenmodes for a broader range of applications. The fact that the potential of localized eigenmodes is restricted in plasmonics by their high ohmic losses has turned the research toward low-loss high-index dielectric resonators, where localized photonic eigenmodes readily demonstrated their unique capabilities to control light-matter interaction. As such, both technologies complete their full cycles for exploration of propagating and localized eigenmodes and meet together. To further utilize their full potentials for synergy of the two fields, the development of a *unified platform* covering both areas is required.

The goal of this book is to provide a unified description of nanophotonics and plasmonics, to highlight their similarities and advantages, and finally to use their synergy for further progress in control of light at nanometer scale. This book overviews the state-of-the-art developments in this field from an *electrodynamics point of view* enabling the joint description of both technologies. The central line of the book is analysis of nanophotonics and plasmonics based on the use of the general theory of electromagnetic eigenmodes. It provides the universal tools for in-depth understanding of localized and propagating eigenmodes in both technologies for further implementation in various applications ranging from chiral and integrated optics to coloration and biosensing.

The book is divided into three parts. Part I presents unified fundamentals of nanophotonics and plasmonics. Chapter 1 introduces the theory of electromagnetic fields in uniform media. Chapter 2 provides an introduction to electromagnetic waves in bounded media. Chapter 3 presents the theory of localized eigenmodes in simple single-particle systems. Chapter 4 discusses the hybridization of localized eigenmodes in complex multi-particle systems.

Part II overviews numerous applications based on the use of localized eigenmodes. Chapter 5 gives an introduction to structural coloration. Chapter 6 explores eigenmodes for fluorescence enhancement. Chapter 7 provides an overview of the developments in chiral optics. Chapter 8 introduces biosensing on localized eigenmodes. Chapter 9 discusses the use of optical metasurfaces made of resonant antennas for efficient light control.

Part III reviews various applications based on electromagnetic waves supported by metallic, dielectric, and semiconductor waveguides. Chapter 10 introduces light guiding with nanoparticle chains. Chapter 11 overviews subwavelength slot waveguides. Chapter 12 discusses photodetectors. Chapter 13 introduces integrated nonlinear photonics. Chapter 14 is on the use of integrated nanophotonics for multi-user quantum key distribution networks.

We gratefully acknowledge the support from the Agency for Science, Technology, and Research (A*STAR) of Singapore and the Institute of High Performance Computing (IHPC). We also acknowledge all the contributors, Drs. Pavlo Rutkevych, Lin Wu, Ping Bai, Song Sun, Ravi Hegde, Eng Huat Khoo, Wee Kee Phua, Yew Li Hor, Yan Jun Liu, Xiaodong Zhou, Ten It Wong, Zhengtong Liu, Hong-Son Chu, Thomas Ang, Jun Rong Ong, Han Chuen Lim, Mao Tong Liu, and Mr. Valerian Chen, without whose significant work this book would never have been published.

We hope this book will serve as a basis for future progress in the field and will be a valuable reference for engineers, researchers, and students in the areas of nanophotonics, plasmonics, and nano-optics in general.

Yuriy Akimov
Ching Eng Png

Part I

Fundamentals

Chapter 1

Electromagnetic fields in uniform media

Yuriy Akimov
Institute of High Performance Computing, Singapore
Pavlo Rutkevych
Institute of High Performance Computing, Singapore

In this chapter, we will give a brief introduction to the classical electrodynamics that constitutes the basis of modern nanophotonics and plasmonics. Section 1.1 will consider the theory of electromagnetic fields in continuous media. In Section 1.2, we will give an optical description of non-magnetic solids within the local response approximation. Section 1.3 will review the theory of electromagnetic eigenwaves in solids. The chapter will conclude with a summary in Section 1.4.

1.1 Electromagnetic field equations

1.1.1 Maxwell's equations in medium

In the classical theory, the concept of *electromagnetic field* is introduced to describe how charges interact with each other [1]. Every charge is assumed to generate electromagnetic field, via which it interacts with other charges. In the case of multiple charges, the overall field is given by the sum of corresponding fields generated by every charge, according to the superposition principle. As a result, full electromagnetic description of a medium requires consideration of all microscopic charges comprising the medium, which is a complicated and troublesome procedure, as the number of such charges in solids is $\sim 10^{30}$ per 1 m^3. To simplify the description, the *macroscopic* approach is introduced [2], where all physical quantities including fields and charges are statistically averaged in space over small volumes, the linear size of which, a_{macro}, is much smaller compared to the wavelength, but much larger than the lattice constant. In other words, we neglect all fluctuations appearing at atomic scales, below a_{macro}, and describe the medium from the macroscopic point of view, where charges and currents are continuously distributed in space. Within this approach, the generation of electromagnetic fields by charges is given by the *macroscopic Maxwell's equations* [3, 4],

$$\nabla \times \mathbf{B} = \mu_0 \left(\varepsilon_0 \frac{\partial \mathbf{E}}{\partial t} + \mathbf{j} \right), \tag{1.1}$$

$$\nabla \times \mathbf{E} = -\frac{\partial \mathbf{B}}{\partial t}, \tag{1.2}$$

where $\varepsilon_0 = 8.85418782 \cdot 10^{-12}$ F/m and $\mu_0 = 4\pi \cdot 10^{-7}$ H/m are the electric and magnetic constants; $\mathbf{j}$ is the macroscopic current density of all charges in the medium; and $\mathbf{E}$ and $\mathbf{B}$ are the macroscopic electric field and magnetic induction, respectively.

Usually, Maxwell's equations are accompanied by two divergence relations,

$$\nabla \cdot \mathbf{E} = \frac{\rho}{\varepsilon_0}, \tag{1.3}$$

$$\nabla \cdot \mathbf{B} = 0, \tag{1.4}$$

with ρ as the macroscopic charge density. These relations are used intensely in classical electrodynamics. Often, they are called the second pair of Maxwell's equations. However, their importance is slightly overestimated. In fact, they are derivatives of Eqs. (1.1) and (1.2) and do not convey additional information. Therefore, they should be considered as auxiliary and solved together with the first pair of Maxwell's equations all the time. Otherwise, their separate solution can lead to substantial errors for time-varying fields.[1]

In Eqs. (1.1)–(1.4), the charge and current densities play the role of sources for the fields $\mathbf{E}$ and $\mathbf{B}$. In general, charge and current densities are related by the charge conservation law that requires

$$\frac{\partial \rho}{\partial t} + \nabla \cdot \mathbf{j} = 0, \tag{1.5}$$

where both ρ and $\mathbf{j}$ should be considered total quantities, i.e., composed by both intrinsic and extrinsic charges. As the intrinsic charges are generally represented by positive and negative ones, it is convenient to split the total charge density ρ into two parts:

$$\rho = \rho_{\text{pol}} + \rho_{\text{ext}},$$

where sign-varying $\rho_{\text{pol}}(\mathbf{r})$ describes fully compensated charges called *polarization charges* and featuring zero total charge

$$\int \rho_{\text{pol}}\, \mathrm{d}V = 0,$$

and the second part, $\rho_{\text{ext}}(\mathbf{r})$, represents uncompensated charges called *external charges* and exhibiting

$$\int \rho_{\text{ext}}\, \mathrm{d}V \neq 0.$$

[1] Auxiliary relations (1.3) and (1.4) can be derived by taking divergence of the first pair of Maxwell's equations. Therefore, they are one order higher than original Eqs. (1.1) and (1.2). As a result, they feature extra solutions, which are unphysical and must be excluded from consideration.

Both polarization and external charges are required to obey the individual continuity equations similar to Eq. (1.5). Following it, we can similarly decompose the total current density $\mathbf{j}$,

$$\mathbf{j} = \mathbf{j}_{\text{pol}} + \mathbf{j}_{\text{ext}}.$$

Of course, such separation of polarization and external charges is relative and ambiguous, as it can be done in different ways if the total charge is uncompensated. But it allows us to represent the overall system as a uniform charge-compensated medium with added external charges. In this description, the external charges and currents introduce initial electromagnetic disturbance to the initially charge-neutral medium with $\rho_{\text{pol}} = 0$, which responds with separation of its charges and $\rho_{\text{pol}} \neq 0$. The concept of polarization charges is the fundamental idea of the classical electrodynamics that enables the elegant and unifying description of electromagnetic response of any charge-neutral medium regardless of its type and types of the charges composing it.

Following the Maxwell's equation (1.2), the density of polarization charges can be represented as the divergence of an arbitrary vector, $-\mathbf{P}$,

$$\rho_{\text{pol}} = -\nabla \cdot \mathbf{P}. \tag{1.6}$$

The condition of full compensation of the polarization charges requires $\mathbf{P}$ to be nonzero inside the medium only and to vanish outside of it. Thus, vector $\mathbf{P}$ describes the polarization of the initially neutral medium and is commonly called the *polarization* field. At the same time, the continuity equation (1.5) gives us a form for the polarization current density,

$$\mathbf{j}_{\text{pol}} = \frac{\partial \mathbf{P}}{\partial t} + \nabla \times \mathbf{M}, \tag{1.7}$$

where $\mathbf{M}$ is another arbitrary vector that describes magnetic induction of the polarization current not accompanied by charge polarization. It is commonly called the *magnetization* field. Introduction of the polarization and magnetization fields allows us to rewrite the macroscopic Maxwell's equations in a more compact form,

$$\nabla \times \mathbf{H} = \frac{\partial \mathbf{D}}{\partial t} + \mathbf{j}_{\text{ext}}, \tag{1.8}$$

$$\nabla \times \mathbf{E} = -\frac{\partial \mathbf{B}}{\partial t}, \tag{1.9}$$

with the divergence relations given by

$$\nabla \cdot \mathbf{D} = \rho_{\text{ext}}, \tag{1.10}$$

$$\nabla \cdot \mathbf{B} = 0, \tag{1.11}$$

where $\mathbf{D}$ and $\mathbf{H}$ are the auxiliary fields called the *electric displacement* and *magnetic field*, introduced to account for the polarization and magnetization of the charge-compensated medium,

$$\mathbf{D} = \varepsilon_0 \mathbf{E} + \mathbf{P}, \tag{1.12}$$

$$\mathbf{H} = \frac{1}{\mu_0} \mathbf{B} - \mathbf{M}. \tag{1.13}$$

In this form, the Maxwell's equations describe generation of electromagnetic field by external currents in terms of two pairs of the electric $\{\mathbf{E},\mathbf{D}\}$ and magnetic $\{\mathbf{B},\mathbf{H}\}$ fields. However, they do not form a closed set of equations, until we provide *material relations* for the response of the charge-neutral medium to electric and magnetic fields. In general, these relations are given by field-dependent functions for the polarization $\mathbf{P} = \mathbf{P}(\mathbf{E})$ and magnetization $\mathbf{M} = \mathbf{M}(\mathbf{B})$ vectors that eventually result in material relations for the auxiliary fields $\mathbf{D} = \mathbf{D}(\mathbf{E})$ and $\mathbf{H} = \mathbf{H}(\mathbf{B})$.

1.1.2 Material equations

Establishing the relations for $\mathbf{D}(\mathbf{E})$ and $\mathbf{H}(\mathbf{B})$ takes the key place in classical electrodynamics, as it describes the charge-neutral media response to electromagnetic fields. In our considerations, we will focus on the high-frequency response in the optical range and above, where most of solids lose their magnetic properties [2], featuring

$$\mathbf{M} = 0, \qquad \mathbf{H} = \frac{1}{\mu_0}\mathbf{B}. \tag{1.14}$$

Below, we will assume that condition (1.14) holds all the time and consider only magnetically inactive materials. The response of such materials is given by dependence $\mathbf{D}(\mathbf{E})$ only, which is generally nonlinear. However, in most interesting cases the electric fields $\mathbf{E}$ are low enough, so the auxiliary field $\mathbf{D}(\mathbf{E})$ can be treated as a linear function. It is the so-called *linear electrodynamics* approach, where the medium's polarization current $\mathbf{j}_{\rm pol}$ at a given point $\mathbf{r}$ and moment t is assumed to be a linear function $\mathbf{J}(\mathbf{E})$, defined as a nonlocal response to electromagnetic fields taken at any point of space $\mathbf{r}'$ at all preceding moments $t' < t$ in accordance with the *causality principle*,

$$J_i(t,\mathbf{r}) = \int_{-\infty}^{t} \mathrm{d}t' \int \mathrm{d}\mathbf{r}' \sigma_{ij}(t,t',\mathbf{r},\mathbf{r}') E_j(t',\mathbf{r}'). \tag{1.15}$$

The tensor $\sigma_{ij}(t,t',\mathbf{r},\mathbf{r}')$ in Eqs. (1.15) characterizes transfer of the material's response from one point of space and time to another. For a homogeneous medium (guaranteed by macroscopic averaging over $a_{\rm macro}$ larger than the lattice constant), the tensor σ_{ij} depends on the differences $t-t'$ and $\mathbf{r}-\mathbf{r}'$ only,

$$\sigma_{ij}(t,t',\mathbf{r},\mathbf{r}') = \sigma_{ij}(t-t',\mathbf{r}-\mathbf{r}'). \tag{1.16}$$

Now, if we perform the Fourier transform,

$$G(t,\mathbf{r}) = \frac{1}{(2\pi)^2} \int\int G(\omega,\mathbf{k})\,\mathrm{e}^{\mathrm{i}(\mathbf{k}\cdot\mathbf{r}-\omega t)}\,\mathrm{d}\omega\,\mathrm{d}\mathbf{k},$$

of $\mathbf{J}$ and $\mathbf{E}$ in the $(t,\mathbf{r})$ space, we can get their relation in the frequency-wavevector space $(\omega,\mathbf{k})$,

$$J_i(\omega,\mathbf{k}) = \sigma_{ij}(\omega,\mathbf{k}) E_j(\omega,\mathbf{k}), \tag{1.17}$$

where $\sigma_{ij}(\omega,\mathbf{k})$ is the *tensor of complex conductivity* given by

$$\sigma_{ij}(\omega,\mathbf{k}) = \int_0^\infty \mathrm{d}\tau \int \mathrm{d}\mathbf{R}\, \sigma_{ij}(\tau,\mathbf{R})\,\mathrm{e}^{-\mathrm{i}(\mathbf{k}\cdot\mathbf{R}-\omega\tau)}, \tag{1.18}$$

with $\tau = t - t'$ and $\mathbf{R} = \mathbf{r} - \mathbf{r}'$.

Now, we can use the relation $\mathbf{J}(\mathbf{E})$ in the $(\omega,\mathbf{k})$ space to get the material equation for $\mathbf{D}(\mathbf{E})$ in the linear approximation,

$$D_i(\omega,\mathbf{k}) = \varepsilon_0 \varepsilon_{ij}(\omega,\mathbf{k}) E_j(\omega,\mathbf{k}). \tag{1.19}$$

Here, $\varepsilon_{ij}(\omega,\mathbf{k})$ is the *tensor of complex permittivity* defined as

$$\varepsilon_{ij}(\omega,\mathbf{k}) = 1 + \mathrm{i}\frac{\sigma_{ij}(\omega,\mathbf{k})}{\varepsilon_0 \omega}. \tag{1.20}$$

Following this definition, $\varepsilon_{ij}(\omega,\mathbf{k})$ characterizes the linear nonlocal response of a charge-neutral medium to electric fields.

For an isotropic medium, properties of which are identical in any direction, $\varepsilon_{ij}(\omega,\mathbf{k})$ can be composed of the unit tensor δ_{ij} and the tensor $k_i k_j$, as they are the only two tensors of the second rank formed of the wavevector $\mathbf{k}$. Thus, we can write

$$\varepsilon_{ij}(\omega,\mathbf{k}) = \left(\delta_{ij} - \frac{k_i k_j}{k^2}\right) \varepsilon_t(\omega,\mathbf{k}) + \frac{k_i k_j}{k^2} \varepsilon_l(\omega,\mathbf{k}). \tag{1.21}$$

Following this expression, among nine components of the tensor ε_{ij}, there are only two independent components, $\varepsilon_t(\omega,\mathbf{k})$ and $\varepsilon_l(\omega,\mathbf{k})$. According to Eq. (1.20), these components have clear physical meaning: $\varepsilon_l(\omega,\mathbf{k})$ gives the medium response to longitudinal electric fields ($\mathbf{E}\times\mathbf{k} = 0$), while $\varepsilon_t(\omega,\mathbf{k})$ describes the response to transverse electric fields ($\mathbf{E}\cdot\mathbf{k} = 0$).

1.1.3 Temporal and spatial dispersion

In general case, tensor ε_{ij} depends on the frequency ω and the wavevector $\mathbf{k}$. Eventually, any electromagnetic pulse disperses by propagating in the medium, as the Fourier components

$$G(\omega,\mathbf{k})\,\mathrm{e}^{\mathrm{i}(\mathbf{k}\cdot\mathbf{r}-\omega t)}$$

with different ω and $\mathbf{k}$ (that comprise the pulse in accordance with the Fourier transform) propagate with different phase velocities ω/k. Thus, the materials with frequency and wavevector dependence are *dispersive*. The frequency dependence of the tensor ε_{ij} describes the *temporal dispersion* of electromagnetic fields, while the wavevector dependence gives the *spatial dispersion*.

Temporal dispersion of solids arises due to the inertia and friction of intrinsic charges that make the polarization inertial to electric field. Thus, the medium response at a given moment t depends on the electric field values at all preceding moments $t' \leq t$.

The time interval $\tau = t - t'$, over which the previous history still has a significant effect, is defined by the characteristic frequencies ω_s. It is obvious that for electromagnetic fields oscillating at a very high frequency $\omega \gg \omega_s$, the intrinsic charges do not have enough time to form any significant polarization. Eventually, the result is very weak temporal dispersion with

$$\varepsilon_{ij}(\omega \to \infty) = \delta_{ij}.$$

However, at frequencies ω below or close to the characteristic frequencies ω_s, the temporal dispersion increases and cannot be ignored anymore.

In contrast to the temporal dispersion, the spatial one comes from nonlocality of the medium response to electric fields. Physically, it is given by the dependence of the polarization vector $\mathbf{P}(t, \mathbf{r})$ on the electric fields $\mathbf{E}(t, \mathbf{r}')$ in the vicinity of the point $\mathbf{r}$. The region over which the nonlocality takes place is defined by the characteristic length $a_s = a_{\text{macro}} + a_{\text{dyn}}$ given by the field averaging and charge dynamics.

The first nonlocal mechanism is purely artificial and introduced through statistical averaging of all fields in space over a_{macro}. It should be considered the inherent error of the macroscopic approach that does not allow us to look at smaller scales, but is necessary to fulfill the condition of the medium's uniformity. The second mechanism is physical, caused by the dynamics of charges in the medium. At high frequencies, it is given by the traveling distance of the electrons at the highest occupied orbital

$$a_{\text{dyn}} = \sqrt{\frac{2E_{\text{HOMO}}}{m\omega^2}},$$

where E_{HOMO} and m are the energy and effective mass of those electrons.

As a_{dyn} decreases with the frequency very fast, the response of typical solids in the optical range is highly localized ($ka_s \approx ka_{\text{macro}} \ll 1$) with negligibly small spatial dispersion. It allows us to treat $\varepsilon_{ij}(\omega, \mathbf{k})$ independent of the wavevector $\mathbf{k}$, when

$$\varepsilon_{ij}(\omega, \mathbf{k}) = \delta_{ij}\varepsilon(\omega) \tag{1.22}$$

with

$$\varepsilon_t(\omega, \mathbf{k}) = \varepsilon_l(\omega, \mathbf{k}) = \varepsilon(\omega).$$

Thus, magnetically inactive uniform solids are well described at optical frequencies and above within the *local response* approximation, where the material relations are completely given by the *scalar complex dielectric permittivity* $\varepsilon(\omega) = \varepsilon'(\omega) + \mathrm{i}\varepsilon''(\omega)$,

$$\mathbf{D}(\omega, \mathbf{k}) = \varepsilon_0\varepsilon(\omega)\mathbf{E}(\omega, \mathbf{k}), \tag{1.23}$$

$$\mathbf{B}(\omega, \mathbf{k}) = \mu_0\mathbf{H}(\omega, \mathbf{k}). \tag{1.24}$$

Note that in Eq. (1.24) we wrote the relation $\mathbf{B}(\mathbf{H})$ instead of $\mathbf{H}(\mathbf{B})$. This formal change is due to the symmetry of the Maxwell's equations, following which it is more natural to describe magnetic properties in terms of $\mathbf{H}$ rather than $\mathbf{B}$. Namely for this reason, the field $\mathbf{H}$ is commonly called the magnetic field by analogy with the electric field $\mathbf{E}$ although it is actually an auxiliary quantity. Hereinafter, electromagnetic fields will be described in terms of vectors $\mathbf{E}$ and $\mathbf{H}$ only.

1.2 Local response approximation

1.2.1 Energy of electromagnetic field in medium

According to the Maxwell's equations, distribution of electromagnetic field in a medium obeys the energy conservation law. For magnetically inactive materials, it can be written as [3, 4]:

$$\frac{1}{2}\frac{\partial}{\partial t}(\varepsilon_0 E^2 + \mu_0 H^2) + \nabla \cdot (\mathbf{E} \times \mathbf{H}) + \mathbf{j}_{\text{pol}} \cdot \mathbf{E} = -\mathbf{j}_{\text{ext}} \cdot \mathbf{E}, \tag{1.25}$$

describing the energy balance between electromagnetic fields and different types of charges composing the medium. The right-hand side of this equation represents the power density of the external currents spent on the polarization of the medium's element dV and the excitation of the electric and magnetic fields inside it,

$$A_{\text{ext}} = -\mathbf{j}_{\text{ext}} \cdot \mathbf{E}, \tag{1.26}$$

while the left-hand side gives us different mechanisms for that energy expenditure. The first one is the increase of the energy density

$$U_0 = \frac{1}{2}(\varepsilon_0 E^2 + \mu_0 H^2) \tag{1.27}$$

of the electric $\mathbf{E}$ and magnetic $\mathbf{H}$ fields over the time dt. The second mechanism is the radiation of the energy out of the volume dV; its contribution is given by the divergence of the Poynting vector

$$\mathbf{S} = \mathbf{E} \times \mathbf{H}, \tag{1.28}$$

which is considered as the energy flux density of the electromagnetic field. The third mechanism is the work done by the polarization charges, described with the power density

$$Q_{\text{med}} = \mathbf{j}_{\text{pol}} \cdot \mathbf{E}. \tag{1.29}$$

In general, Q_{med} is comprised of the dissipation power density Q_{dis} and the change of the energy density U_{pol} stored in the medium in the form of the polarization field $\mathbf{P}$,

$$Q_{\text{med}} = \frac{\partial U_{\text{pol}}}{\partial t} + Q_{\text{dis}}. \tag{1.30}$$

Thus, the conservation law for the overall electromagnetic field in the medium can be written as

$$\frac{\partial U}{\partial t} + \nabla \cdot S + Q_{\text{dis}} = A_{\text{ext}}, \tag{1.31}$$

where $U = U_0 + U_{\text{pol}}$ is the total energy density of the three fields $\mathbf{E}$, $\mathbf{H}$, and $\mathbf{P}$.

To derive the energy characteristics in the local response approximation, we calculate the time-averaged value of Q_{med},

$$\overline{Q_{\text{med}}} = \frac{1}{2}\text{Re}(\mathbf{j}_{\text{pol}} \cdot \mathbf{E}^*) = \frac{1}{2}\text{Re}\left(\frac{\partial \mathbf{P}}{\partial t} \cdot \mathbf{E}^*\right), \tag{1.32}$$

assuming that the electromagnetic field is not purely monochromatic, but has a small frequency width,

$$\mathbf{E} = \mathbf{E}_0(t)\,\mathrm{e}^{-\mathrm{i}\omega t}, \quad \mathbf{H} = \mathbf{H}_0(t)\,\mathrm{e}^{-\mathrm{i}\omega t},$$

where $\mathbf{E}_0(t)$ and $\mathbf{H}_0(t)$ are slowly varying functions, as compared to $\mathrm{e}^{-\mathrm{i}\omega t}$. By doing that, we can capture the effect of temporal dispersion that gives a significant contribution to the energy stored in a medium [3, 5],

$$\overline{Q_{\text{med}}} = \frac{\partial}{\partial t}\frac{\varepsilon_0|E|^2}{4}\left[\frac{\partial(\omega\varepsilon')}{\partial\omega} - 1\right] + \frac{\omega}{2}\varepsilon_0\varepsilon''|E|^2. \tag{1.33}$$

By comparing with Eq. (1.30), we find that the time-averaged energy density of the overall electromagnetic field (including the polarization one) is

$$\overline{U} = \frac{1}{4}\left[\frac{\partial(\omega\varepsilon')}{\partial\omega}\varepsilon_0|E|^2 + \mu_0|H|^2\right] \tag{1.34}$$

with the dispersion effect given by the function $\partial(\omega\varepsilon')/\partial\omega$, where ε' is the real part of the complex dielectric permittivity $\varepsilon(\omega)$. For the time-averaged dissipation rate, we obtain

$$\overline{Q_{\text{dis}}} = \frac{\omega}{2}\varepsilon_0\varepsilon''|E|^2, \tag{1.35}$$

where ε'' is the imaginary part of $\varepsilon(\omega)$. In this way, we conclude that the real part of complex dielectric permittivity defines the energy of electromagnetic field, while the imaginary part describes its dissipation.

1.2.2 Properties of complex dielectric permittivity

As we have already seen, scalar dielectric permittivity plays an essential role in description of the electrodynamic field interaction with a medium. Therefore, it is vital to understand the main properties of $\varepsilon(\omega)$.

Some common qualities of the scalar dielectric permittivity can be obtained from the general definition of $\sigma_{ij}(\omega,\mathbf{k})$ given by Eq. (1.18). Note the difference between the tensors $\sigma_{ij}(t,\mathbf{r})$ and $\sigma_{ij}(\omega,\mathbf{k})$ in that equation — the former is a purely real function, as it relates two real vectors $\mathbf{J}(t,\mathbf{r})$ and $\mathbf{E}(t,\mathbf{r})$, while the latter is generally complex. Following this peculiarity, we can write the symmetry relation for $\sigma_{ij}(\omega,\mathbf{k})$,

$$\sigma_{ij}(-\omega,-\mathbf{k}) = \sigma_{ij}^*(\omega,\mathbf{k}),$$

which can be translated to the identical relation for $\varepsilon_{ij}(\omega,\mathbf{k})$ with Eq. (1.20),

$$\varepsilon_{ij}(-\omega,-\mathbf{k}) = \varepsilon_{ij}^*(\omega,\mathbf{k}).$$

Thus, for isotropic materials in the high-frequency range, we have

$$\varepsilon(-\omega) = \varepsilon^*(\omega).$$

In terms of the real $\varepsilon'(\omega)$ and imaginary $\varepsilon''(\omega)$ parts of $\varepsilon(\omega)$, it can be rewritten as follows:

$$\varepsilon'(-\omega) = \varepsilon'(\omega), \quad \varepsilon''(-\omega) = -\varepsilon''(\omega), \tag{1.36}$$

exhibiting evenness for the real part and oddness for the imaginary part of scalar permittivity.

Another property can be obtained if we consider the limiting case of $\omega \to \infty$. This is the case of highly localized response with $a_s \to 0$, when all intrinsic charges oscillate at such a high frequency and short scale, so they do not feel each other. In this regime, nuclei are immobile due to their high masses, so the medium's response is given by light electrons only driven by the electric field,

$$\frac{\partial \mathbf{v}(t,\mathbf{r})}{\partial t} = -\frac{e}{m}\mathbf{E}(t,\mathbf{r}) = -\frac{e}{m}\mathbf{E}_0(\mathbf{r})\,\mathrm{e}^{-\mathrm{i}\omega t},$$

where $\mathbf{v}$, $-e$, and m are the directional velocity, charge, and mass of the electrons. Oscillating in the electric field $\mathbf{E}$, electrons create the polarization current $\mathbf{j}_{\mathrm{pol}}$ given by

$$\mathbf{j}_{\mathrm{pol}}(t,\mathbf{r}) = -en_0\mathbf{v}(t,\mathbf{r}),$$

where n_0 is the total density of the medium's electrons. Performing the Fourier transform for $\mathbf{v}, \mathbf{E}$, and $\mathbf{j}$, we derive

$$\mathbf{j}_{\mathrm{pol}}(\omega) = \mathrm{i}\frac{e^2 n_0}{m\omega}\mathbf{E}(\omega).$$

At the same time, according to the definition of the polarization vector $\mathbf{P}$, the internal current is given by

$$\mathbf{j}_{\mathrm{pol}}(\omega) = -\mathrm{i}\omega\mathbf{P}(\omega) = -\mathrm{i}\omega\varepsilon_0[\varepsilon(\omega) - 1]\mathbf{E}(\omega).$$

Thus, we get

$$\varepsilon(\omega) = 1 - \frac{e^2 n_0}{\varepsilon_0 m\omega^2}$$

for the asymptotic behavior of scalar dielectric permittivity at extremely high frequencies. In the limit of $\omega \to \infty$, it gives us

$$\lim_{\omega\to\infty} \varepsilon(\omega) = 1,$$

or alternatively

$$\lim_{\omega\to\infty} \varepsilon'(\omega) = 1, \quad \lim_{\omega\to\infty} \varepsilon''(\omega) = 0. \tag{1.37}$$

Other, thermodynamic, properties of the scalar dielectric permittivity can be obtained if we consider the energy density of the overall electromagnetic field $\overline{U}$ and the energy dissipation rate $\overline{Q_{\mathrm{dis}}}$ given by Eqs. (1.34) and (1.35), respectively. Following them, under the thermodynamic equilibrium, $\varepsilon'(\omega)$ and $\varepsilon''(\omega)$ should feature

$$\frac{\partial}{\partial\omega}\left[\omega\varepsilon'(\omega)\right] \geq 0, \quad \omega\varepsilon''(\omega) \geq 0. \tag{1.38}$$

The former condition imposes a restriction on temporal dispersion, while the latter defines the sign of $\varepsilon''(\omega)$ for optically passive media.

1.2.3 Medium response modeling

Up to this point, we gave the description of a charge-compensated medium without mentioning its nature. It was the unified theory covering magnetically inactive materials ranging from dielectrics to semiconductors and metals. Now, let us discuss different groups of materials and review common models developed for their description in the local response approximation.

Within the local response approximation, the scalar complex dielectric permittivity can be written in the following form

$$\varepsilon(\omega) = 1 + \chi(\omega), \tag{1.39}$$

where $\chi(\omega)$ is the scalar susceptibility of the neutral-charge medium that gives us the material relation for the polarization field

$$\mathbf{P}(\omega) = \varepsilon_0 \chi(\omega) \mathbf{E}(\omega). \tag{1.40}$$

If there are independent mechanisms for polarization of the medium, we can write the susceptibility as a sum over them

$$\chi(\omega) = \sum_{\alpha} \chi_{\alpha}(\omega). \tag{1.41}$$

Following it, understanding of typical dependences of $\chi_{\alpha}(\omega)$ given by polarization of different groups α of intrinsic charges is of critical importance for medium response modeling.

Below we overview most commonly used classical models developed for descriptions of different media. They all are based on the relation of the polarization field and current,

$$\frac{\partial \mathbf{P}}{\partial t} = \mathbf{j}_{\text{pol}} = \sum_{\alpha} q_{\alpha} n_{\alpha} \mathbf{v}_{\alpha}, \tag{1.42}$$

where the polarization current is contributed by different species of the polarization charges α characterized with the charge q_{α}, volume density n_{α}, and directional velocity $\mathbf{v}_{\alpha}$. After additional differentiation, we get

$$\frac{\partial^2 \mathbf{P}}{\partial t^2} = \sum_{\alpha} q_{\alpha} n_{\alpha} \frac{\partial \mathbf{v}_{\alpha}}{\partial t}, \tag{1.43}$$

where $\partial \mathbf{v}_{\alpha}/\partial t$ is given by the motion equation different for every species.

Within the local response approximation (valid at high frequencies only), the polarization current is mainly contributed by conduction and bound electrons due to their light weight. Heavy nuclei can be considered immobile at these frequencies without significant loss of accuracy, as their contribution to the polarization current is at least a thousand times smaller than those of conduction and bound electrons.

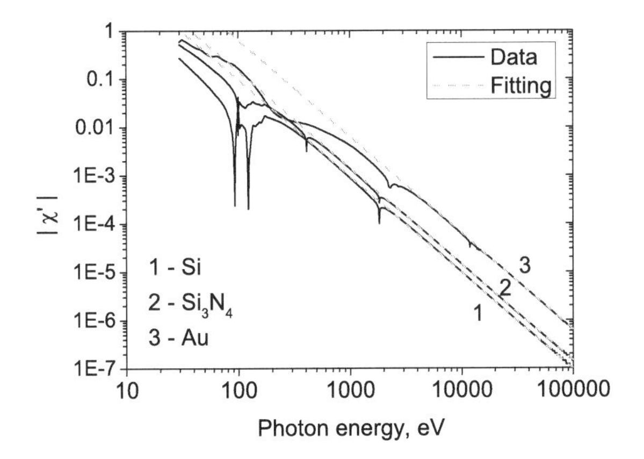

Figure 1.1: Fitting of $|\chi'|$ with the dissipationless Drude model for silicon (Si), silicon nitride (Si_3N_4), and gold (Au). The fitted values of ω_{p0} are 31, 38, and 80 eV, respectively. The obtained total electron density is in excellent agreement with the mass densities of the corresponding materials, which are 2.329, 3.44, and 19.32 g/cm^3. The optical data fitted are taken from Refs. [6, 7].

Dissipationless Drude model

The dissipationless Drude model has already been used in the previous section when we considered properties of the scalar dielectric permittivity at extremely high frequencies. In this case, all electrons (regardless of whether they are conduction or bound) behave as free: they oscillate with the directional velocity driven by the high-frequency electric field only,

$$\frac{\partial \mathbf{v}}{\partial t} = -\frac{e}{m}\mathbf{E}, \tag{1.44}$$

without feeling each other and nuclei. As a result, the polarization of the medium is given by the differential equation

$$\frac{\partial^2 \mathbf{P}}{\partial t^2} = \omega_{p0}^2 \varepsilon_0 \mathbf{E}, \tag{1.45}$$

where $\omega_{p0} = \sqrt{n_0 e^2/(m\varepsilon_0)}$ is the plasma frequency of all electrons in the medium, characterized with the total electron density n_0. The modeled polarization field and scalar susceptibility have the following form

$$\mathbf{P}(\omega) = -\frac{\omega_{p0}^2 \varepsilon_0}{\omega^2}\mathbf{E}, \qquad \chi(\omega) = -\frac{\omega_{p0}^2}{\omega^2}. \tag{1.46}$$

This model describes the limiting case of highly localized response of any medium, regardless of its electronic structure. The applicability of this model is restricted to extremely high frequencies, above all characteristic frequencies of the medium, $\omega \gg \omega_s$. Fitting of experimental data with this model allows us to get ω_{p0} and estimate the total density of electrons in the medium, as shown in Fig. 1.1.

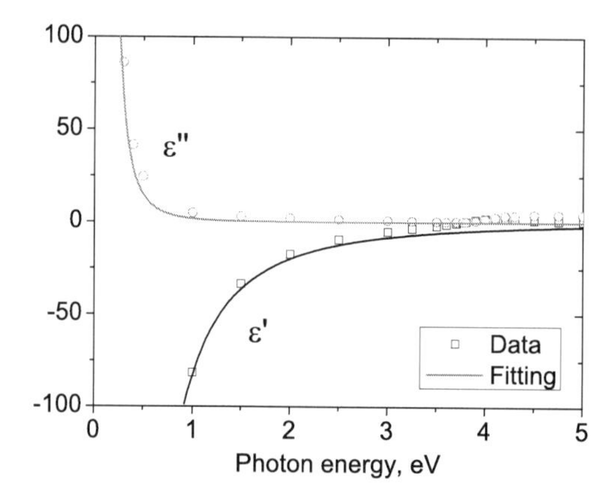

Figure 1.2: Fitting of the scalar dielectric permittivity of silver (Ag) with the dissipative Drude model. The estimated ω_p and γ correspond to 9.19 and 0.02 eV. The fitted data are taken from Ref. [8].

Dissipative Drude model

In the dissipationless Drude model, we ignored interactions between electrons. If we add friction to the equation of motion, we can get the dissipative Drude model that describes the response of free electrons with significant contribution of electron-electron scattering,

$$\frac{\partial \mathbf{v}}{\partial t} = -\frac{e}{m}\mathbf{E} - \gamma \mathbf{v}, \tag{1.47}$$

where γ is the scattering rate. In this case, the equation for polarization field $\mathbf{P}$ has the following form,

$$\frac{\partial^2 \mathbf{P}}{\partial t^2} + \gamma \frac{\partial \mathbf{P}}{\partial t} = \omega_p^2 \varepsilon_0 \mathbf{E}, \tag{1.48}$$

resulting in

$$\mathbf{P}(\omega) = -\frac{\omega_p^2 \varepsilon_0}{\omega(\omega + i\gamma)}\mathbf{E}, \qquad \chi(\omega) = -\frac{\omega_p^2}{\omega(\omega + i\gamma)}, \tag{1.49}$$

where $\omega_p = \sqrt{ne^2/(m\varepsilon_0)}$ is the plasma frequency of the free electrons with volume density n.

This model was first proposed by Paul Drude in 1900 to explain electrical conduction of metals. It provides an adequate description of good conductors with dominating contribution of conduction electrons. As a rule, Drude-like response is observed at lower frequencies, below the electron inter-band transitions where the polarization by conduction electrons dominates (Fig. 1.2). At higher frequencies, additional terms accounting for the inter-band transitions should be considered as well. Through fitting of ω_p and γ, this model allows us to determine density of the conduction electrons together with their relaxation time.

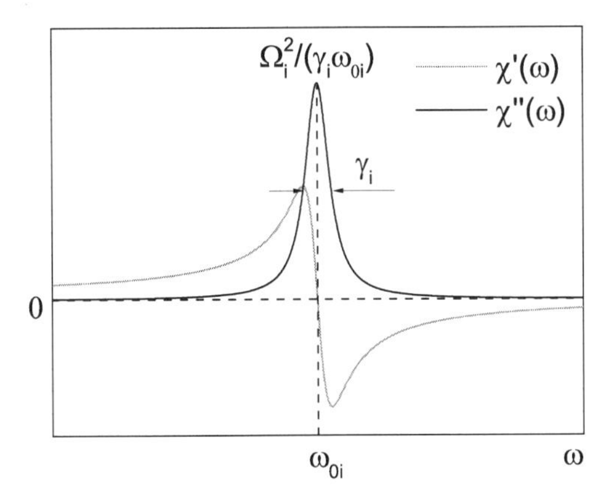

Figure 1.3: Characteristics of the oscillator susceptibility $\chi_i(\omega)$ in the Lorentz model.

Lorentz oscillators model

Although the dissipative Drude model accounts for the electron-electron scattering, it describes free electrons only and does not consider electron-nucleus bonds. Of course, a full description of bound electrons response is very complicated; it requires a quantum-mechanical consideration of all possible electron transitions. Therefore, it is common to use the methods of classical mechanics to model the quantum responses of bound electrons. The most widely used model is of Lorentz oscillators. It describes bound electrons as a system of N decoupled oscillators. Every oscillator represents a set of identical bound electrons of density n_i featuring the same bond potential U_i and scattering rate γ_i, where $i = 1, 2, ..., N$. Then, the electron dynamics of the i-th oscillator is given by

$$\frac{\partial \mathbf{v}_i}{\partial t} = -\frac{e}{m}\mathbf{E} - \gamma_i \mathbf{v}_i - \frac{\nabla U_i}{m}, \tag{1.50}$$

so the corresponding partial polarization $\mathbf{P}_i$ obeys the equation

$$\frac{\partial^2 \mathbf{P}_i}{\partial t^2} + \gamma_i \frac{\partial \mathbf{P}_i}{\partial t} + \omega_{0i}^2 \mathbf{P}_i = \Omega_i^2 \varepsilon_0 \mathbf{E}, \tag{1.51}$$

where $\Omega_i^2 = e^2 n_i (m\varepsilon_0)^{-1}$ and $\omega_{0i}^2 = (\mathbf{E} \cdot \nabla)^2 U_i|_{r_{\min}} (|E|^2 m)^{-1}$. By solving Eq. (1.51) we get the polarization field

$$\mathbf{P}_i(\omega) = -\frac{\Omega_i^2 \varepsilon_0}{\omega^2 - \omega_{0i}^2 + \mathrm{i}\gamma_i \omega}\mathbf{E}, \tag{1.52}$$

and susceptibility of the i-th Lorentz oscillator,

$$\chi_i(\omega) = -\frac{\Omega_i^2}{\omega^2 - \omega_{0i}^2 + \mathrm{i}\gamma_i \omega}. \tag{1.53}$$

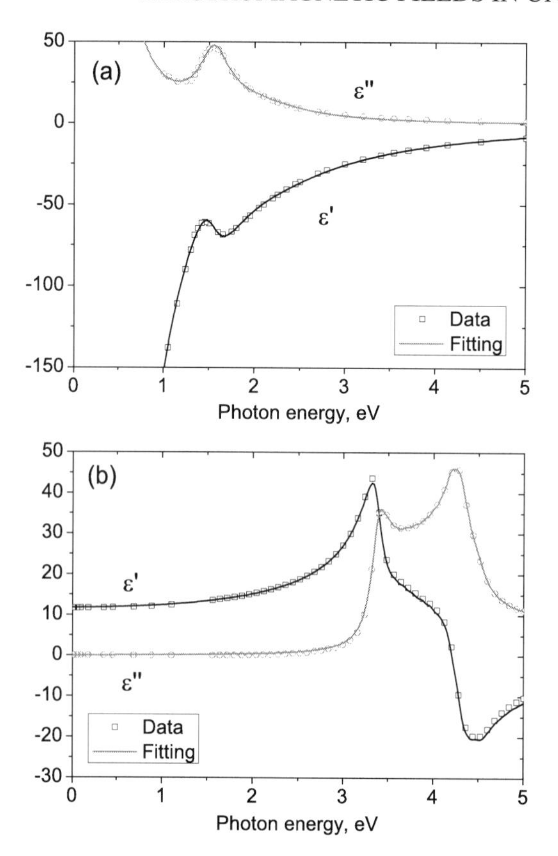

Figure 1.4: Fitting of the scalar dielectric permittivity of (a) aluminium (Al) and (b) silicon (Si) with the model of Lorentz oscillators. Optical data of Al are fitted well with $N = 3$ oscillators, while Si data require $N > 10$ for satisfactory matching. The fitted data are taken from Ref. [9].

Finally, to get the total polarization field and susceptibility, we need to sum the respective partial contributions over the oscillator index i

$$\mathbf{P}(\omega) = \sum_{i=1}^{N} \mathbf{P}_i(\omega), \qquad \chi(\omega) = \sum_{i=1}^{N} \chi_i(\omega). \tag{1.54}$$

Although the Lorentz oscillators model is based on classical mechanics principles, it provides a very good fitting for bound electron responses. More accurate quantum-mechanical considerations confirm the Lorentz shape of scalar susceptibility caused by electron transitions [10], although the model's parameters Ω_i, ω_{0i}, γ_i bear different meanings. In addition, the Lorentz oscillators model covers the response of conduction electrons as well. If we choose $\omega_{0i} = 0$ and $N = 1$, we get the Drude model where $\omega_p = \Omega_i$. All this makes $\chi_i(\omega)$ a universal spectroscopic function that can be used in composition of any type of dielectric permittivity. Indeed, the parameters Ω_i, ω_{0i}, γ_i bear a clear spectroscopic meaning for $\chi''(\omega)$ — they are strength, central frequency, and the width of the spectral line, as shown in Fig. 1.3.

In general, oscillator susceptibility (1.53) works well for both conductors and insulators, as shown in Fig. 1.4. However, as real spectral lines in $\chi''(\omega)$ are often asymmetric due to quasi-continuous electron transitions, the procedure may require a higher number N of the harmonic oscillators to provide better matching [see Fig. 1.4(b)].

1.3 Electromagnetic fields in medium

1.3.1 Electromagnetic field generation

Traditionally, excitation of electromagnetic fields in magnetically inactive materials is described by two Maxwell's equations:

$$\nabla \times \mathbf{E} = -\mu_0 \frac{\partial \mathbf{H}}{\partial t}, \qquad \nabla \times \mathbf{H} = \frac{\partial \mathbf{D}}{\partial t} + \mathbf{J},$$

with sources given by $\mathbf{J} = \mathbf{j}_{\text{ext}}$. In the general case of a uniform isotropic medium with spatial dispersion, the field excitation in the $(\omega, \mathbf{k})$ space is given by

$$\mathbf{k} \times \mathbf{E}(\omega, \mathbf{k}) = \omega \mu_0 \mathbf{H}(\omega, \mathbf{k}), \tag{1.55}$$

$$\mathbf{k} \times \mathbf{H}(\omega, \mathbf{k}) = -\omega \mathbf{D}(\omega, \mathbf{k}) - \mathrm{i}\mathbf{J}(\omega, \mathbf{k}), \tag{1.56}$$

where $D_i(\omega, \mathbf{k}) = \varepsilon_0 \varepsilon_{ij}(\omega, \mathbf{k}) E_j(\omega, \mathbf{k})$ with

$$\varepsilon_{ij}(\omega, \mathbf{k}) = \left(\delta_{ij} - \frac{k_i k_j}{k^2} \right) \varepsilon_t(\omega, \mathbf{k}) + \frac{k_i k_j}{k^2} \varepsilon_l(\omega, \mathbf{k}).$$

Following the Helmholtz theorem, all vectors $\mathbf{F} = \{\mathbf{E}, \mathbf{H}, \mathbf{J}\}$ can be uniquely decomposed into the longitudinal (curl-free) and transverse (divergence-free) fields,

$$\mathbf{F}(\omega, \mathbf{k}) = \mathbf{F}_t(\omega, \mathbf{k}) + \mathbf{F}_l(\omega, \mathbf{k}), \tag{1.57}$$

where

$$\mathbf{F}_t(\omega, \mathbf{k}) = \mathbf{k} \times [\mathbf{F}(\omega, \mathbf{k}) \times \mathbf{k}]/k^2, \qquad \mathbf{F}_l(\omega, \mathbf{k}) = \mathbf{k} \cdot (\mathbf{F}(\omega, \mathbf{k}) \cdot \mathbf{k})/k^2.$$

Then, we can get expressions for the excited transverse,

$$\mathbf{E}_t(\omega, \mathbf{k}) = -\mathrm{i} \frac{\omega \mu_0 \mathbf{J}_t(\omega, \mathbf{k})}{k_0^2 \varepsilon_t(\omega, \mathbf{k}) - k^2}, \qquad \mathbf{H}_t(\omega, \mathbf{k}) = -\mathrm{i} \frac{\mathbf{k} \times \mathbf{J}_t(\omega, \mathbf{k})}{k_0^2 \varepsilon_t(\omega, \mathbf{k}) - k^2}, \tag{1.58}$$

and longitudinal fields,

$$\mathbf{E}_l(\omega, \mathbf{k}) = -\mathrm{i} \frac{\omega \mu_0 \mathbf{J}_l(\omega, \mathbf{k})}{k_0^2 \varepsilon_l(\omega, \mathbf{k})}, \qquad \mathbf{H}_l(\omega, \mathbf{k}) = 0, \tag{1.59}$$

where $k_0 = \omega \sqrt{\varepsilon_0 \mu_0}$ is the free-space wavenumber. Notably the generation of electromagnetic fields in an isotropic medium occurs in two independent ways. First is a response of the medium to the transverse external currents given by $\mathbf{J}_t$, while the second is excitation of electric fields by the longitudinal currents with $\mathbf{J}_l$. These processes occur independently and do not affect each other.

1.3.2 Bulk eigenwaves

Another interesting feature of field excitation is that both transverse and longitudinal fields exhibit resonances; the transverse fields have singularity at

$$k^2 - k_0^2 \varepsilon_t(\omega, \mathbf{k}) = 0, \tag{1.60}$$

while the longitudinal electric field diverges at

$$\varepsilon_l(\omega, \mathbf{k}) = 0. \tag{1.61}$$

At these conditions, media resonantly respond to external currents by generation of strong fields. These resonances are attributed to excitation of self-consistent oscillations of the coupled polarization and electromagnetic fields supported by the medium. Indeed, if we solve the *eigenvalue problem* composed by Maxwell's equations without external sources ($\mathbf{J} = 0$),

$$\mathbf{k} \times \mathbf{E}(\omega, \mathbf{k}) = \omega \mu_0 \mathbf{H}(\omega, \mathbf{k}), \tag{1.62}$$

$$\mathbf{k} \times \mathbf{H}(\omega, \mathbf{k}) = -\omega \varepsilon_0 \varepsilon_t(\omega, \mathbf{k}) \mathbf{E}(\omega, \mathbf{k}), \tag{1.63}$$

we find that the self-consistent solutions exist only for certain dependencies of $\omega(\mathbf{k})$, given by Eqs. (1.60) and (1.61). The corresponding solutions are called *bulk eigenwaves* and are of extreme importance for understanding of resonant light-matter interactions. They represent a special type of electromagnetic fields that can freely propagate in a medium without any scattering. In this sense, bulk eigenwaves are the medium's intrinsic characteristics describing its ability to respond resonantly to external excitations.

At high frequencies, when the difference between two components of dielectric permittivity vanishes and the spatial dispersion becomes negligible, $\varepsilon_t(\omega, \mathbf{k}) = \varepsilon_l(\omega, \mathbf{k}) = \varepsilon(\omega)$, the dispersion equations reduce to

$$k^2 - k_0^2 \varepsilon(\omega) = 0, \tag{1.64}$$

$$\varepsilon(\omega) = 0. \tag{1.65}$$

These relations give us the dependencies $\omega(\mathbf{k})$ for bulk eigenwaves in the local response approximation. In this regime, $\omega(\mathbf{k})$ are complex due to finite dissipation in real materials. In addition, they feature independence of $\mathbf{k}$ for longitudinal eigenfields, when eigenwaves with different $\mathbf{k}$ propagate at the same frequency $\omega(\mathbf{k}) = \omega$. This comes from the neglect of spatial dispersion in Eq. (1.61), restricted by the applicability conditions of the local response approximation.

1.3.3 Quasi-particle classification

Elementary excitations

As we mentioned, bulk eigenwaves represent special, self-consistent electromagnetic oscillations. They can freely propagate in a medium, similar to photons in vacuum.

Indeed, if we consider the case of free space with $\varepsilon(\omega) = 1$, Eq. (1.65) for longitudinal eigenwaves has no solutions, while Eq. (1.64) for transverse eigenwaves reduces to

$$k^2 = k_0^2. \tag{1.66}$$

This is the dispersion relation of pure electromagnetic fields traveling in vacuum. It gives us the well-known linear relation for $\omega(\mathbf{k})$ [1, 4],

$$\omega = \pm ck, \tag{1.67}$$

where $c = 1/\sqrt{\varepsilon_0\mu_0}$ is the speed of light in vacuum. Thus, pure electromagnetic fields propagate in vacuum at constant velocity c and bear the energy

$$\overline{U}_{EMF} = \frac{1}{2}\varepsilon_0|E|^2. \tag{1.68}$$

In the quasi-particle classification, quanta of such oscillations are known as *photons* and considered elementary electromagnetic quasi-particles. At the same time, following the electromagnetic field classification, photons are the transverse bulk eigenwaves of unbounded free space.

Now, let us consider how the polarization charges modify the free-space eigenwaves. First of all, they give rise to a new class of longitudinal bulk eigenwaves. Following Eq. (1.65), they require zero dielectric permittivity that can be achieved only in the presence of the polarization charges and therefore are not applicable to vacuum. Hence, they are inherent characteristics of the medium. In the quasi-particle classification, they are treated as *elementary polarization excitations*, similar to photons, which are *elementary electromagnetic excitations*. However, in contrast to the latter, they feature longitudinal fields with $\mathbf{k} \times \mathbf{E}(\omega, \mathbf{k}) = 0$ and $\mathbf{H}(\omega, \mathbf{k}) = 0$.

For non-magnetic materials, there are three types of elementary polarization excitations, given by different kinds of polarization charges. In quasi-particle classification, they are: i) *phonons* resulting from oscillations of the crystal lattice, ii) *excitons* appearing due to bound electrons and their inter-band transitions, and iii) *plasmons* caused by collective oscillation of the conduction electrons. Since those charge groups react to electromagnetic fields differently, the corresponding excitations exist at distinct energies. Phonons typically appear in low-frequency range, far below 1 eV; excitons are commonly seen in the high-frequency range, above 1 eV; while plasmons may appear at either low or high frequencies, depending on the density of conduction electrons (for most metals, plasmons occur at visible and ultraviolet wavelengths).

Regardless of their types, all elementary polarization excitations exhibit the same optical properties. They feature longitudinal electric fields and absence of magnetic. In the local response approximation, elementary polarization excitations are not accompanied by any energy transfer, as their Poynting vectors are always zero. In fact, they transfer energy at a very low rate defined by the spatial dispersion ignored in the local response approximation.

The quasi-particle differentiation of longitudinal eigenwaves is given by distinct dynamics of all groups of polarization charges. However, in real materials, we never have a single type of charges. Instead, we always deal with a combination of two or three. As a result, longitudinal bulk eigenwaves in real materials are *hybrid* quasi-particles. Nonetheless, it is common to call them phonons, excitons, or plasmons, depending on which group of the polarization charges dominates in the medium's response.

Composite excitations

Another major change caused by the presence of polarization charges is modification of transverse eigenwaves. Their dependence $\omega(\mathbf{k})$ is no longer linear and real, as it is for photons,

$$\omega = \pm ck/\sqrt{1+\chi(\omega)}, \tag{1.69}$$

given by varying susceptibility $\chi(\omega) = \chi'(\omega) + \mathrm{i}\chi''(\omega)$ of the polarization charges. It means that transverse eigenwaves propagate in a medium with different phase velocities defined by their frequencies ω and exhibit attenuation.

In addition, pure electromagnetic fields (photons) become so strongly coupled to the polarization charges, that we cannot separate their energies,

$$\overline{U} = \frac{1}{4}\left[1 + |1+\chi(\omega)| + \frac{\partial(\omega\chi')}{\partial\omega}\right]\varepsilon_0|E|^2. \tag{1.70}$$

Here, the coupling is given by the function $|1+\chi(\omega)|$ which is generally non-additive, as $\chi(\omega)$ is complex. Mathematically, this function becomes additive only if $\chi''(\omega) = 0$. Then, the energy of pure electromagnetic field $\overline{U}_{EMF}$ can be distinguished. However, physically, the condition $\chi''(\omega) = 0$ is met in vacuum only and is never reached in real materials.[2]

In view of the above-mentioned considerations, the transverse bulk eigenwaves in a medium are the coupled states of electromagnetic fields and polarization charges. In the quasi-particle classification, they are considered *composite excitations* named *polaritons*, the quanta of coupled elementary electromagnetic and polarization excitations. From elementary electromagnetic excitations (photons), polaritons take over the transverse fields, $\mathbf{k}\cdot\mathbf{E}(\omega,\mathbf{k}) = 0$ and $\mathbf{k}\cdot\mathbf{H}(\omega,\mathbf{k}) = 0$. At the same time, their behaviors strictly follow the response of the polarization charges.

In fact, polaritons dependence on polarization charges determines their main differences from photons. First of all, it results in attenuation of polaritons. During their propagation, a part of the guided energy stored in the polarization charges dissipates through different mechanisms. As a result, polaritons can transmit power over finite lengths only, which vary depending on the material dissipation level. Second, polaritons usually have limited frequency bands to propagate, in contrast to photons that can freely travel at any frequency. These bands are given by the condition $k' > k''$ and

[2]In insulators, $\chi''(\omega)$ is always positive due to finite optical losses; even in the transparency regime, insulators feature negligibly small but nonzero losses given by incomplete occupancy of the valence band at finite temperatures. For conductors, the condition of zero losses is reached in the regime of superconductivity, but restricted to $\omega = 0$ only, where the local response approximation considered here does not hold.

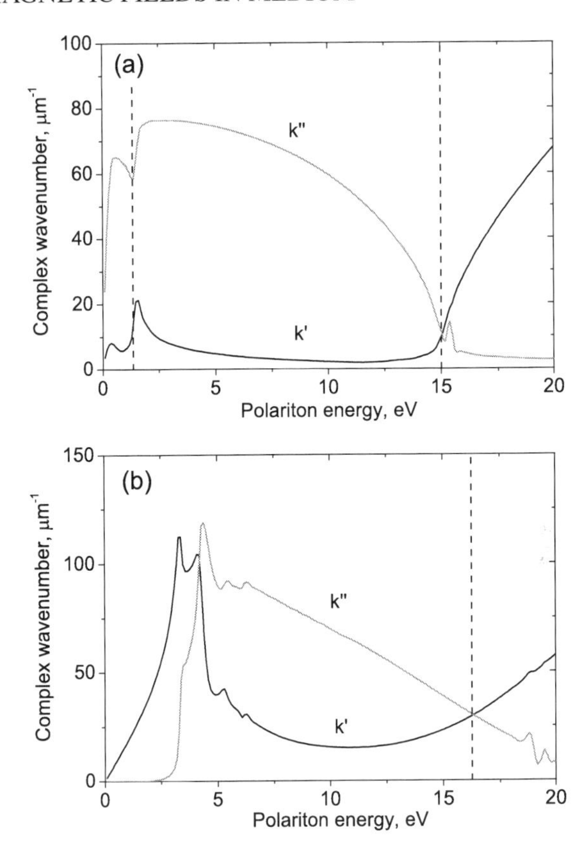

Figure 1.5: Complex wavenumbers k of bulk polaritons as a function of their energy $\hbar\omega$ in (a) Al and (b) Si. The dashed lines depict the energy of plasmons and excitons. At frequencies, where $k' > k''$, polaritons are weakly attenuating, while $k' < k''$ corresponds to highly attenuating polaritons. The optical data are taken from Refs. [9, 11].

clearly seen in Fig. 1.5 for both Al and Si. Beyond these propagation bands, polaritons are highly attenuating, possessing $k' < k''$, so their excitation can be neglected.

Notably the energy of elementary polarization excitations often coincides with the edges of the forbidden and propagation bands. It is caused by the fact that longitudinal bulk eigenwaves define zeros of polaritons' wavenumbers, as follows from their dispersion relation (1.64). Thus, at these energies we can observe changes of polariton behavior, as seen in Fig. 1.5.

Similar to elementary polarization excitations, polaritons in real materials have a hybrid nature. Nonetheless, specification of the dominating polarization charges is a common practice in the literature. Thus, phonon-polaritons, exciton-polaritons, and plasmon-polaritons are common terms in different areas of wave optics, although

<table>
<tr><td colspan="2" rowspan="2"></td><td colspan="2">Fields</td></tr>
<tr><td>Transverse</td><td>Longitudinal</td></tr>
<tr><td rowspan="2">Quasi-particles</td><td>Elementary</td><td>Photons</td><td>Phonons
Excitons
Plasmons</td></tr>
<tr><td>Composite</td><td>Polaritons</td><td>—</td></tr>
</table>

Table 1.1: Quasi-particle and field classification of bulk eigenwaves.

such differentiation is ambiguous and sometimes impugnable.[3] However, despite different polarization nature they all belong to the same class of transverse composite excitations and exhibit similar properties. Complete classification of bulk eigenwaves is presented in Table 1.1.

1.4 Summary

In this chapter, we considered the theory of electromagnetic fields in uniform media. We have discussed the main electromagnetic properties of such media within the local response approximation. We have demonstrated that the resonant character of their response is attributed to the self-consistent eigen solutions of the Maxwell's equations, the bulk eigenwaves. We have reviewed the basic types of the bulk eigenwaves supported in uniform media and considered their hierarchy from both the quasi-particle and field perspectives.

Bibliography

[1] L. D. Landau and E. M. Lifshitz, *The Classical Theory of Fields.* Pergamon Press, 2nd ed., 1962.

[2] L. D. Landau and E. M. Lifshitz, *Statistical Physics.* Pergamon Press, 2nd ed., 1969.

[3] L. D. Landau, E. M. Lifshitz, and L. P. Pitaevskii, *Electrodynamics of Continuous Media.* Pergamon Press, 2nd ed., 1984.

[4] J. D. Jackson, *Classical Electrodynamics.* Wiley, 2nd ed., 1975.

[3]In the literature, the type of polaritons is often based not on the dominating charge contribution, but on the medium type, that causes additional miscommunication between different subfields of optics.

[5] Y. A. Akimov, "Fundamentals of plasmonics," in *Plasmonic Nanoelectronics and Sensing* (E. P. Li and H. S. Chu, eds.), pp. 1–19, Cambridge University Press, 2014.

[6] CXRO. `http://www-cxro.lbl.gov`.

[7] LLNL. `http://www-phys.llnl.gov/V_Div/scattering/asf.html`.

[8] H. J. Hagemann, W. Gudat, and C. Kunz, "Optical constants from the far infrared to the X-ray region: Mg, Al, Cu, Ag, Au, Bi, C, and Al_2O_3," *J. Opt. Soc. Am.*, vol. 65, pp. 742–744, 1975.

[9] E. D. Palik, *Handbook of Optical Constants of Solids*. Academic Press, Inc., 1985.

[10] J. M. Ziman, *Principles of the Theory of Solids*. Cambridge University Press, 1972.

[11] E. Shiles, T. Sasaki, M. Inokuti, and D. Y. Smith, "Self-consistency and sum-rule tests in the Kramers-Kronig analysis of optical data: applications to aluminum," *Phys. Rev. B*, vol. 22, pp. 1612–1628, 1980.

Chapter 2

Electromagnetic waves in bounded media

Yuriy Akimov
Institute of High Performance Computing, Singapore

In this chapter, we will discuss peculiarities of electromagnetic fields in bounded media. Section 2.1 will consider electromagnetic fields in generally inhomogeneous media. In Section 2.2, we will demonstrate how boundaries affect the wave excitation in homogeneous media. Section 2.3 will study eigenfields of bounded media with the main focus on piecewise homogeneous waveguides. Summary will appear in Section 2.4.

2.1 Electromagnetic fields in bounded media

The theory of electromagnetic fields presented in Chapter 1 was developed for unbounded homogeneous media. The results obtained there are generally valid until the distance between the source (localized external currents) and the medium's boundary is much larger than the characteristic scale of electromagnetic energy confinement given by the bulk eigenwave propagation length, $L_{\mathrm{EMF}} = 1/(2k'')$. In other words, it assumes that the excited electromagnetic waves never reach boundaries of the medium. If this condition does not hold, we must account for the medium's boundaries and adjoining materials. To give a consistent description of electromagnetic waves in bounded media, we start from generalization of the theory of electromagnetic fields to the case of a generally inhomogeneous medium.

2.1.1 Material relations for inhomogeneous media

Recall that macroscopic Maxwell's equations (1.1) and (1.2) were derived after averaging all microscopic quantities such as fields, charges, and currents in space over the length a_{macro} larger than the lattice constant. The intent was to suppress all microscopic inhomogeneities appearing at scales below a_{macro}, including effects of the lattice in order to treat the medium's response as homogeneous. However, this step does not eliminate inhomogeneities that can appear at scales above a_{macro}. In a case of macroscopically inhomogeneous medium, consideration of the spatial dispersion becomes generally impossible [1, 2]. The material relation $\mathbf{D}(\mathbf{E})$ for inhomogeneous nonmagnetic media can be written within the local response approximation only, where the polarization current $\mathbf{j}_{\text{pol}}$ is given by the linear function $\mathbf{J}(\mathbf{E})$ defined in accordance with the causality principle,

$$J_i(t,\mathbf{r}) = \int_{-\infty}^{t} \mathrm{d}t' \sigma_{ij}(t-t',\mathbf{r}) E_j(t',\mathbf{r}). \tag{2.1}$$

By performing the Fourier transform of $\mathbf{J}(t,\mathbf{r})$ and $\mathbf{E}(t,\mathbf{r})$ to the $(\omega,\mathbf{r})$ space, we derive

$$J_i(\omega,\mathbf{r}) = \sigma_{ij}(\omega,\mathbf{r}) E_j(\omega,\mathbf{r}), \tag{2.2}$$

where

$$\sigma_{ij}(\omega,\mathbf{r}) = \int_{0}^{\infty} \mathrm{d}\tau \, \sigma_{ij}(\tau,\mathbf{r}) \mathrm{e}^{\mathrm{i}\omega\tau} \tag{2.3}$$

is the tensor of complex conductivity. Now, the material relation for inhomogeneous media can be written in the form of

$$D_i(\omega,\mathbf{r}) = \varepsilon_0 \varepsilon_{ij}(\omega,\mathbf{r}) E_j(\omega,\mathbf{r}), \tag{2.4}$$

with

$$\varepsilon_{ij}(\omega,\mathbf{r}) = 1 + \mathrm{i}\frac{\sigma_{ij}(\omega,\mathbf{r})}{\varepsilon_0 \omega} \tag{2.5}$$

being the tensor of complex dielectric permittivity.

Eventually, for magnetically inactive isotropic media, the material equations are given by

$$\mathbf{D}(\omega,\mathbf{r}) = \varepsilon_0 \varepsilon(\omega,\mathbf{r}) \mathbf{E}(\omega,\mathbf{r}), \tag{2.6}$$

$$\mathbf{B}(\omega,\mathbf{r}) = \mu_0 \mathbf{H}(\omega,\mathbf{r}), \tag{2.7}$$

where $\varepsilon(\omega,\mathbf{r})$ is the inhomogeneous scalar complex dielectric permittivity in the local response approximation. In this view, the homogeneous scalar permittivity $\varepsilon(\omega)$ studied in Chapter 1 is the limit of $\varepsilon(\omega,\mathbf{r})$ far away from any macroscopic inhomogeneities.

2.1.2 Electromagnetic fields in inhomogeneous media

Following the material relations obtained, a macroscopic inhomogeneity does not allow us to describe the medium response in the $(\omega,\mathbf{k})$ space, as in the case of uniform media. Instead, we should solve the Maxwell's equations in the $(\omega,\mathbf{r})$ space,

$$\nabla \times \mathbf{E}(\omega,\mathbf{r}) = \mathrm{i}\omega\mu_0 \mathbf{H}(\omega,\mathbf{r}), \tag{2.8}$$

$$\nabla \times \mathbf{H}(\omega,\mathbf{r}) = -\mathrm{i}\omega \mathbf{D}(\omega,\mathbf{r}) + \mathbf{J}(\omega,\mathbf{r}), \tag{2.9}$$

where by $\mathbf{J}(\omega,\mathbf{r})$ we assume the external current density $\mathbf{j}_{\text{ext}}(\omega,\mathbf{r})$. For the rest of this chapter, we omit the arguments $(\omega,\mathbf{r})$, assuming that all quantities are frequency- and space-dependent.

According to the Helmholtz theorem in vector analysis, all fields $\mathbf{F} = \{\mathbf{E}, \mathbf{H}, \mathbf{J}\}$ can be uniquely decomposed into longitudinal $\mathbf{F}_l$ and transverse $\mathbf{F}_t$ components,

$$\mathbf{F} = \mathbf{F}_l + \mathbf{F}_t, \tag{2.10}$$

where

$$\nabla \times \mathbf{F}_l = 0, \qquad \nabla \cdot \mathbf{F}_t = 0.$$

By using this decomposition, we can separate the Maxwell's equations for longitudinal

$$-\mathrm{i}\omega\varepsilon_0\varepsilon\mathbf{E}_l + \mathbf{J}_l = 0,$$
$$\mathrm{i}\omega\mu_0\mathbf{H}_l = 0,$$

and transverse fields

$$\nabla \times \mathbf{E}_t = \mathrm{i}\omega\mu_0\mathbf{H}_t,$$
$$\nabla \times \mathbf{H}_t = -\mathrm{i}\omega\varepsilon_0\varepsilon\mathbf{E}_t + \mathbf{J}_t.$$

The solution for excited longitudinal fields is then straightforward,

$$\mathbf{E}_l = -\mathrm{i}\frac{\omega\mu_0\mathbf{J}_l}{k_0^2\varepsilon}, \qquad \mathbf{H}_l = 0. \tag{2.11}$$

It is given by the local dependence of $\mathbf{E}_l$ on $\mathbf{J}_l$ and represents the exact solution universal for any profile of $\varepsilon = \varepsilon(\omega, \mathbf{r})$. For transverse fields, the situation is more complicated, as the existence of an exact analytical solution is defined by the spatial profile of $\varepsilon(\omega, \mathbf{r})$.

In general, the solution for transverse fields can be sought as a sum of two polarizations: transverse-electric (TE) and transverse-magnetic (TM). For the TE polarization, the Maxwell's equations are reduced to the wave-like equation for the electric field $\mathbf{E}_t$,

$$\nabla^2\mathbf{E}_t + k_0^2\varepsilon\mathbf{E}_t = -\mathrm{i}\omega\mu_0\mathbf{J}_t, \tag{2.12}$$

while the magnetic field is given by

$$\mathbf{H}_t = -\frac{\mathrm{i}}{\omega\mu_0}\nabla \times \mathbf{E}_t. \tag{2.13}$$

For the TM polarization, the Maxwell's equations are reduced to the differential equation for the magnetic field $\mathbf{H}_t$,

$$\nabla^2\mathbf{H}_t + \frac{\nabla\varepsilon}{\varepsilon} \times (\nabla \times \mathbf{H}_t) + k_0^2\varepsilon\mathbf{H}_t = -\nabla \times \mathbf{J}_t + \frac{\nabla\varepsilon}{\varepsilon} \times \mathbf{J}_t, \tag{2.14}$$

while the electric field can be written as

$$\mathbf{E}_t = \frac{\mathrm{i}}{\omega\varepsilon_0\varepsilon}(\nabla \times \mathbf{H}_t - \mathbf{J}_t). \tag{2.15}$$

In both cases, calculation of electromagnetic fields requires numerical computation, as analytical solutions of Eqs. (2.12) and (2.14) exist only for certain spatial profiles of $\varepsilon = \varepsilon(\mathbf{r})$ [1]. Despite that, we can make some general conclusions about peculiarities of the transverse field excitation directly from the equations obtained. Namely, Eq. (2.15) demonstrates that a TM-polarized electric field $\mathbf{E}_t$ may experience a resonant growth at the points $\mathbf{r} = \mathbf{r}_0$, where $\varepsilon \approx 0$. Indeed, at these points, the magnetic field exhibits $\nabla \times \mathbf{H}_t \approx \mathbf{J}_t$, following Eq. (2.14). If, in addition to that, $\mathbf{J}_t$ is zero at $\mathbf{r} = \mathbf{r}_0$, the transverse fields become quasi-static with $\nabla \times \mathbf{H}_t \approx 0$. Such degeneracy of the TM fields makes them similar to the longitudinal ones and causes coupling to the longitudinal eigenwaves at $\mathbf{r} = \mathbf{r}_0$ with a local enhancement of the TM-polarized electric field featuring $|\mathbf{E}_t| \gg |\mathbf{H}_t|$. The localized resonance of degenerated TM fields on longitudinal eigenwaves is a common feature of the field generation in inhomogeneous media that results in a range of qualitatively new effects extrinsic to uniform materials [1].

2.1.3 *Piecewise homogeneous media*

Now, let us consider the most frequent and fundamental macroscopic inhomogeneity — an interface between two media. We assume that every medium is homogeneous and isotropic alone. It means that far away from the interface, every medium can be described with bulk scalar permittivities $\varepsilon_{1,2}(\omega)$. However, close to the interface, ε is inhomogeneous. It smoothly changes from $\varepsilon_1(\omega)$ to $\varepsilon_2(\omega)$ within a finite transition layer, properties of which are defined by microscopic composition of the interface. The width of such transition is generally given by $a_{\text{trans}} = a_{\text{micro}} + a_{\text{macro}}$, where a_{micro} is the microscopic size of the inhomogeneity, which typically varies from one to tens of nanometers, depending on the type of interface [3].

Note that the minimal size of transition layer is given by a_{macro}, which is the smallest recognizable length in the macroscopic description. Thus, any microscopic interface is macroscopically continuous. Even an ideal sharp interface with an abrupt step-like microscopic profile and $a_{\text{micro}} = 0$ exhibits macroscopically continuous transition with $a_{\text{trans}} = a_{\text{micro}}$. Additional factors such as interface composition, roughness, and diffuseness only increase a_{trans} and make the transition layers thicker [3].

At the same time, the typical values of a_{trans} are often much below the wavelength, $a_{\text{trans}}k \ll 1$. It allows us to model such transitions as macroscopic discontinuities between piecewise uniform media. Following this criterion, more complex structures composed of several uniform domains can be considered piecewise homogeneous if the size of every domain is much larger than a_{trans}. In this case, the Maxwell's equations can be solved separately for every homogeneous domain with the uniform material equations,

$$\mathbf{D} = \varepsilon_0 \varepsilon_j(\omega) \mathbf{E}, \qquad \mathbf{B} = \mu_0 \mathbf{H},$$

where $\varepsilon_j(\omega)$ is the scalar permittivity of the j-th domain. Finally, the fields in adjoining domains can be seamed with the respective boundary conditions.

2.2 Boundary effects

2.2.1 *Boundary conditions*

The boundary conditions for piecewise homogeneous structures can be derived from the macroscopic Maxwell's equations directly. To obtain them, let us consider an arbitrary field $\mathbf{F}(t,\mathbf{r})$ and derive auxiliary equations for its jumps at a discontinuous interface. If this field describes a real physical quantity, it experiences only finite jumps that can be calculated with the following integrals:

$$\mathbf{n}\times(\mathbf{F}_2-\mathbf{F}_1)|_\Omega = \int_{\mathbf{n}\cdot\mathbf{r}_\Omega-0}^{\mathbf{n}\cdot\mathbf{r}_\Omega+0} \nabla\times\mathbf{F}\,\mathrm{d}(\mathbf{n}\cdot\mathbf{r}),$$

$$\mathbf{n}\cdot(\mathbf{F}_2-\mathbf{F}_1)|_\Omega = \int_{\mathbf{n}\cdot\mathbf{r}_\Omega-0}^{\mathbf{n}\cdot\mathbf{r}_\Omega+0} \nabla\cdot\mathbf{F}\,\mathrm{d}(\mathbf{n}\cdot\mathbf{r}),$$

where the integration is performed across the interface, with $\mathbf{n}$ being the normal vector to the interface Ω pointing from medium 1 to medium 2 at $\mathbf{r}=\mathbf{r}_\Omega$. Now, if we perform similar integration of the Maxwell's equations and divergence relations (1.8) through (1.11), we get

$$\mathbf{n}\times(\mathbf{H}_1-\mathbf{H}_2)|_\Omega = \mathbf{i}_{\text{ext}}, \tag{2.16}$$

$$\mathbf{n}\times(\mathbf{E}_1-\mathbf{E}_2)|_\Omega = 0, \tag{2.17}$$

and

$$\mathbf{n}\cdot(\mathbf{D}_1-\mathbf{D}_2)|_\Omega = \sigma_{\text{ext}}, \tag{2.18}$$

$$\mathbf{n}\cdot(\mathbf{H}_1-\mathbf{H}_2)|_\Omega = 0, \tag{2.19}$$

where σ_{ext} and $\mathbf{i}_{\text{ext}}$ are the surface densities of external charges and currents,

$$\sigma_{\text{ext}} = \int_{\mathbf{n}\cdot\mathbf{r}_\Omega-0}^{\mathbf{n}\cdot\mathbf{r}_\Omega+0} \rho_{\text{ext}}\,\mathrm{d}(\mathbf{n}\cdot\mathbf{r}), \qquad \mathbf{i}_{\text{ext}} = \int_{\mathbf{n}\cdot\mathbf{r}_\Omega-0}^{\mathbf{n}\cdot\mathbf{r}_\Omega+0} \mathbf{j}_{\text{ext}}\,\mathrm{d}(\mathbf{n}\cdot\mathbf{r}). \tag{2.20}$$

Thus, the fields' discontinuities are given by external surface charges and currents at the interface, if any. In the case of finding eigenwaves, the boundary conditions transform to the respective continuity equations with $\sigma_{\text{ext}}=0$ and $\mathbf{i}_{\text{ext}}=0$.

2.2.2 *Bulk eigenfields in real space*

Bulk eigenwaves and resonances on them were studied in Section 1.3 in the $(\omega,\mathbf{k})$ space. Now, we investigate their fields in the $(\omega,\mathbf{r})$ space in order to get deeper understanding of eigenfields and their role in the medium's response. For the sake of simplicity, we assume that the external current exciting electromagnetic fields is one-dimensional with the point-like spatial dependence,

$$\mathbf{j}_{\text{ext}} = \mathbf{i}_0\delta(x). \tag{2.21}$$

Below, we investigate two cases of $\mathbf{i}_0$ for which the external current is purely longitudinal and transverse.

Longitudinal eigenfields

First, we consider the response of an unbounded medium to a pure longitudinal current, $\nabla \times \mathbf{j}_{\mathrm{ext}} = 0$. It is realized when $\mathbf{i}_0 = i_0 \mathbf{e}_x$. In this case, we can write the excited fields by using Eqs. (2.11),

$$\mathbf{E}_l = -\mathrm{i}\frac{\omega \mu_0 i_0}{k_0^2 \varepsilon}\delta(x)\mathbf{e}_x, \qquad \mathbf{H}_l = 0. \tag{2.22}$$

This represents the localized excitation of the longitudinal fields by $\mathbf{j}_{\mathrm{ext}} = i_0 \delta(x)\mathbf{e}_x$, with zero fields excited everywhere except $x = 0$. Such a localized interaction is a consequence of the spatial dispersion neglect that forbids any energy transfer by longitudinal fields.

At the same time, if we solve the eigenvalue problem for longitudinal eigenwaves,

$$-\mathrm{i}\omega\varepsilon_0\varepsilon\mathbf{E}_l = 0, \qquad \mathrm{i}\omega\mu_0\mathbf{H}_l = 0, \tag{2.23}$$

we find that their fields are given by

$$\mathbf{E}_l = -\nabla\Phi(\mathbf{r}), \qquad \mathbf{H}_l = 0, \tag{2.24}$$

where $\Phi(\mathbf{r})$ is an arbitrary function smooth on the a_{macro} scale. These fields are generally defined in the whole space of $\mathbf{r}$, but only for certain complex frequencies $\omega = \omega' + \mathrm{i}\omega''$ satisfying the bulk dispersion relation $\varepsilon(\omega) = 0$. Since their ω'' are nonzero, the longitudinal bulk eigenwaves are *virtual* in the local response approximation. They exist in the complex ω space and hence can never appear as the excited fields for any $\mathbf{j}_{\mathrm{ext}}$. What we see in Eqs. (2.22) is just a localized resonance of the medium on the longitudinal eigenfields. In this view, the longitudinal eigenfields should be treated as mathematical tools for description and characterization of the polarization-induced resonances rather than the real electromagnetic fields.

Transverse eigenfields

The second case we consider is the field generation by a pure transverse external current $\nabla \cdot \mathbf{j}_{\mathrm{ext}} = 0$ given by $\mathbf{i}_0(z) = i_0 \mathbf{e}_y$. Calculation of the electromagnetic fields excited by such a current is simple too, if we take into consideration that $\mathbf{j}_{\mathrm{ext}}$ is actually a surface current with

$$\mathbf{i}_{\mathrm{ext}} = \int_{-0}^{+0} \mathbf{j}_{\mathrm{ext}}\,\mathrm{d}x = i_0 \mathbf{e}_y, \tag{2.25}$$

and hence can be easily accounted through boundary condition (2.16). In this case, we need to solve Eqs. (2.12) and (2.13) separately in the currentless domains $x < 0$ and $x > 0$, and then apply boundary conditions (2.16) and (2.17) at $x = 0$.

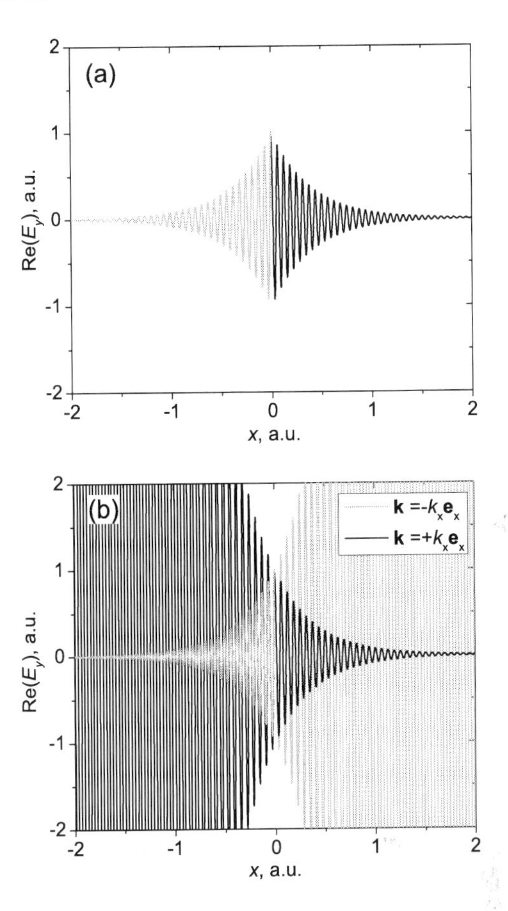

Figure 2.1: Spatial distribution of (a) the electric field of bulk polaritons excited by $\mathbf{j}_{\text{ext}} = i_0\delta(x)\mathbf{e}_y$ and (b) the non-truncated eigenfields with $\mathbf{k} = \pm k_x\mathbf{e}_x$.

Solutions of Eqs. (2.12) and (2.13) in the currentless domains are generally given by

$$\mathbf{E}_t = \mathbf{E}_+ \exp(+\mathrm{i}k_x x) + \mathbf{E}_- \exp(-\mathrm{i}k_x x), \tag{2.26}$$

$$\mathbf{H}_t = \frac{k_x}{\omega\mu_0}\mathbf{e}_x \times [\mathbf{E}_+ \exp(+\mathrm{i}k_x x) - \mathbf{E}_- \exp(-\mathrm{i}k_x x)], \tag{2.27}$$

where $k_x = (k_0^2\varepsilon)^{1/2}$. As k_x'' is always positive in real materials due to finite dissipation, the field finiteness at $x \to \pm\infty$ requires

$$\mathbf{E}_+(x<0) \equiv 0, \qquad \mathbf{E}_-(x>0) \equiv 0. \tag{2.28}$$

Eventually, after application of boundary conditions (2.16) and (2.17) at $x = 0$, we get the amplitudes of the excited waves

$$\mathbf{E}_{-}(x<0) = \mathbf{E}_{+}(x>0) = \frac{\omega\mu_0 \mathrm{i}_0}{2k_x}\mathbf{e}_y. \tag{2.29}$$

This solution describes two outgoing plane waves excited by $\mathbf{j}_{\mathrm{ext}} = i_0\delta(x)\mathbf{e}_y$, which propagate with the complex wavenumbers $\pm k_x$ away from the source. The excited waves experience the exponential decay given by k_x'' and penetrate into the medium over the length $L_{\mathrm{EMF}} = 1/(2k_x'')$, as shown in Fig. 2.1(a).

Now, let us consider the distribution of the transverse bulk eigenfields in the real space. For that, we should solve the eigenvalue problem, i.e., Eqs. (2.12) and (2.13) in a uniform medium with $\mathbf{j}_{\mathrm{ext}} = 0$. Formally, the eigenfields can be written as

$$\mathbf{E}_t = \mathbf{E}_0(\mathbf{k})\exp(\mathrm{i}\mathbf{k}\cdot\mathbf{r}), \tag{2.30}$$

$$\mathbf{H}_t = \frac{1}{\omega\mu_0}\mathbf{k}\times\mathbf{E}_0(\mathbf{k})\exp(\mathrm{i}\mathbf{k}\cdot\mathbf{r}), \tag{2.31}$$

where $k^2 = k_0^2\varepsilon$ and $\mathbf{E}_0(\mathbf{k})$ is an arbitrary eigenfield amplitude. If we compare these eigenfields with the fields excited by $\mathbf{j}_{\mathrm{ext}}$ [see Eqs. (2.26) and (2.28)], we find that the latter are composed of the eigenfields with $\mathbf{k} = \pm k_x\mathbf{e}_x$ only. In addition to that, this composition is piecewise; eigenfields with $\mathbf{k} = -k_x\mathbf{e}_x$ and $\mathbf{k} = +k_x\mathbf{e}_x$ appear as the excited fields in the limited space only, for $x<0$ and $x>0$, respectively (see Fig. 2.1). This is a common feature of any eigenfields; since they represent the eigensolutions of the current-free Maxwell's equations, they naturally appear in a *piecewise truncated* form, limited by the space of *current-free domains.*

The considered examples show the fundamental difference between longitudinal and transverse bulk eigenfields in the local response approximation. While the former are unphysical fields describing the inherent polarization-related resonances, the latter are real electromagnetic waves that can transfer energy through the medium's current-free domains.

2.2.3 Eigenfields of piecewise homogeneous media

Finding of eigenfields in bounded systems is generally more complicated than finding them in unbounded bulk materials. It requires eigenfields to satisfy not only the bulk conditions (the Maxwell's equations with zero external currents), but also the boundary conditions given by the medium's shape and adjacent materials. As a result, eigenfields of bounded media become dependent on the geometry and composition of the whole structure. Below, we consider common properties of the longitudinal and transverse eigenwaves of piecewise homogeneous structures.

Longitudinal eigenfields

In any uniform domain with $\varepsilon = \varepsilon_j$, longitudinal eigenfields are the solutions of Eqs. (2.23). Formally, they still can be written in the form of Eqs. (2.24),

$$\mathbf{E}_l = -\nabla\Phi_j(\mathbf{r}), \qquad \mathbf{H}_l = 0, \qquad \mathbf{r}\in V_j, \tag{2.32}$$

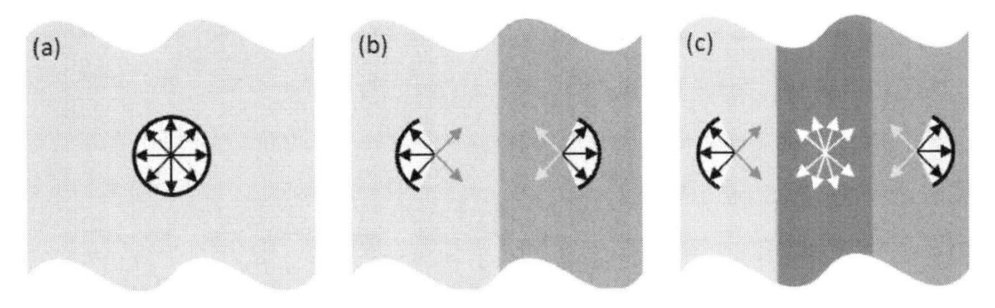

Figure 2.2: Eigenfields excited by $\mathbf{j}_{\text{ext}}$ localized (a) in the external medium far away from the boundary, (b) in the external medium close to the boundary, and (c) in the internal medium. Arrows depict the directions of eigenwave propagation.

where $\Phi_j(\mathbf{r})$ is a function smooth on the a_{macro} scale given by

$$\Phi_j(\mathbf{r}) = \begin{cases} \text{any}, & \varepsilon_j(\omega) = 0, \\ \text{const}, & \varepsilon_j(\omega) \neq 0. \end{cases} \tag{2.33}$$

Following the bulk conditions, the longitudinal eigenfields are nonzero only within the domains of the corresponding resonant material. At the same time, boundary conditions (2.16) through (2.19) require $\Phi_j(\mathbf{r})$ to satisfy $\nabla \times [\mathbf{n}\Phi_j(\mathbf{r})] = 0$ at the edges of all resonant domains. Similar to bulk materials, the longitudinal eigenfields of bounded media oscillate at complex frequencies $\omega = \omega' + \mathrm{i}\omega''$ with nonzero ω'', making their realization (excitation) impossible in principle. Therefore, they are virtual and should be considered only as the inherent characteristics of the structure's resonant response to longitudinal currents.

Transverse eigenfields

In a similar way, we can consider transverse eigenfields. For the j-th uniform domain, they can be written as

$$\mathbf{E}_t = \mathbf{E}_j(\mathbf{k}_j)\exp(\mathrm{i}\mathbf{k}_j \cdot \mathbf{r}), \qquad \mathbf{H}_t = \frac{1}{\omega\mu_0}\mathbf{k}_j \times \mathbf{E}_t, \qquad \mathbf{r} \in V_j, \tag{2.34}$$

where $k_j^2 = k_0^2\varepsilon_j$. Thus, the transverse eigenfields are generally piecewise. Application of the boundary conditions imposes restrictions on the amplitudes of eigenfields $\mathbf{E}_j(\mathbf{k})$ in every domain according to the boundary geometry and the adjoining material. We should remember that the boundary conditions are applicable only for those eigenfields that propagate toward the corresponding boundaries and physically can reach them. In view of that, the types of the excited waves strictly depend on the external current localization.

For instance, if the external currents are completely localized in an external domain and placed far away from the boundary, they excite the bulk eigenwaves propagating in all possible directions with negligible effect of the boundary [see Fig. 2.2(a)]. But if they are moved close to the interface, most of the bulk eigenwaves propagating toward it will experience reflection, while those propagating away from the interface will not be affected by the boundary at all [see Fig. 2.2(b)].

Physically, the fields reflected by an interface correspond to the outgoing bulk eigenwaves of the external medium. Thus, the total solution in the external medium is generally given by the coupled incoming and outgoing bulk eigenwaves. Such coupled fields cannot be treated as eigenwaves, as *the true eigensolution must be of pure incoming or outgoing type in the external medium.* In other words, among all incoming bulk eigenwaves originally excited in the external medium only those represent the true eigenfields of the structure, which feature *zero reflection* from the interface. In general, this condition is fulfilled only for very limited waves of the initially excited spectrum of the incoming bulk polaritons.

Thus, the external currents localized in external domains always excite a continuum of the outgoing bulk polaritons together with a discrete set of the incoming polaritons of the whole structure, as shown in Fig. 2.2(b). At the same time, the external currents located within an internal part can excite only discrete eigenwaves of the respective subsystems [see Fig. 2.2(c)].

In view of the above-mentioned difference in the excitation of the longitudinal and transverse eigenfields, in what follows we will discuss properties of the transverse eigenfields only with the focus on their use for resonant energy transfer.

2.3 Polaritons of bounded media

In general, a discrete set of transverse eigenwaves appears in any piecewise uniform system as a result of hybridization of the bulk polaritons supported by different domains. Their fields are given by Eqs. (2.34), while the boundary conditions define how they intercouple each other. As a result, the net eigenfields become *hybrid*. They are no longer bulk eigenwaves of either medium; they become equally dependent on properties of all materials in the system and describe it as a whole.

2.3.1 Hybridization of bulk polaritons

To demonstrate the hybridization of bulk polaritons, let us consider a flat interface between two semi-infinite media characterized with ε_1 and ε_2. This is a general case that permits us to understand the underlying processes at the basic level. Following Eqs. (2.34), each medium supports its own polaritons fulfilling the bulk conditions,

$$k_1^2 = k_0^2 \varepsilon_1, \qquad k_2^2 = k_0^2 \varepsilon_2,$$

which define k_1 and k_2 for a given ω, but not the directions of $\mathbf{k}_1$ and $\mathbf{k}_2$. These are the bulk polaritons propagating in any direction within the respective media.

In the case of the interface, the bulk polaritons of each medium are no longer independent. Their fields (2.34) are coupled at the interface by boundary conditions (2.16) through (2.19). These conditions impose the additional restrictions on the tangential and normal wavevectors,

$$\mathbf{n} \times (\mathbf{k}_1 - \mathbf{k}_2) = 0, \qquad \mathbf{n} \cdot \left(\frac{\mathbf{k}_1}{\varepsilon_1} - \frac{\mathbf{k}_2}{\varepsilon_2} \right) = 0. \tag{2.35}$$

Eventually, only polaritons with

$$(\mathbf{n}\times\mathbf{k}_{1,2})^2 = k_0^2\frac{\varepsilon_1\varepsilon_2}{\varepsilon_1+\varepsilon_2}, \qquad (\mathbf{n}\cdot\mathbf{k}_{1,2})^2 = k_0^2\frac{\varepsilon_{1,2}^2}{\varepsilon_1+\varepsilon_2}$$

are allowed to propagate in the system. These are the *interface polaritons* that satisfy all bulk and boundary conditions and thus represent the hybrid states of the transverse bulk eigenwaves of two media. They propagate along the interface in all possible directions with the same tangential wavenumber given by

$$k_{\parallel} = k_0\sqrt{\frac{\varepsilon_1\varepsilon_2}{\varepsilon_1+\varepsilon_2}}. \tag{2.36}$$

Thus, among the original continuum of the bulk polaritons, we get a smaller continuum of the interface polaritons featuring the same tangential wavenumber $k_{\parallel}$. This new continuum is of pure TM (transverse-magnetic) polarization, when the magnetic field lies out of the propagation plane. Another, TE (transverse-electric), polarization with the electric field lying out of the propagation plane is not supported by a single flat interface. It is in contrast to the bulk polaritons, which are of both TM and TE polarizations. However, it does not mean that TE polaritons are impossible in piecewise homogeneous structures at all. Such polaritons are generally supported in more complex systems consisting of at least two interfaces. We will discuss them and their properties later, in Section 2.3.3.

2.3.2 *Polaritons of interface*

Now, let us investigate the polaritons' eigenfields of a single interface in more details. The obtained solution suggests that they feature different normal wavenumbers in two media,

$$\mathbf{n}\cdot\mathbf{k}_{1,2} = \pm i\kappa_{1,2} \qquad \kappa_{1,2} = k_0\sqrt{-\varepsilon_{1,2}^2/(\varepsilon_1+\varepsilon_2)}, \tag{2.37}$$

where the signs $\pm$ in $\mathbf{k}_1$ and $\mathbf{k}_2$ are generally independent and characterize whether the corresponding fields are exponentially growing ($-$) or decaying ($+$) along the direction of the interface normal vector $\mathbf{n}$. Physically, it means that we deal with four groups of the interface polaritons exhibiting different combinations of the normal wavenumbers in the media 1 and 2. The first two groups feature the same field behavior in two media, $(\mathbf{n}\cdot\mathbf{k}_1'')(\mathbf{n}\cdot\mathbf{k}_2'') > 0$, while the other two groups represent the opposite field behaviors, $(\mathbf{n}\cdot\mathbf{k}_1'')(\mathbf{n}\cdot\mathbf{k}_2'') < 0$.

Transmission polaritons

We start our consideration from the polaritons featuring either growing or decaying fields in both media, $(\mathbf{n}\cdot\mathbf{k}_1'')(\mathbf{n}\cdot\mathbf{k}_2'') > 0$. These are the so-called *transmission polaritons* representing the hybrid states of the bulk polaritons of the media 1 and 2, which by falling on the interface do not experience any reflection from it. These polaritons propagate through the interface in a way similar to the bulk polaritons in an unbounded medium. The only difference is the refraction of the transmission polaritons after they pass through the interface.

Depending on the source localization, the transmission polaritons propagate either from the medium 1 to the medium 2 and exhibit exponential decay,

$$\mathbf{n}\cdot\mathbf{k}_1 = +\mathrm{i}\kappa_1, \qquad \mathbf{n}\cdot\mathbf{k}_2 = +\mathrm{i}\kappa_2,$$

or in the opposite direction, from the medium 2 to the medium 1, exhibiting exponential growth in the $\mathbf{n}$ direction,

$$\mathbf{n}\cdot\mathbf{k}_1 = -\mathrm{i}\kappa_1, \qquad \mathbf{n}\cdot\mathbf{k}_2 = -\mathrm{i}\kappa_2.$$

Among these solutions, only one is the true polariton with fully physical fields, while the other is a virtual eigenwave featuring fields violating the energy conservation law.[1] Following boundary conditions (2.35), the solution corresponding to the transmission polaritons realizes only if $\varepsilon_1'\varepsilon_2' > 0$, while the condition of their propagation along the interface, $k_\parallel' > k_\parallel''$, requires $\varepsilon_1' + \varepsilon_2' > 0$. Thus, their propagation band is given by

$$\varepsilon_1'(\omega) > 0, \qquad \varepsilon_2'(\omega) > 0. \tag{2.38}$$

Under these conditions, we have $|\mathbf{n}\cdot\mathbf{k}_{1,2}'| > |\mathbf{n}\cdot\mathbf{k}_{1,2}''|$, which gives us the propagating eigenfields along the normal direction for both media.

Properties of the transmission polaritons are in agreement with Fresnel's equations for plane wave reflection by a flat interface. In particular, Fresnel's equations confirm that only TM-polarized light can exhibit zero reflection [4, 5]. For nearly transparent materials, it takes place when the incident angle equals the Brewster's angle given by

$$\tan\theta_{B1} = \sqrt{\varepsilon_2/\varepsilon_1}, \qquad \tan\theta_{B2} = \sqrt{\varepsilon_1/\varepsilon_2},$$

if the incident TM wave falls on the interface from the medium 1 or the medium 2, respectively. Now, if we calculate the tangential wavenumbers corresponding to those angles,

$$k_{\parallel 1}^2 = k_0^2\varepsilon_1\sin^2\theta_{B1} = k_0^2\frac{\varepsilon_1\varepsilon_2}{\varepsilon_1+\varepsilon_2},$$
$$k_{\parallel 2}^2 = k_0^2\varepsilon_2\sin^2\theta_{B2} = k_0^2\frac{\varepsilon_1\varepsilon_2}{\varepsilon_1+\varepsilon_2},$$

we get the same values coinciding with the tangential wavenumber of the transmission polaritons $k_\parallel$.

To quantify the reflection in materials with significant absorption, we use Fresnel's equations to calculate the amplitude ratio of the TM reflected $|E_-|$ and incident $|E_+|$ plane waves at the interface as a function of their real tangential wavenumber $k_\parallel$,

$$\frac{|E_-|}{|E_+|} = \left|\frac{\mathbf{n}\cdot(\varepsilon_1\mathbf{k}_2 - \varepsilon_2\mathbf{k}_1)}{\mathbf{n}\cdot(\varepsilon_1\mathbf{k}_2 + \varepsilon_2\mathbf{k}_1)}\right|, \qquad \mathbf{n}\cdot\mathbf{k}_{1,2} = \sqrt{k_0^2\varepsilon_{1,2} - k_\parallel^2}. \tag{2.39}$$

[1] Virtual polaritons are common solutions of the eigenvalue problem, which can never appear as the fields excited by external currents, but define the resonant response of the system.

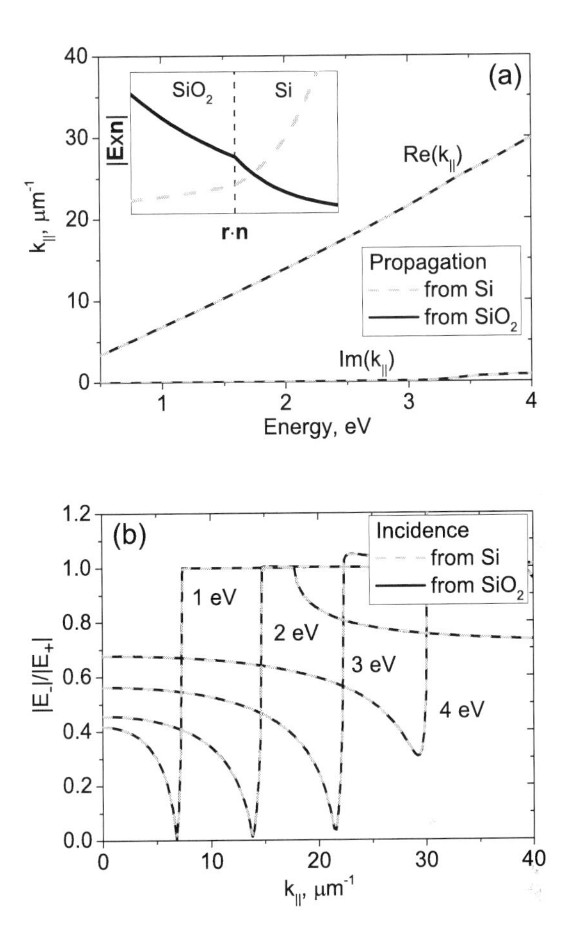

Figure 2.3: (a) Complex tangential wavenumber $k_{\|}$ of the transmission polaritons as a function of their energy $\hbar\omega$ for a planar Si/SiO_2 interface. The inset shows typical distributions of the electric field amplitude of the transmission polaritons across the interface. Note that the transmission polaritons propagating from Si are the true polaritons, while those propagating from SiO_2 are virtual. (b) Amplitude ratios $|E_-|/|E_+|$ of the reflected and incident plane waves at the Si/SiO_2 interface as functions of the real tangential wavenumbers for different energies.

Note that this ratio is invariant to the medium index change. Thus, it is an inherent characteristic of the interface and does not depend on the medium from which the plane wave falls on the interface. Figure 2.3 shows comparison of the complex tangential wavenumber $k_{\|}$ of the transmission polaritons supported by a planar Si/SiO_2 interface with the amplitude ratio of the TM waves reflected from the interface. It demonstrates that at tangential wavenumbers $k_{\|}$ close to those of the transmission polaritons, the amplitude ratio of the reflected and incident waves exhibits a local minimum.

Absorption and guided polaritons

Now let us consider two remaining groups of the interface polaritons with $(\mathbf{n}\cdot\mathbf{k}_1'')(\mathbf{n}\cdot\mathbf{k}_2'') < 0$. Following boundary conditions (2.35), these types of solutions require $\varepsilon_1'\varepsilon_2' < 0$, while the condition of their propagation along the interface $k_\parallel' > k_\parallel''$ imposes the restriction $\varepsilon_1' + \varepsilon_2' < 0$. Thus, the propagation band of these groups of polaritons can be defined as

$$0 < \varepsilon_1'(\omega) < -\varepsilon_2'(\omega) \quad \text{or} \quad 0 < \varepsilon_2'(\omega) < -\varepsilon_1'(\omega). \tag{2.40}$$

Under these conditions, the eigenfields are evanescent in both media with $|\mathbf{n}\cdot\mathbf{k}_{1,2}'| < |\mathbf{n}\cdot\mathbf{k}_{1,2}''|$.

The solution given by

$$\mathbf{n}\cdot\mathbf{k}_1 = +i\kappa_1, \qquad \mathbf{n}\cdot\mathbf{k}_2 = -i\kappa_2,$$

represents the eigenfields exponentially growing from the interface into both media. These are the so-called *absorption polaritons*, which are virtual eigenwaves featuring unphysical fields forbidden by the energy conservation law.

The fourth type of the interface polaritons, given by

$$\mathbf{n}\cdot\mathbf{k}_1 = -i\kappa_1, \qquad \mathbf{n}\cdot\mathbf{k}_2 = +i\kappa_2,$$

exhibits the fields exponentially decaying from the interface into two media. These are the *guided polaritons*[2] that possess evanescent fields enabling strong energy confinement across the interface and efficient transmission along it. Physically, this confinement is provided by high tangential wavenumbers of the guided polaritons, $k_\parallel'^2 > k_0^2\varepsilon_{1,2}'$, within their propagation band.

Similar to the transmission polaritons, the guided polaritons can be seen in the reflection spectrum of incident plane waves. Since their existence requires $\varepsilon_1'\varepsilon_2' < 0$, the guided polaritons can be observed at metal interfaces and traced for high tangential wavenumbers only, given by $k_\parallel^2 > k_0^2\varepsilon_{1,2}'$. In this case, both incident and reflected waves are evanescent. The incident wave is exponentially decaying with the surface amplitude $|E_+|$, while the reflected wave is exponentially growing with $|E_-|$ at the interface. Thus, coupling of the evanescent plane waves to the guided polaritons of a single interface is accompanied with a local maximum of $|E_-|/|E_+|$ given by Eq. (2.39). This effect is clearly seen in Fig. 2.4 that shows complex $k_\parallel$ of the polaritons guided by a planar Ag/SiO_2 interface and the amplitude ratio $|E_-|/|E_+|$ as a function of the real $k_\parallel$ of the incident waves.

[2]Guided polaritons of a single interface are often referred to as *surface polaritons* due to their field confinement at the interface. They are the subject of intensive research in the field of plasmonics, where they are commonly called surface plasmon-polaritons. Another popular term for them is surface plasmons [6], which is misleading, as it confuses two definitions of plasmons and polaritons (for details, see Section 1.3.3), and is not used in this book.

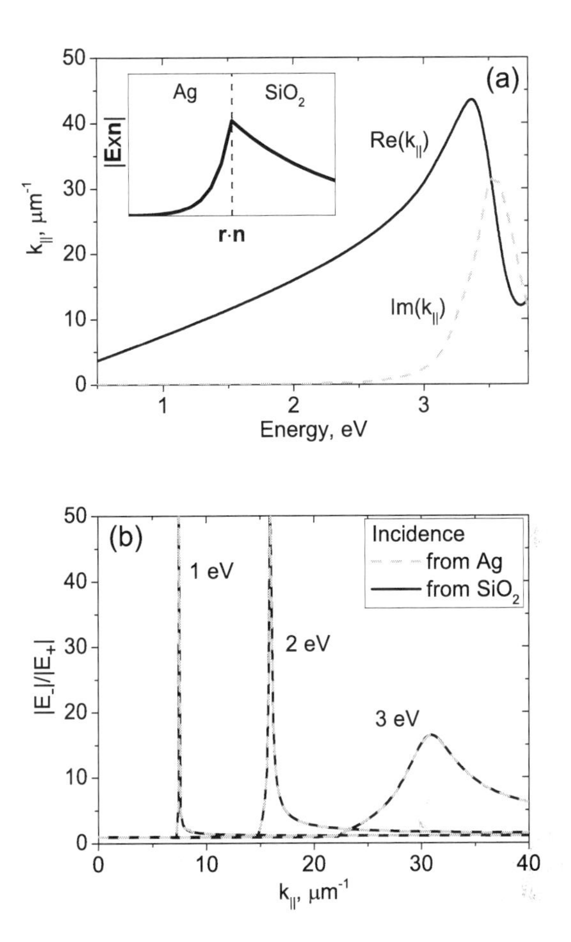

Figure 2.4: (a) Complex tangential wavenumber $k_{||}$ of the guided surface polaritons as a function of their energy $\hbar\omega$ for a planar Ag/SiO_2 interface. The inset shows a typical distribution of the electric field amplitude of the guided polaritons across the interface. (b) Amplitude ratios $|E_-|/|E_+|$ of the reflected and incident plane waves at the Ag/SiO_2 interface as functions of the real tangential wavenumbers for different energies.

2.3.3 *Polaritons of slab*

To study polaritons of more complex bounded systems, we consider a slab of uniform material embedded in a homogeneous medium. The slab is assumed to be of thickness L in one direction and infinitely large in the other directions. The dielectric permittivities of the slab and embedding medium are ε_i and ε_e, respectively. First, we write general expressions for the polaritons' eigenfields inside the slab and external medium and then study different types of the supported polaritons.

Planar eigenfields

As we have seen for polaritons of a single interface, their eigenfields do not possess a fixed tangential wavevector $\mathbf{k}_{\parallel} = \mathbf{n} \times \mathbf{k}_{1,2}$. They are generally allowed to propagate in any direction along the interface with a fixed wavenumber $k_{\parallel}$ being the function of the structure. This principle remains valid for any planar system. For the sake of simplicity, we direct the x axis perpendicularly to the slab's interfaces. Then, the wavevector of eigenfields in every homogeneous domain can be written as

$$\mathbf{k}_j = \mathbf{k}_{\parallel} + \mathbf{e}_x k_{xj}, \qquad \mathbf{k}_{\parallel} \cdot \mathbf{e}_x = 0,$$

where according to the polariton's bulk conditions,

$$k_{xj} = \pm \mathrm{i}\kappa_j, \qquad \kappa_j = \sqrt{k_{\parallel}^2 - k_0^2 \varepsilon_j}.$$

Thus, the general solution for planar eigenfields in the homogeneous domains can be written with the use of Eqs. (2.34),

$$\mathbf{E}_t = [\mathbf{E}_{j-} \exp(-\kappa_j x) + \mathbf{E}_{j+} \exp(+\kappa_j x)] \exp(\mathrm{i}\mathbf{k}_{\parallel} \cdot \mathbf{r}), \tag{2.41}$$

$$\mathbf{H}_t = \frac{1}{\omega \mu_0} \mathbf{k}_j \times [\mathbf{E}_{j-} \exp(-\kappa_j x) + \mathbf{E}_{j+} \exp(+\kappa_j x)] \exp(\mathrm{i}\mathbf{k}_{\parallel} \cdot \mathbf{r}). \tag{2.42}$$

For the considered structure,

$$j = \begin{cases} 1, & x \in (-\infty; -L/2), \\ 2, & x \in [-L/2; L/2], \\ 3, & x \in (L/2; \infty). \end{cases}$$

Following these notations,

$$\varepsilon_1 = \varepsilon_3 = \varepsilon_e, \qquad \varepsilon_2 = \varepsilon_i,$$
$$\kappa_1 = \kappa_3 = \kappa_e, \qquad \kappa_2 = \kappa_i.$$

The eigenfields above represent the general solution of Maxwell's equations with zero external currents in the planar geometry, where the vectors $\mathbf{E}_{j\pm}$ strictly depend on the type of the polaritons studied. Below, we consider the transmission, absorption, and guided polaritons of a slab waveguide.

Transmission polaritons

In the external domains, eigenfields of the transmission polaritons should be $\exp(-\kappa_{1,3} x)$ (if the polaritons propagate in the positive direction of $\mathbf{e}_x$) or $\exp(+\kappa_{1,3} x)$ (if they travel in the opposite direction). In the former case, we should require

$$\mathbf{E}_{1+} = \mathbf{E}_{3+} = 0,$$

while the latter is given by

$$\mathbf{E}_{1-} = \mathbf{E}_{3-} = 0.$$

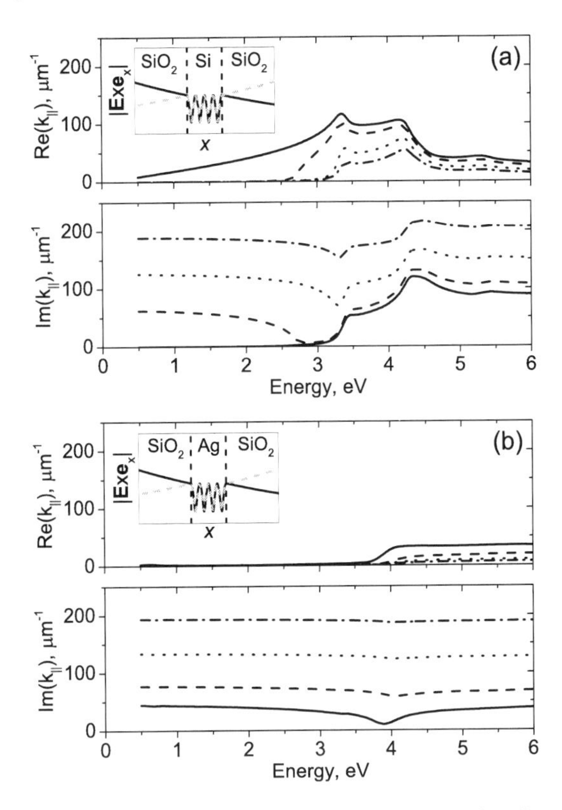

Figure 2.5: Complex tangential wavenumbers $k_{\parallel}$ of the Fabry-Perot transmission polaritons as a function of their energy $\hbar\omega$ for the (a) $SiO_2/Si/SiO_2$ and (b) $SiO_2/Ag/SiO_2$ waveguiding structures with film thickness $L = 50$ nm. The insets show typical distributions of the electric field amplitude of the Fabry-Perot transmission polaritons across the structure.

By applying boundary conditions (2.16) through (2.19) at $x = \pm L/2$ and solving them for the unknown coefficients $\mathbf{E}_{j\pm}$, we can obtain the dispersion relations of transmission polaritons for the TM and TE polarizations,

$$(\varepsilon_i^2 \kappa_e^2 - \varepsilon_e^2 \kappa_i^2)\sinh(\kappa_i L) = 0, \tag{2.43}$$

$$\sinh(\kappa_i L) = 0. \tag{2.44}$$

Following these dispersion relations, both the TM and TE polaritons exist under

$$\sinh(\kappa_i L) = 0.$$

This is the condition of the *Fabry-Perot resonances* for zero reflection of the incident plane waves falling on a film [4, 5]. Their tangential wavenumbers can be written in the following form

$$k_{\parallel} = \sqrt{k_0^2 \varepsilon_i - (n\pi/L)^2}, \qquad n = 1, 2, 3.... \tag{2.45}$$

Figure 2.5 shows the complex wavenumbers of the low-order Fabry-Perot transmission polaritons of the Si and Ag films embedded in SiO_2. They possess similar non-attenuating standing wave fields across the slabs, but with very different propagation characteristics: losses of these polaritons are generally higher in the Ag film and lower for the Si slab.

In addition to the Fabry-Perot transmission polaritons, the TM polarization features another dispersion relation given by

$$\varepsilon_i^2 \kappa_e^2 - \varepsilon_e^2 \kappa_i^2 = 0.$$

This relation defines the interface transmission polaritons with $k_{\parallel}$ given by Eq. (2.36) and studied in details in Section 2.3.2. The dispersion curves $k_{\parallel}(\omega)$ of the interface polaritons for the considered cases of Si and Ag films embedded in SiO_2 are shown in Figs. 2.3 and 2.4. In contrast to the Fabry-Perot transmission polaritons, these eigenwaves exhibit decaying (or growing) fields across the slabs.

For the considered cases of the Si and Ag slabs embedded in SiO_2, the transmission polaritons including the Fabry-Perot and interface polaritons are virtual.

Absorption polaritons

An additional solution supported by a slab is the *absorption polaritons*. In general, they require the external eigenfields exponentially growing from the slab, realizing when

$$\mathbf{E}_{1+} = \mathbf{E}_{3-} = 0.$$

By applying the boundary conditions for eigenfields at $x = \pm L/2$, we again get two dispersion relations for the TM and TE polarizations:

$$\frac{\varepsilon_i \kappa_e - \varepsilon_e \kappa_i}{\varepsilon_i \kappa_e + \varepsilon_e \kappa_i} = \pm \exp(-\kappa_i L), \tag{2.46}$$

$$\frac{\kappa_e - \kappa_i}{\kappa_e + \kappa_i} = \pm \exp(-\kappa_i L), \tag{2.47}$$

where the sign $\pm$ corresponds to the *even* and *odd* distributions of the tangential electric field across the slab. The dispersion curves of these polaritons for the Si and Ag films embedded in SiO_2 are shown in Fig. 2.6. In contrast to the Fabry-Perot transmission polaritons, the absorption eigenwaves exhibit attenuating standing wave field profiles inside the slabs. These are the so-called *volume absorption polaritons* that require $\varepsilon_i' > 0$. In addition, there exist *surface absorption polaritons* that occur at $\varepsilon_i' < 0$ and possess the internal fields highly confined at the slab interfaces.

In general, absorption polaritons may be either true or virtual. The true absorption polaritons can be excited with two coherent sources located in media 1 and 3. Depending on the symmetry of the polaritons to be excited, the sources should be adjusted to have odd or even field distribution. Under this matching, we can efficiently pump the sources' power into the system with no reflection produced.

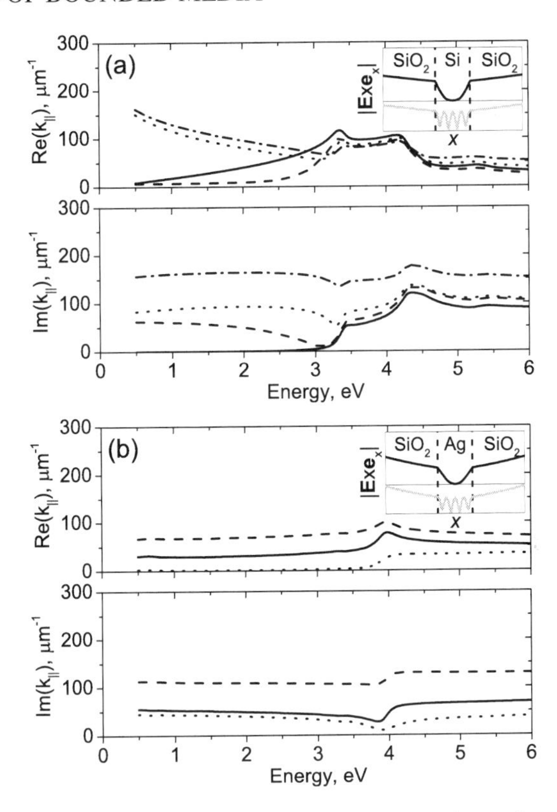

Figure 2.6: Complex tangential wavenumbers $k_{\parallel}$ of the absorption polaritons as a function of their energy $\hbar\omega$ for the (a) SiO_2/Si/SiO_2 and (b) SiO_2/Ag/SiO_2 waveguiding structures with film thickness $L = 50$ nm. The insets show typical distributions of the tangential electric field amplitude across the structure for the surface (black lines) and volume (gray lines) polaritons.

Guided polaritons

The *guided polaritons* of bounded systems always feature the fields decaying into the external domains. To find their dispersion relations for a slab, we should require

$$\mathbf{E}_{1-} = \mathbf{E}_{3+} = 0.$$

By applying the boundary conditions, we get the following dispersion relations of the TM and TE guided polaritons [7]

$$\frac{\varepsilon_i \kappa_e - \varepsilon_e \kappa_i}{\varepsilon_i \kappa_e + \varepsilon_e \kappa_i} = \pm \exp(\kappa_i L), \tag{2.48}$$

$$\frac{\kappa_e - \kappa_i}{\kappa_e + \kappa_i} = \pm \exp(\kappa_i L), \tag{2.49}$$

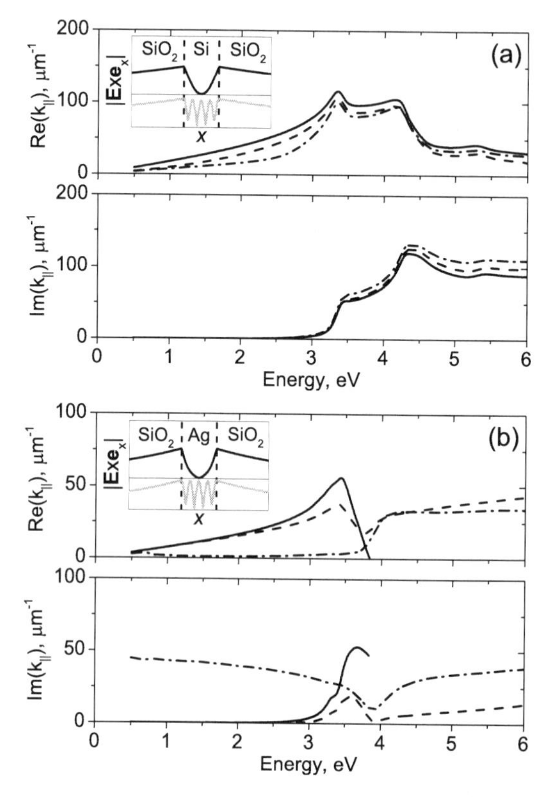

Figure 2.7: Complex tangential wavenumbers $k_{\parallel}$ of the guided polaritons as a function of their energy $\hbar\omega$ for the (a) SiO_2/Si/SiO_2 and (b) SiO_2/Ag/SiO_2 waveguiding structures with film thickness $L = 50$ nm. The insets show typical distributions of the tangential electric field amplitude across the structure for the surface (black lines) and volume (gray lines) polaritons.

where similarly to the absorption polaritons, the sign $\pm$ corresponds to the even and odd distributions of the guided eigenfields. Figure 2.7 shows the dispersion curves $k_{\parallel}(\omega)$ of the guided polaritons for the Si and Ag thin films embedded in SiO_2 cladding. Their fields inside the slabs look very similar to those of the absorption polaritons. They are either volume if $\varepsilon_i' > 0$ or surface when $\varepsilon_i' < 0$. The first type is commonly studied in photonics for nearly transparent high-index slabs with $\varepsilon_i' > \varepsilon_e'$. Within their propagation band, the volume guided polaritons feature the eigenfields confined inside the slab provided by the total internal reflection. The second type is of intensive study in plasmonics for metal slabs[3] with $\varepsilon_i' < -\varepsilon_e'$. Within their propagation bands, the surface guided polaritons exhibit the eigenfields confined at two interfaces. In general, the volume and surface guided polaritons are inherent to both photonic and plasmonic systems, but characterized with different propagation bands.

[3]In plasmonics, surface guided polaritons are traditionally called surface plasmon-polaritons.

2.4 Summary

In this chapter, we demonstrated that media boundaries strongly affect the excitation and propagation of electromagnetic waves. Nevertheless, the responses of bounded media remain resonant. Similar to unbounded media, they feature longitudinal and transverse eigenwaves that characterize resonant light-matter interactions. Among them, only polaritons (the transverse eigenwaves) can efficiently transmit energy through the system. In general, bounded systems support infinite number of polaritons. They are inherent to both photonic and plasmonic structures, but feature slightly different field and propagation characteristics. Regardless of the structural composition, all polaritons can be divided into transmission, absorption, and guided polaritons. The first group features the eigenfields traveling across the structure, the second one possesses the absorption fields, while the third group exhibits the confined eigenfields guided along the structure. All three groups can be used in various applications for resonant energy transfer through and within a system.

Bibliography

[1] V. L. Ginzburg, *The Propagation of Electromagnetic Waves in Plasmas*. Pergamon Press, 1970.

[2] A. F. Aleksandrov, L. S. Bogdankevich, and A. A. Rukhadze, *Principles of Plasma Electrodynamics*. Springer, 1984.

[3] V. M. Agranovich and D. L. Mills, eds., *Surface Polaritons: Electromagnetic Waves at Surfaces and Interfaces*. North Holland, 1982.

[4] J. D. Jackson, *Classical Electrodynamics*. Wiley, 2nd ed., 1975.

[5] L. D. Landau, E. M. Lifshitz, and L. P. Pitaevskii, *Electrodynamics of Continuous Media*. Pergamon Press, 2nd ed., 1984.

[6] H. Raether, *Surface Plasmons on Smooth and Rough Surfaces and on Gratings*. Springer, 1988.

[7] Y. A. Akimov, "Plasmonic properties of metal nanostructures," in *Plasmonic Nanoelectronics and Sensing* (E. P. Li and H. S. Chu, eds.), pp. 20–66, Cambridge University Press, 2014.

Chapter 3

Localized polaritons of single-particle systems

Yuriy Akimov
Institute of High Performance Computing, Singapore

In this chapter, we will study localized polaritons of basic nanostructures used in the areas of photonics and plasmonics. First, in Section 3.1, we will discuss the main differences of localized and propagating transverse eigenmodes. Then, we will consider the localized polaritons of single-particle nanostructures of the planar (Section 3.2), cylindrical (Section 3.3) and spherical (Section 3.4) geometry. Summary will appear in Section 3.5.

3.1 Localized polaritons

The polaritons considered in Chapter 2 were studied under the assumption of infinite lengths of the waveguiding structures. The study allowed us to describe transverse eigenfields in terms of the waves traveling with complex parallel wavenumbers $k_{\parallel}(\omega)$. In general, this approach works until the excited fields do not experience significant reflection from the ends of the waveguiding structure, i.e., when the external currents exciting the eigenwaves are located far away from the waveguide's ends.

If the distance between localization of the external currents and the waveguide's ends is comparable to the polaritons' propagation length $1/(2k''_{\parallel})$, we no longer can describe eigenfields in terms of $k_{\parallel}$ and must account for boundaries and respective diffraction of the fields at the waveguide's ends. Finally, it leads to quantization of the wavenumber $k_{\parallel}$ and formation of the standing eigenfields inside the waveguide. Thus, polaritons of three-dimensional objects represent a discrete set of three-dimensionally quantized eigenmodes of the whole system in addition to a continuum of unquantized outwardly propagating bulk eigenwaves of the external media.

3.1.1 Eigenfield quantization

To better understand polariton eigenfield quantization, let us consider a material bar with edges a_x, a_y, and a_z embedded in a uniform medium. In this case, the *geometry-*

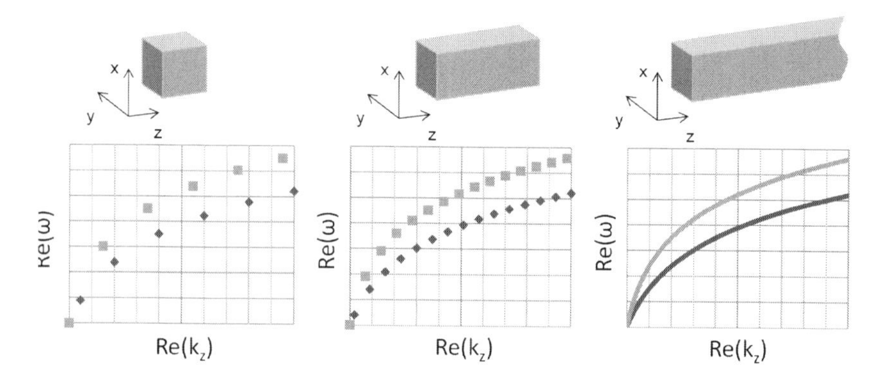

Figure 3.1: Transformation of localized eigenoscillations into propagating eigenwaves with increasing structure size.

induced quantization of the eigenfields' wavevector appears with the steps of

$$\Delta k_x \sim \pi/a_x, \quad \Delta k_y \sim \pi/a_y, \quad \Delta k_z \sim \pi/a_z,$$

while the frequencies of eigenfields are restricted to a discrete set $\omega = \omega_{nml}$ in accordance with the bulk and boundary conditions, where the three integer indices n, m, and l characterize the three-dimensional quantization of the wavevector $\mathbf{k}$. These are the transverse electromagnetic eigenoscillations representing fully quantized *localized polaritons*, also known as *cavity modes*.

Now, if we fix two dimensions, a_x and a_y, of the bar and start to gradually increase the third one, a_z, then Δk_z will decrease until it reaches zero for infinitely large a_z, as shown in Fig. 3.1. Mathematically, such an increase of a_z weakens the boundary conditions in the z direction. Finally, when the bar becomes infinitely long, the boundary conditions in the z direction lose their strength and the geometry-induced quantization of k_z disappears. It gives us the polaritons that are bound in two dimensions (along the x and y axes), but can freely move in the third one (along the z axis). Thus, we get the transverse electromagnetic eigenwaves called *propagating polaritons*.

The main difference between the propagating and localized polaritons is that the former can freely travel at least in one direction (by accepting arbitrary wavenumber), while the latter are bound in all three dimensions (given by discrete values for the wavevector). For the case considered, the eigenwaves propagate in the z direction with wavenumber k_z that can take any arbitrary value.[1] As a result, the eigenwave's frequency becomes a function of k_z, when $\omega = \omega_{nm}(k_z)$.

In a similar way, we can extend another dimension a_y of the bar and obtain an infinite slab of thickness a_x that supports two-dimensional polaritons propagating

[1]Applicability of the infinitely long waveguide model to real structures of finite length is restricted by the condition of weak diffraction of the excited polaritons at the structure's ends. It requires spatial attenuation of the eigenwaves during their propagation. Hence, the polariton wavenumber k_z can accept only complex values with $k_z'' \neq 0$.

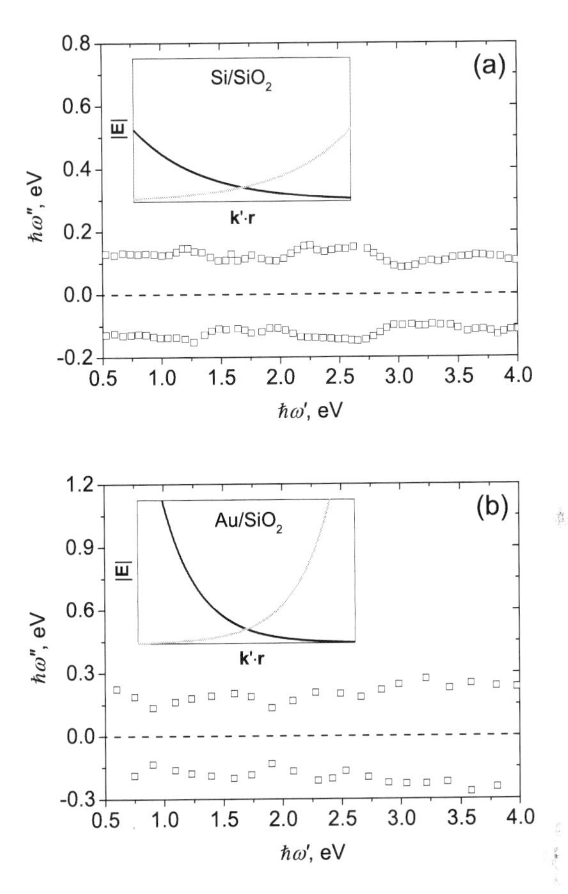

Figure 3.2: Elementary transmission polaritons of (a) Si/SiO_2 and (b) Au/SiO_2 bi-material structures. The insets show typical distributions of the electric field amplitude across the structure for polaritons propagating in the positive (black curve) and negative (gray curve) directions.

both in the y and z directions. Such one-dimensionally bound eigenwaves are characterized by two independent wavenumbers, k_y and k_z, and frequency $\omega = \omega_n(k_y, k_z)$, where the index n denotes quantization of the wavenumber k_x. These are the slab polaritons studied in Section 2.3.3. At the end, if we make the remaining bar's dimension a_x infinitely large, we reduce the one-dimensionally bound eigenwaves to the bulk polaritons studied in Section 1.3.2, with completely free wavevector $\mathbf{k}$ and the frequency $\omega = \omega(\mathbf{k})$ given by the bulk conditions only.

Note that the materiality of eigenfields requires their frequencies to be real; otherwise they cannot be realized in dispersive media by any means. It gives us an additional peculiarity of the eigenfield quantization for propagating and localized polaritons. In accordance with the wavevector quantization, there exist two groups of propagating polaritons designated $\omega''_{nm} = 0$ and $\omega''_{nm} \neq 0$. The first group represents *physical* solutions, which may appear as the fields excited by external currents, while the second group gives purely *virtual* eigenfields, which can never be realized. For localized polaritons, the situation is different, as the full wavevector quantization generally leads to $\omega''_{nml} \neq 0$. As a result, such polaritons are virtual, featuring unphysical electromagnetic fields.[2]

As eigenoscillations, virtual polaritons play an important role in resonant interactions of light and matter. They define the frequencies, at which we can observe resonant power transfer in a system. For example, in a structure composed of two materials, the simplest polaritons are given by $\varepsilon_1(\omega) = \varepsilon_2(\omega)$. These are the *elementary transmission polaritons* that define the condition of zero reflection or scattering in the structure regardless of its geometry. As their eigenfrequencies ω_{nml} are complex, they are virtual and cannot appear as the excited fields. However, at the resonant condition $\omega = \omega'_{nml}$, we can observe minimal reflection or scattering of the incident power flux. In this case, the excited fields are said to resonate on virtual polaritons leading to the enhanced power transfer.

Eigenfrequencies of the elementary transmission polaritons for bi-material Si/SiO_2 and Au/SiO_2 structures are shown in Fig. 3.2. In both cases, the elementary polaritons are virtual, featuring the eigenfield profiles similar to the bulk eigenwaves of unbounded media. However, in contrast to the bulk eigenwaves, the elementary transmission eigenoscillations are quantized. This is the so-called *material-induced wavevector quantization* based on the materials' permittivity match. As another fundamental mechanism, the material-induced quantization is common to any composite structure and generally occurs in addition to the geometry-induced quantization of the eigenfields considered in the beginning of this section.

3.1.2 *Wavefunctions of polariton eigenfields*

In Section 2.2.3, we have shown the general solution for transverse eigenfields in homogeneous domains. However, the practical implementation of that solution for piecewise homogeneous structures is troublesome due to the requirement to satisfy the boundary conditions at interfaces of different topologies. An alternative way is to solve Maxwell's equations in terms of the functions inherent to the structure geometry that simplify fulfilling of the boundary conditions. Such inherent functions are called *wavefunctions* and can be found by considering transverse eigenfields in a homogeneous domain.

Following the Helmholtz theorem, all transverse fields $\mathbf{F}_t = \{\mathbf{E}_t, \mathbf{H}_t\}$ can be written in terms of the two independent divergence-free vectors [1–3]:

$$\mathbf{F}_t = a\mathbf{M} + b\mathbf{N}, \qquad \nabla \cdot \mathbf{M} = \nabla \cdot \mathbf{N} = 0, \tag{3.1}$$

where a and b are considered coordinates of the field $\mathbf{F}_t$ in the basis of vectors $\mathbf{M}$

[2]The condition of complex frequency of fully quantized eigenoscillations is generally kept for all dispersive materials and structures. However, in rare singular cases that require a particular matching of the structure's material and geometrical parameters, the localized eigenmodes can exhibit $\omega''_{nml} = 0$. Such polaritons are physical and can appear as the fields excited by external currents.

and **N**. The latter can be arbitrary non-collinear vectors featuring zero divergence. It is convenient to compose them with an arbitrary constant vector **c** and a space-dependent scalar function $\Psi(\mathbf{r})$,

$$\mathbf{M} = \nabla \times \mathbf{c}\Psi(\mathbf{r}), \qquad \mathbf{N} = \frac{1}{k_0\sqrt{\varepsilon}}\nabla \times \nabla \times \mathbf{c}\Psi(\mathbf{r}). \tag{3.2}$$

The basis vectors defined in this way are divergence-free and orthogonal. Now, if we require such $\mathbf{F}_t$ to satisfy Maxwell's equations, $\Psi(\mathbf{r})$ must fulfill the Helmholtz equation,

$$\nabla^2\Psi(\mathbf{r}) + k_0^2\varepsilon\Psi(\mathbf{r}) = 0, \tag{3.3}$$

in the whole domain.

In other words, if we can get a set of solutions $\Psi_n(\mathbf{r})$ that fulfill Eq. (3.3) for a fixed vector **c**, we can generate a number of wavefunctions $\mathbf{M}_n$ and $\mathbf{N}_n$ that satisfy the homogeneous Maxwell's equation. Eventually, the transverse electromagnetic fields can be written in the form [1–3],

$$\mathbf{E}_t = \sum_n (a_n\mathbf{M}_n + b_n\mathbf{N}_n), \tag{3.4}$$

$$\mathbf{H}_t = -\mathrm{i}\sqrt{\frac{\varepsilon_0\varepsilon}{\mu_0}}\sum_n (b_n\mathbf{M}_n + a_n\mathbf{N}_n), \tag{3.5}$$

where the summation is performed over all the wavefunctions. In this way, we can get the distribution of transverse eigenfields in every homogeneous domain for subsequent application of the boundary conditions.

3.2 Localized polaritons of planar systems

We will start our consideration of localized polaritons from the transverse eigenmodes oscillating across a slab of uniform material embedded in a homogeneous medium that represent the simplest one-dimensional case of eigenoscillations. The slab is assumed to be of thickness L in the x direction and infinitely large in the other directions. Dielectric permittivities of the slab and embedding medium are ε_i and ε_e, respectively. First, we will write general expressions for the polaritons' eigenfields in terms of the planar wavefunctions and then study one-dimensional eigenoscillations supported in photonic and plasmonic structures.

3.2.1 Planar eigenfields

The planar wavefunctions featuring eigenfield propagation in the xz plane can be generated with

$$\mathbf{c} = \mathbf{e}_y, \qquad \Psi(\mathbf{r}) = X(x). \tag{3.6}$$

Following Eq. (3.3), the function $X(x)$ should satisfy the one-dimensional Helmholtz equation

$$\frac{\mathrm{d}^2X}{\mathrm{d}x^2} + k_0^2\varepsilon X = 0. \tag{3.7}$$

Solution of this equation is generally given by two exponential functions:

$$X_{\pm}(x) = \exp(\pm \mathrm{i}kx), \qquad k = k_0\sqrt{\varepsilon}, \tag{3.8}$$

which represent two elementary plane waves propagating in the positive and negative directions of the axis x. By using this generating function, the planar wavefunctions can be written with Eqs. (3.2) as

$$\mathbf{M}_{\pm} = \pm \mathrm{i}\mathbf{e}_z kX_{\pm}(x), \tag{3.9}$$

$$\mathbf{N}_{\pm} = k\mathbf{e}_y X_{\pm}(x). \tag{3.10}$$

Now, the derived wavefunctions can be used for eigenfield calculation.

For eigenfields in the external domains, we can write

$$\mathbf{E}_t = \sum_{\pm}(a_{\pm}\mathbf{M}^e_{\pm} + b_{\pm}\mathbf{N}^e_{\pm}), \tag{3.11}$$

$$\mathbf{H}_t = -\mathrm{i}\sqrt{\frac{\varepsilon_0\varepsilon_e}{\mu_0}}\sum_{\pm}(b_{\pm}\mathbf{M}^e_{\pm} + a_{\pm}\mathbf{N}^e_{\pm}) \tag{3.12}$$

for $x \in (-\infty; -L/2]$ and

$$\mathbf{E}_t = \sum_{\pm}(c_{\pm}\mathbf{M}^e_{\pm} + d_{\pm}\mathbf{N}^e_{\pm}), \tag{3.13}$$

$$\mathbf{H}_t = -\mathrm{i}\sqrt{\frac{\varepsilon_0\varepsilon_e}{\mu_0}}\sum_{\pm}(d_{\pm}\mathbf{M}^e_{\pm} + c_{\pm}\mathbf{N}^e_{\pm}) \tag{3.14}$$

for $x \in [L/2; \infty)$. The summation in the eigenfields is performed over the $\pm$ index, while the superscript e denotes the wavefunctions of the external domain with $\varepsilon = \varepsilon_e$. Following this notation, the fields inside the slab, $x \in [-L/2; L/2]$, can be written as

$$\mathbf{E}_t = \sum_{\pm}(e_{\pm}\mathbf{M}^i_{\pm} + f_{\pm}\mathbf{N}^i_{\pm}), \tag{3.15}$$

$$\mathbf{H}_t = -\mathrm{i}\sqrt{\frac{\varepsilon_0\varepsilon_i}{\mu_0}}\sum_{\pm}(f_{\pm}\mathbf{M}^i_{\pm} + e_{\pm}\mathbf{N}^i_{\pm}), \tag{3.16}$$

where the superscript i denotes the wavefunctions of the internal domain with $\varepsilon = \varepsilon_i$.

The eigenfields written above represent a general solution of Maxwell's equations with zero external currents in the planar geometry, where the decomposition coefficients $a_{\pm}$, $b_{\pm}$, $c_{\pm}$, $d_{\pm}$, $e_{\pm}$, $f_{\pm}$ strictly depend on the type of the polaritons studied. We will separately consider transmission, radiation, and absorption localized polaritons of metal and dielectric planar slabs below.

3.2.2 *Planar eigenoscillations*

Transmission polaritons

For transmission polaritons, the eigenfields in external domains should comprise either $\mathbf{M}^e_{+}$, $\mathbf{N}^e_{+}$ (if the polaritons propagate in the positive direction of $\mathbf{e}_x$) or $\mathbf{M}^e_{-}$, $\mathbf{N}^e_{-}$

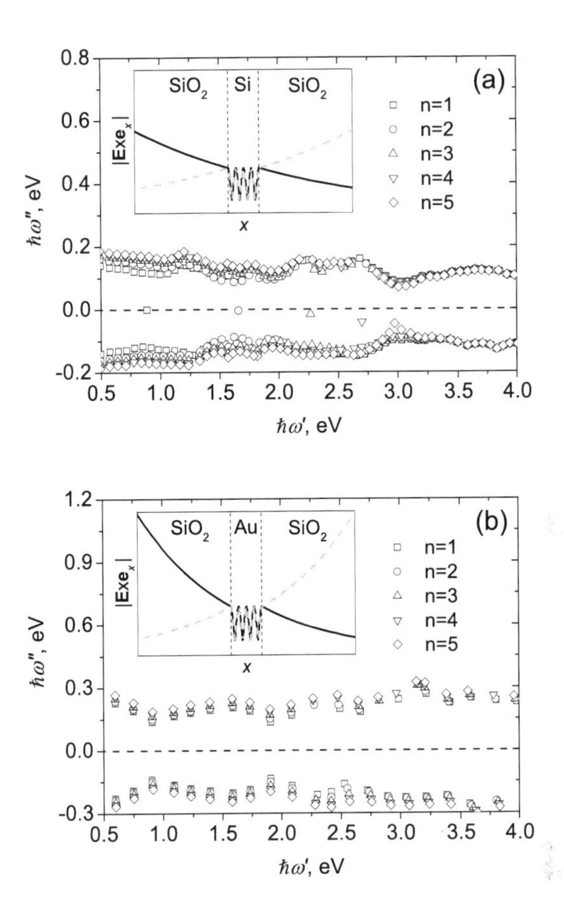

Figure 3.3: Localized Fabry-Perot transmission polaritons of (a) $SiO_2/Si/SiO_2$ and (b) $SiO_2/Au/SiO_2$ planar structures with slab thickness of $L = 200$ nm. The insets show typical distributions of the tangential electric field amplitude across the structure for the polaritons propagating in the positive (black curve) and negative (gray curve) directions of the x axis.

(if they travel in the opposite direction). However, due to the symmetry of the considered structure in the x direction, both types of transmission polaritons are given by the same dispersion relation. We assume below that the transmission polaritons propagate in the positive direction of $\mathbf{e}_x$. In this case, we should require

$$a_- = b_- = c_- = d_- = 0.$$

By applying boundary conditions (2.16) through (2.19) at $x = \pm L/2$ and solving them for the unknown decomposition coefficients, we obtain the dispersion relation of transmission polaritons,

$$(\varepsilon_i - \varepsilon_e)\sin(k_i L) = 0. \tag{3.17}$$

The obtained dispersion relation gives us the *elementary transmission polaritons* with

$$\varepsilon_i = \varepsilon_e,$$

as well as the *localized Fabry-Perot polaritons* defined by the slab properties only,

$$k_i L = n\pi,$$

where n is a positive integer.

Figure 3.3 shows eigenfrequencies of the localized Fabry-Perot transmission polaritons for single Si and Au planar slabs embedded in SiO_2. In both cases, the polaritons are virtual, featuring similar non-attenuating standing wave field patterns inside the slabs with n denoting the number of half-wavelengths packed. However, properties of these eigenoscillations are slightly different. First of all, the density of localized eigenmodes is higher for the Si slab due to the denser material-induced quantization of the eigenfields in the Si/SiO_2 bi-material system, as shown in Fig. 3.2. Another difference is given by ω_n''. For the Au film, $|\omega_n''|$ are generally higher compared to those in the Si slab where they can become negligibly small, especially in the transparency regime of silicon [see Fig. 3.3(a)]. This is a common feature of all nearly transparent materials with $\varepsilon' \gg \varepsilon''$, for which ω_n'' can approach zero, causing the pronounced Farby-Perot resonances with nearly zero reflection for plane waves incident on such films [3, 4].

Radiation polaritons

For radiation polaritons, the external eigenfields propagate outward from the slab. In terms of the planar wavefunctions, they are given by $\mathbf{M}_-^e$, $\mathbf{N}_-^e$ at $x \in (-\infty; -L/2]$ and $\mathbf{M}_+^e$, $\mathbf{N}_+^e$ at $x \in [L/2; \infty)$. Hence, they feature

$$a_+ = b_+ = c_- = d_- = 0.$$

Application of the boundary conditions for such fields results in the following dispersion relation:

$$\frac{k_e - k_i}{k_e + k_i} = \pm\exp(-\mathrm{i}k_i L), \tag{3.18}$$

where the sign $\pm$ corresponds to the odd $(+)$ and even $(-)$ distribution of the electric field across the structure.

Figure 3.4 shows eigenfrequencies of the radiation polaritons for single Si and Au films embedded in SiO_2 cladding. Eigenfields of these polaritons feature similar odd and even distributions, following the geometry-induced wavevector quantization. In general, radiation polaritons are purely virtual, possessing $\omega_n'' \neq 0$. Regardless of the structure's material and geometric parameters, radiation polaritons never reach $\omega_n'' = 0$, as it violates the energy conservation law.

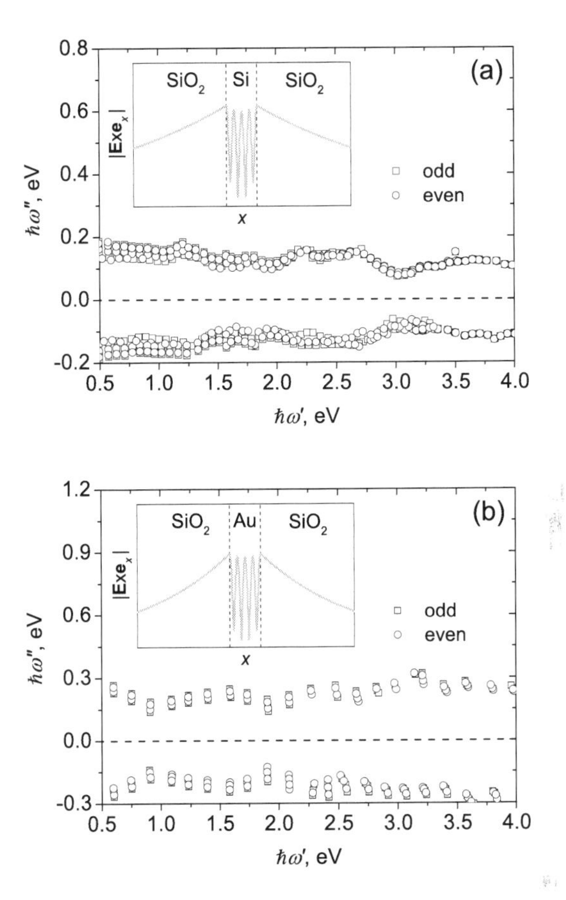

Figure 3.4: Localized radiation polaritons of (a) $SiO_2/Si/SiO_2$ and (b) $SiO_2/Au/SiO_2$ planar structures with slab thickness of $L = 200$ nm. The insets show typical distributions of the polaritons' tangential electric field amplitude across the structure.

Absorption polaritons

To find the dispersion relation of the last group of localized excitations that feature incoming external fields, we should take into account that their eigenfields are given by $\mathbf{M}_+^e$, $\mathbf{N}_+^e$ at $x \in (-\infty; -L/2]$ and $\mathbf{M}_-^e$, $\mathbf{N}_-^e$ at $x \in [L/2; \infty)$. In other words, we use

$$a_- = b_- = c_+ = d_+ = 0.$$

After application of boundary conditions (2.16) through (2.19) at $x = \pm L/2$, we get the following dispersion relation of the localized absorption polaritons:

$$\frac{k_e - k_i}{k_e + k_i} = \pm \exp(\mathrm{i}k_i L), \tag{3.19}$$

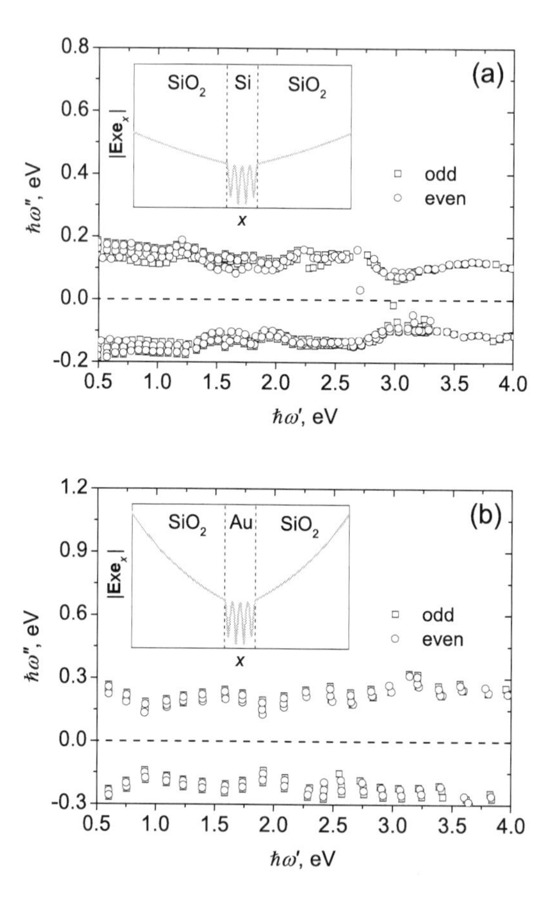

Figure 3.5: Localized absorption polaritons of (a) SiO_2/Si/SiO_2 and (b) SiO_2/Au/SiO_2 planar structures with slab thickness of $L = 200$ nm. The insets show typical distributions of the polaritons' tangential electric field amplitude across the structure.

where the sign $\pm$ denotes the odd (+) and even ($-$) distributions of the electric eigenfield across the structure.

Figure 3.5 shows eigenfrequencies of the localized absorption polaritons for single Si and Au films embedded in SiO_2. Eigenfields of these polaritons feature similar attenuating standing wave patterns of odd and even distributions inside the slabs. However, the Si film exhibits a higher density of eigenmodes caused by denser material-induced quantization [see Fig. 3.2(a)].

In general, absorption eigenoscillations are virtual ($\omega_n'' \neq 0$) for all structures. However, at certain thicknesses (dependent on the material properties), absorption polaritons can become true with $\omega_n'' = 0$. At these conditions, the coherent radiations incident from two sides of the structure can be completely absorbed in the system with no reflection from the slab if the amplitudes and phases of two sources are matched properly [5].

3.3 Localized polaritons of cylindrical systems

In this section, we will consider a more complicated case of two-dimensional polaritons oscillating across an infinite cylinder of the radius R and permittivity ε_i embedded in a homogeneous medium with $\varepsilon = \varepsilon_e$. First, we will introduce the cylindrical wavefunctions for polariton eigenfields, and then study two-dimensional eigenoscillations of photonic and plasmonic structures.

3.3.1 Cylindrical eigenfields

To study the localized eigenmodes oscillating across an infinite cylinder, we will introduce the cylindrical wavefunctions [1, 3]. They can be generated with the use of

$$\mathbf{c} = \mathbf{e}_z, \qquad \Psi(\mathbf{r}) = \Psi(r,\phi), \tag{3.20}$$

in the cylindrical system (r,ϕ,z). In general, the generating function $\Psi(r,\phi)$ can be written in the following form:

$$\Psi(r,\phi) = Z_m(r)\exp(im\phi), \tag{3.21}$$

with m denoting the integer azimuthal number. Then, following Eq. (3.3), the radial function $Z_m(r)$ should obey the differential equation

$$r^2\frac{\mathrm{d}^2 Z_m}{\mathrm{d}r^2} + r\frac{\mathrm{d}Z_m}{\mathrm{d}r} + (r^2k_0^2\varepsilon - m^2)Z_m = 0. \tag{3.22}$$

This is the cylindrical Bessel equation that has two linearly independent solutions that can be written in terms of the Hankel functions of the first and second kinds:

$$Z_{+,m}(r) = H_m^{(1)}(kr), \quad Z_{-,m}(r) = H_m^{(2)}(kr), \quad k = k_0\sqrt{\varepsilon}. \tag{3.23}$$

The chosen functions $Z_{\pm,m}(r)$ represent the elementary outgoing ($Z_{+,m}$) and incoming ($Z_{-,m}$) cylindrical waves [2, 6].

Using the cylindrical generating functions, we can compose the cylindrical wavefunctions fulfilling Maxwell's equations,

$$\mathbf{M}_{\pm,m} = \left[\mathrm{i}\frac{m}{r}Z_{\pm,m}(r)\mathbf{e}_r - Z'_{\pm,m}(r)\mathbf{e}_\phi\right]\exp(im\phi), \tag{3.24}$$

$$\mathbf{N}_{\pm,m} = k\mathbf{e}_z Z_{\pm,m}(r)\exp(im\phi). \tag{3.25}$$

Then, the eigenfields in the external medium can be generally written as

$$\mathbf{E}_t = \sum_{m=-\infty}^{\infty}\sum_{\pm}(a_{\pm,m}\mathbf{M}^e_{\pm,m} + b_{\pm,m}\mathbf{N}^e_{\pm,m}), \tag{3.26}$$

$$\mathbf{H}_t = -\mathrm{i}\sqrt{\frac{\varepsilon_0\varepsilon_e}{\mu_0}}\sum_{m=-\infty}^{\infty}\sum_{\pm}(b_{\pm,m}\mathbf{M}^e_{\pm,m} + a_{\pm,m}\mathbf{N}^e_{\pm,m}). \tag{3.27}$$

The summation in the eigenfields is performed over the $\pm$ and m indices, while the superscript e is to denote the wavefunctions of the external domain with $\varepsilon = \varepsilon_e$. In a similar way, we can write the fields inside the cylinder,

$$\mathbf{E}_t = \sum_{m=-\infty}^{\infty} \sum_{\pm} (c_{\pm,m}\mathbf{M}^i_{\pm,m} + d_{\pm,m}\mathbf{N}^i_{\pm,m}), \tag{3.28}$$

$$\mathbf{H}_t = -\mathrm{i}\sqrt{\frac{\varepsilon_0\varepsilon_i}{\mu_0}} \sum_{m=-\infty}^{\infty} \sum_{\pm} (d_{\pm,m}\mathbf{M}^i_{\pm,m} + c_{\pm,m}\mathbf{N}^i_{\pm,m}), \tag{3.29}$$

where the superscript i denotes the wavefunctions of the internal domain with $\varepsilon = \varepsilon_i$. Since all the internal wavefunctions diverge at $r = 0$, the coefficients $c_{\pm,m}$ and $d_{\pm,m}$ should be chosen as [2, 6]

$$\frac{c_{+,m}}{c_{-,m}} = \frac{d_{+,m}}{d_{-,m}} = (-1)^{m+1} \tag{3.30}$$

to ensure finiteness of the total fields at $r = 0$.

The obtained fields represent a general solution of Maxwell's equations in the cylindrical system. For localized polaritons, we should consider only partial solutions corresponding to either incoming or outgoing cylindrical waves in the external domains. Below, we will consider them for photonic and plasmonic structures.

3.3.2 Cylindrical eigenoscillations

Radiation polaritons

In contrast to planar structures, systems with cylindrical symmetry do not support transmission polaritons. They exhibit radiation and absorption eigenmodes only. For the radiation polaritons, the external eigenfields propagate outward from the cylinder, given by the $\mathbf{M}^e_{+,m}$ and $\mathbf{N}^e_{+,m}$ cylindrical wavefunctions. Therefore, they can be obtained from the general solution, if we use

$$a_{-,m} = b_{-,m} = 0.$$

After application of the boundary conditions at $r = R$, we can get two independent dispersion relations for the TE- and TM-polarized eigenoscillations possessing nonzero E_z and H_z eigenfields, respectively,

$$q_e J_m(q_i) H_m^{(1)\prime}(q_e) - q_i J_m'(q_i) H_m^{(1)}(q_e) = 0, \tag{3.31}$$

$$q_i J_m(q_i) H_m^{(1)\prime}(q_e) - q_e J_m'(q_i) H_m^{(1)}(q_e) = 0, \tag{3.32}$$

where J_m are the Bessel functions of the first kind with $q_i = k_i R$ and $q_e = k_e R$.

Note that both the TE and TM radiation polaritons' eigenfrequencies depend on the azimuthal number m. It defines the wavevector quantization along the azimuthal direction $\mathbf{e}_\phi$. Another quantization appears along the radial direction $\mathbf{e}_r$ given by the oscillating behavior of the cylindrical functions J_m and $H_m^{(1)}$. This is the so-called

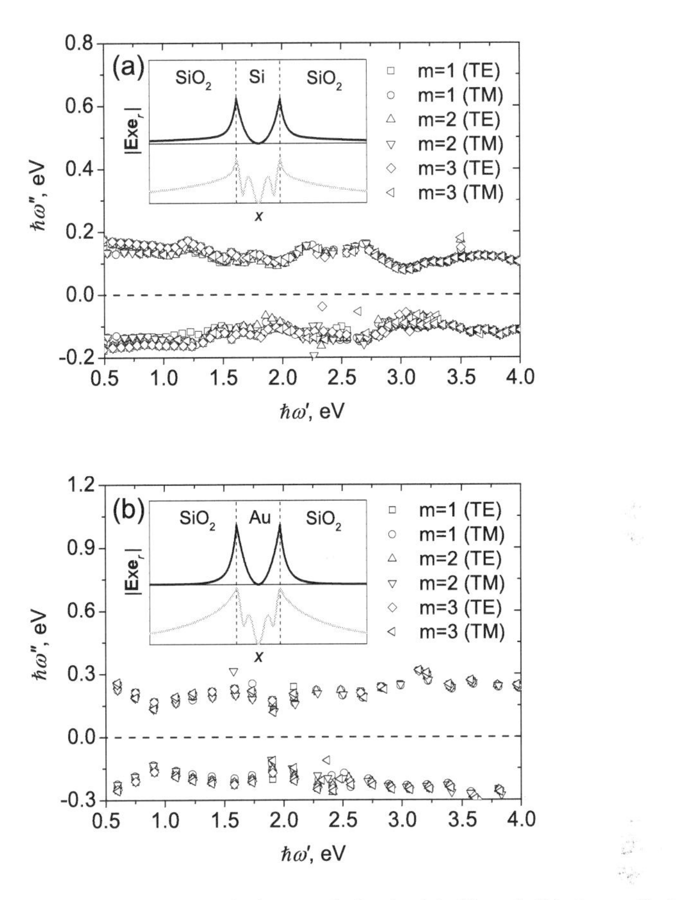

Figure 3.6: Localized radiation polaritons of single (a) Si and (b) Au cylinders of radius $R = 100$ nm embedded in SiO_2 cladding. The insets show typical distributions of the tangential electric field amplitude across the structure for the surface (black lines) and volume (gray lines) polaritons.

two-dimensional geometry-induced quantization. In addition, all eigenfields experience the material-induced quantization discussed in Section 3.1.1.

Figure 3.6 shows eigenfrequencies of the localized radiation polaritons for single Si and Au cylinders embedded in SiO_2 cladding. Eigenmodes of these structures exhibit similar fields oscillating at complex frequencies ω_{nm}, where n is the index accounting for the radial quantization of the wavevector. Similar to planar radial polaritons, the Si cylinder features higher density of the radiation eigenmodes packing due to the denser material-induced quantization of the eigenfields [see Fig. 3.2(a)].

In general, two groups of cylindrical eigenoscillations exist featuring different fields. The first group known as *volume polaritons* exhibits the attenuating standing wave fields inside the cylinder given by $\varepsilon_i'(\omega_{nm}) > 0$. They can be of both the TE and

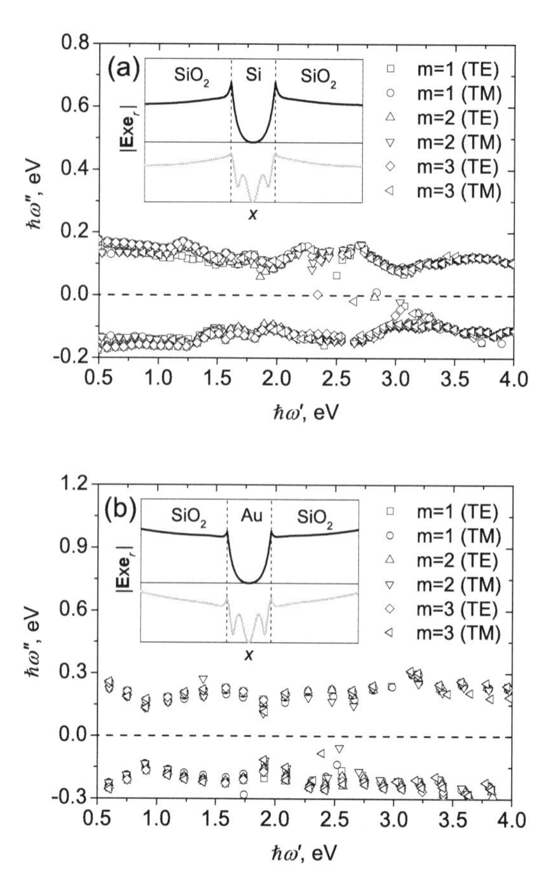

Figure 3.7: Localized absorption polaritons of single (a) Si and (b) Au cylinders of radius $R = 100$ nm embedded in SiO_2 cladding. The insets show typical distributions of the tangential electric field amplitude across the structure for the surface (black lines) and volume (gray lines) polaritons.

TM polarizations. The second group is *surface polaritons* that possess the internal fields confined at the cylinder's interface. They are of the TM polarization only and exist at $\varepsilon_i'(\omega_{nm}) < 0$.

Both volume and surface cylindrical radiation polaritons are inherent to photonic and plasmonic systems. Regardless of the structure's material and geometrical parameters, they all are virtual. The condition of the true polaritons, $\omega_{nm}'' = 0$, is generally forbidden for cylindrical radiation polaritons, as it violates the energy conservation law for real electromagnetic fields.

Absorption polaritons

Another type of localized cylindrical polaritons is the absorption eigenoscillations exhibiting the incoming external eigenfields for all azimuthal angles. In terms of the wavefunctions, they are given by $\mathbf{M}^e_{-,m}$ and $\mathbf{N}^e_{-,m}$. In other words, to find the absorption polariton fields, we should use

$$a_{+,m} = b_{+,m} = 0.$$

After application of the boundary conditions at $r = R$, we can derive two independent dispersion relations,

$$q_e J_m(q_i) H_m^{(2)\prime}(q_e) - q_i J_m'(q_i) H_m^{(2)}(q_e) = 0, \quad (3.33)$$

$$q_i J_m(q_i) H_m^{(2)\prime}(q_e) - q_e J_m'(q_i) H_m^{(2)}(q_e) = 0, \quad (3.34)$$

for the TE- and TM-polarized absorption polaritons.

The wavenumber quantization of the absorption polaritons is very similar to that of the radiation counterparts. Their fields are two-dimensionally quantized, with the attenuating standing wave patterns inside the cylinder for the *volume* eigenmodes and confined at the cylinder surface for the *surface* ones.[3] Figure 3.7 shows typical localized absorption polaritons observed for Si and Au cylinders embedded in SiO_2. Among these two structures, the Si cylinder again demonstrates higher density of eigenmodes packing due to the denser material-induced quantization of the eigenfields in Si/SiO_2 bi-material systems. In addition, the Si cylinder supports more absorption polaritons featuring $|\omega''_{nm}| \ll |\omega'_{nm}|$ than the Au cylinder, as it holds only for the TM eigenmodes of the Au cylinder and for both the TE and TM polaritons of the Si cylinder. Following this difference, metal cylinders exhibit pronounced resonances mainly on the TM absorption polaritons, while dielectric and semiconductor structures can equally resonate on both TM and TE eigenmodes.

Cylindrical absorption polaritons are generally virtual, with $\omega''_{nm} \neq 0$. However, in some cases their eigenfrequencies can be tuned (e.g., by changing the cylinder size) to get $\omega''_{nm} = 0$. Under these conditions, a particular polariton becomes the true solution describing the total absorption of the incident cylindrical waves at the respective frequency $\omega = \omega'_{nm}$.

3.4 Localized polaritons of spherical systems

In this section, we will consider three-dimensional localized polaritons oscillating in a sphere of radius R and permittivity ε_i embedded in a homogeneous medium with $\varepsilon = \varepsilon_e$. We will introduce the spherical wavefunctions for polaritons' eigenfields and then study three-dimensional eigenoscillations of single metal and dielectric spheres.

[3]Surface absorption polaritons provide strong optical resonances in metal nanoparticles. They are the subjects of intensive research in the field of plasmonics, where they are commonly known as *localized surface plasmon-polaritons*. Another popular term for them is localized surface plasmons, which is misleading, as it confuses two definitions of plasmons and polaritons (for details, see Section 1.3.3), and therefore is not in use in this book.

3.4.1 Spherical eigenfields

To study localized eigenoscillations of a sphere, we will compose spherical wavefunctions [1, 3]. They can be generated with the use of

$$\mathbf{c} = \mathbf{e}_r, \qquad \Psi(\mathbf{r}) = \Psi(r, \theta, \phi) \tag{3.35}$$

in the spherical system (r, θ, ϕ). In general, the three-dimensional generating function $\Psi(r, \theta, \phi)$ can be written in terms of the spherical harmonics $Y_{lm}(\theta, \phi)$, as

$$\Psi(r, \theta, \phi) = F_l(r) Y_{lm}(\theta, \phi), \tag{3.36}$$

with $Y_{lm}(\theta, \phi)$ given by

$$Y_{lm}(\theta, \phi) = \sqrt{\frac{2l+1}{4\pi} \frac{(l-m)!}{(l+m)!}} P_l^m(\cos\theta)\, \mathrm{e}^{\mathrm{i}m\phi},$$

where $P_l^m(\cos\theta)$ represent the associated Legendre polynomials, with the integers m and l indicating the azimuthal and orbital indices, respectively. Following Eq. (3.3), the radial function $F_l(r)$ must satisfy

$$r^2 \frac{\mathrm{d}^2 F_l}{\mathrm{d}r^2} + 2r \frac{\mathrm{d}F_l}{\mathrm{d}r} + [r^2 k_0^2 \varepsilon - l(l+1)] F_l = 0. \tag{3.37}$$

This is the spherical Bessel equation that has two linearly independent solutions that can be written in terms of the spherical Hankel functions of the first and second kinds:

$$F_{+,l}(r) = h_l^{(1)}(kr), \quad F_{-,l}(r) = h_l^{(2)}(kr), \quad k = k_0\sqrt{\varepsilon}. \tag{3.38}$$

The radial functions $F_{\pm,l}(r)$ are chosen to provide the elementary outgoing ($F_{+,l}$) and incoming ($F_{-,l}$) spherical waves [2, 6].

Now, we can use Eqs. (3.2) to generate the wavefunctions fulfilling Maxwell's equations in the spherical coordinates,

$$\mathbf{M}_{\pm,lm} = \frac{F_{\pm,l}}{r} \left[\frac{\mathbf{e}_\theta}{\sin\theta} \frac{\partial Y_{lm}}{\partial \phi} - \mathbf{e}_\phi \frac{\partial Y_{lm}}{\partial \theta} \right], \tag{3.39}$$

$$\mathbf{N}_{\pm,lm} = \frac{1}{kr} \left[\mathbf{e}_r \frac{l(l+1)}{r} F_{\pm,l} Y_{lm} + \frac{\mathrm{d}F_{\pm,l}}{\mathrm{d}r} \left(\mathbf{e}_\theta \frac{\partial Y_{lm}}{\partial \theta} - \frac{\mathbf{e}_\phi}{\sin\theta} \frac{\partial Y_{lm}}{\partial \phi} \right) \right]. \tag{3.40}$$

With these wavefunctions, the external eigenfields at $r \geq R$ can be written as

$$\mathbf{E}_t = \sum_{l=0}^{\infty} \sum_{m=-l}^{l} \sum_{\pm} (a_{\pm,lm} \mathbf{M}_{\pm,lm}^e + b_{\pm,lm} \mathbf{N}_{\pm,lm}^e), \tag{3.41}$$

$$\mathbf{H}_t = -\mathrm{i}\sqrt{\frac{\varepsilon_0 \varepsilon_e}{\mu_0}} \sum_{l=0}^{\infty} \sum_{m=-l}^{l} \sum_{\pm} (b_{\pm,lm} \mathbf{M}_{\pm,lm}^e + a_{\pm,lm} \mathbf{N}_{\pm,lm}^e). \tag{3.42}$$

The summation in eigenfields is performed over the $\pm$, l, and m indices, while the superscript e is to denote the wavefunctions of the external domain with $\varepsilon = \varepsilon_e$. Similarly, we can write the fields inside the sphere at $r \leq R$,

$$\mathbf{E}_t = \sum_{l=0}^{\infty} \sum_{m=-l}^{l} \sum_{\pm} (c_{\pm,lm}\mathbf{M}^i_{\pm,lm} + d_{\pm,lm}\mathbf{N}^i_{\pm,lm}), \tag{3.43}$$

$$\mathbf{H}_t = -\mathrm{i}\sqrt{\frac{\varepsilon_0\varepsilon_i}{\mu_0}} \sum_{l=0}^{\infty} \sum_{m=-l}^{l} \sum_{\pm} (d_{\pm,lm}\mathbf{M}^i_{\pm,lm} + c_{\pm,lm}\mathbf{N}^i_{\pm,lm}), \tag{3.44}$$

where the superscript i denotes the wavefunctions of the domain with $\varepsilon = \varepsilon_i$. Since all the internal wavefunctions diverge at $r = 0$, the coefficients $c_{\pm,m}$ and $d_{\pm,m}$ should be chosen by

$$\frac{c_{+,lm}}{c_{-,lm}} = \frac{d_{+,lm}}{d_{-,lm}} = \mathrm{i}(-1)^{l+1} \tag{3.45}$$

to ensure finiteness of the total fields at $r = 0$ [2, 6].

The fields above represent the general solution of eigenfields inside and outside the sphere. For eigenoscillations though, we should use the particular solutions corresponding to either outgoing or incoming spherical waves in the external domains based on the type of the eigenmodes considered. We will discuss them for photonic and plasmonic structures below.

3.4.2 Spherical eigenoscillations

Radiation polaritons

Spherical radiation polaritons feature the outgoing eigenfields in the external medium given by the $\mathbf{M}^e_{+,lm}$ and $\mathbf{N}^e_{+,lm}$ wavefunctions. These polaritons exhibit

$$a_{-,lm} = b_{-,lm} = 0.$$

After application of boundary conditions (2.16) through (2.19) at $r = R$, we can derive the dispersion relations for the TM and TE radiation polaritons [1, 3] exhibiting zero H_r and E_r fields,

$$q_i\psi_l(q_i)\xi_l'(q_e) - q_e\xi_l(q_e)\psi_l'(q_i) = 0, \tag{3.46}$$

$$q_e\psi_l(q_i)\xi_l'(q_e) - q_i\xi_l(q_e)\psi_l'(q_i) = 0, \tag{3.47}$$

where $\psi_l(q) = qj_l(q)$ and $\xi_l(q) = qh_l^{(1)}(q)$ with $q_i = k_iR$ and $q_e = k_eR$.

Note that the TE and TM radiation polaritons are three-dimensionally quantized. Their azimuthal and orbital wavevector quantizations are given by the m and l indices, while the radial quantization is denoted with the index n. Despite the three-dimensional quantization of spherical polaritons, their eigenfrequencies are invariant of the azimuthal index, $\omega_{nml} \equiv \omega_{nl}$, clearly seen from their dispersion relations, which are the explicit functions of l and implicit functions of n (given by the oscillating behavior of ψ_l and ξ_l).

Figure 3.8 shows eigenfrequencies of the localized radiation polaritons for single Si and Au spheres embedded in SiO_2 cladding. Eigenmodes of these structures

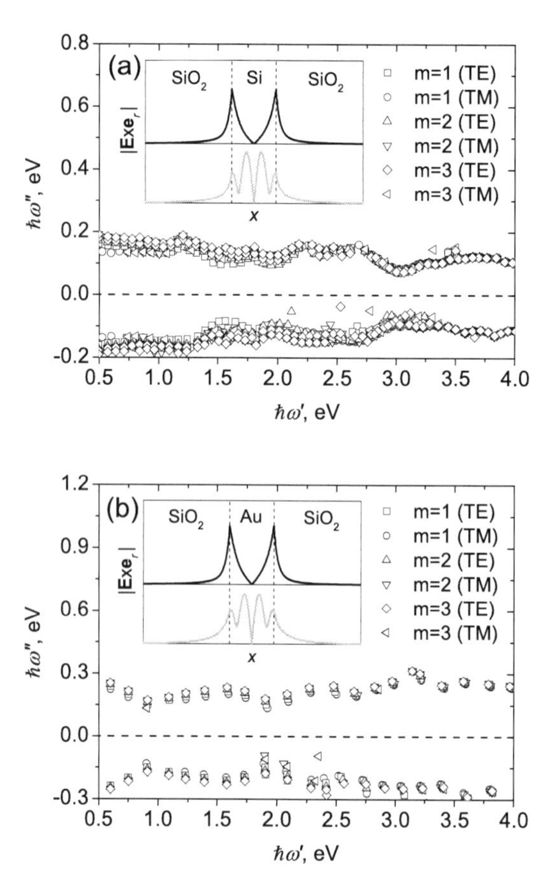

Figure 3.8: Localized radiation polaritons of (a) Si and (b) Au spheres of radius $R = 100$ nm embedded in SiO_2. The insets show typical distributions of the tangential electric field amplitude across the structure for the surface (black lines) and volume (gray lines) polaritons.

exhibit similar attenuating standing wave fields inside the spheres for the *volume* TE and TM polaritons and fields confined at the sphere surface for the *surface* TM oscillations. Similar to the planar and cylindrical radiation polaritons, the Si sphere features higher density of the radiation eigenmode packing due to the denser material-induced quantization of the eigenfields [see Fig. 3.2(a)].

All spherical radiation polaritons are purely virtual. Regardless of the structure's material and geometrical parameters, they never achieve $\omega''_{nl} = 0$, as it violates the energy conservation law for true electromagnetic fields.

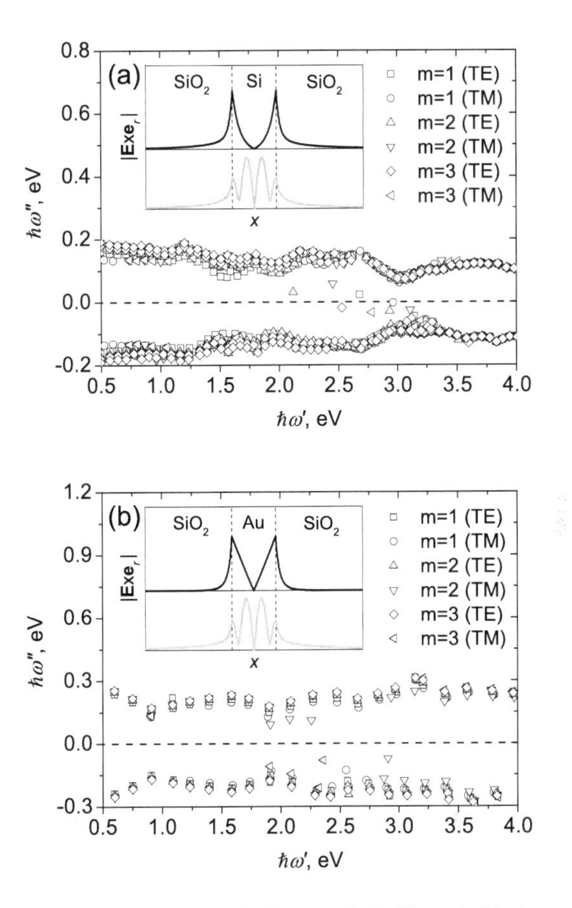

Figure 3.9: Localized absorption polaritons of (a) Si and (b) Au spheres of radius $R = 100$ nm embedded in SiO_2. The insets show typical distributions of the tangential electric field amplitude across the structure for the surface (black lines) and volume (gray lines) polaritons.

Absorption polaritons

Eigenfields of the absorption polaritons feature the spherically incoming external fields given by $\mathbf{M}^e_{-,lm}$ and $\mathbf{N}^e_{-,lm}$ wavefunctions. In other words, they exhibit

$$a_{+,lm} = b_{+,lm} = 0.$$

After application of boundary conditions (2.16) through (2.19) to the internal and external fields at $r = R$, we find two independent dispersion relations for the TM and TE absorption polaritons [7],

$$q_i \psi_l(q_i) \zeta_l'(q_e) - q_e \zeta_l(q_e) \psi_l'(q_i) = 0, \tag{3.48}$$

$$q_e \psi_l(q_i) \zeta_l'(q_e) - q_i \zeta_l(q_e) \psi_l'(q_i) = 0, \tag{3.49}$$

where $\zeta_l(q) = qh_l^{(2)}(q)$.

Similar to the spherical radiation polaritons, the eigenfrequencies of absorption eigenmodes are degenerated with respect to the azimuthal index m for both the polarizations, whereas their eigenfields are given by all three indices according to the complete quantization of the wavevector. The implicit radial quantization is given by oscillating behavior of the spherical functions ψ_l and ζ_l in the dispersion relations. The resultant radial profiles of the eigenfields represent attenuating standing waves formed inside the sphere for the TE and TM *volume* polaritons, while for the TM *surface* modes, they are confined at the sphere surface, as shown in Fig. 3.9 for Si and Au nanospheres embedded in SiO_2. Note that the Si sphere supports more absorption polaritons featuring $|\omega''_{nl}| \ll |\omega'_{nl}|$ than the Au sphere, as this condition holds only for the TM eigenmodes of the Au particle and for both the TE and TM polaritons of the Si sphere. This is a general difference in dielectric and metallic particles, according to which dielectric particles exhibit higher numbers of pronounced resonances on absorption polaritons compared to their metal counterparts.

In general, spherical absorption polaritons are virtual, with complex ω_{nl}. However, by tuning the eigenfrequencies (e.g., by changing the sphere's radius) some of the absorption eigenmodes can become true with $\omega''_{nl} = 0$. They give us the resonant frequency of the spherical waves that can be completely absorbed by the structure with no scattering produced.

3.5 Summary

In this chapter, we reviewed localized polaritons of the basic single-particle systems of the planar, cylindrical, and spherical geometries. We have shown that regardless of structures' material and geometrical parameters, all localized polaritons can be classified as the transmission, radiation, and absorption eigenmodes featuring different distributions of the external eigenfields. These polaritons are inherent to both photonic and plasmonic structures but possess slightly different eigenfields. In general, localized polaritons are virtual; they cannot appear as the excited fields and should be considered as resonant response characteristics only. However, under certain conditions that require a particular matching of the material and geometrical parameters, localized polaritons can become true eigenmodes with fully physical fields that enable pronounced energy transfer in a system.

Bibliography

[1] J. A. Stratton, *Electromagnetic Theory*. McGraw-Hill, 1941.

[2] P. M. Morse and H. Feshbach, *Methods of Theoretical Physics*. McGraw-Hill, 1953.

[3] C. F. Bohren and D. R. Huffman, *Absorption and Scattering of Light by Small Particles*. Wiley-Interscience, 2nd ed., 1998.

[4] J. D. Jackson, *Classical Electrodynamics*. Wiley, 2nd ed., 1975.

[5] S. Longhi, "Backward lasing yields a perfect absorber," *Physics Online Journal*, vol. 3, p. 61, July 2010.

[6] A. N. Tikhonov and A. A. Samarskii, *Equations of Mathematical Physics*. Macmillan, 1963.

[7] Y. A. Akimov, "Plasmonic properties of metal nanostructures," in *Plasmonic Nanoelectronics and Sensing* (E. P. Li and H. S. Chu, eds.), pp. 20–66, Cambridge University Press, 2014.

Chapter 4

Localized polaritons of multi-particle systems

Lin Wu
Institute of High Performance Computing, Singapore

Valerian Hongjie Chen
Institute of High Performance Computing, Singapore

Ping Bai
Institute of High Performance Computing, Singapore

Song Sun
Institute of High Performance Computing, Singapore

This chapter refers to systems of more than one nanoparticle. We will begin by discussing the inter-particle hybridization of localized polaritons in Section 4.1 for the case of a laterally coupled dimer. We will then analyze various hybridization effects in Section 4.2 with focus on the commonly studied multi-particle systems such as trimers, hexamers, asymmetric dimers, stereo-metamaterials, and nanospheres-in-nanoshells. This will be followed by examination of more complex multi-particle systems in Section 4.3. Summary will be presented in Section 4.4.

4.1 Inter-particle polariton hybridization

The simplest and most commonly studied multi-particle system is a laterally coupled nanoparticle dimer. Using such a system as an example, we examine the inter-particle eigenmode hybridization, i.e., capacitive (Section 4.1.1) and conductive (Section 4.1.2) couplings, and the links between them (Section 4.1.3).

4.1.1 Capacitive coupling: eigenmode hybridization theory

A coupling is said to be *capacitive* when the oscillations of intrinsic charges occur within the individual nanoparticles. In order to understand capacitive coupling of arbitrarily shaped nanostructures, P. Nordlander's group developed the *hybridization method* [1] to describe the capacitive coupling of polaritons between two metallic nanoparticles. They showed that the transverse eigenmodes of a complex system can

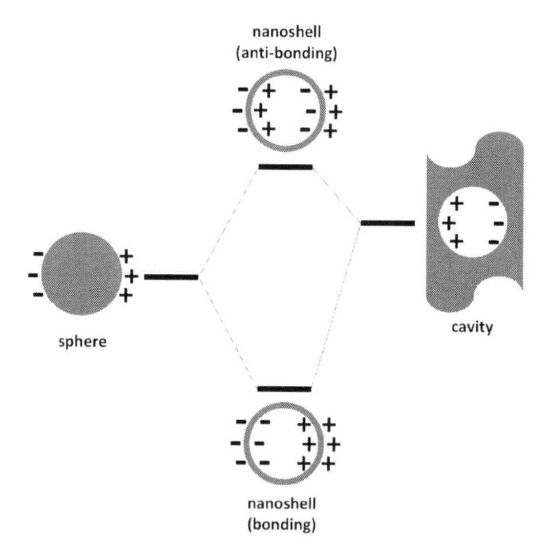

Figure 4.1: The energy diagram of the polariton hybridization in a nanoshell arising from the interaction between the sphere and cavity eigenmodes (where + and − signs represent typical distribution of surface polarization charges).

be understood as the hybridizations of polaritons of all subsystems. For instance, the polariton hybridization in nanoshells arises from the interaction between the sphere and cavity eigenmodes, as shown in Fig. 4.1. The two nanoshell polaritons are the symmetrically coupled (bonding, ω_-) and anti-symmetrically coupled (anti-bonding, ω_+) eigenmodes of the sphere and cavity. This can be qualitatively understood by analogy with the well-known hybridization of atomic orbitals.

The hybridization method was then employed to understand nanoparticle dimers [2]. The dimer polaritons can be understood as the bonding and anti-bonding couplings of the individual nanoparticle eigenmodes. Here, bonding corresponds to the two dipole moments moving in phase (symmetric fields), and the anti-bonding corresponds to anti-phase (anti-symmetric fields). As the net dipole moment of the anti-bonding polaritons is close to zero for identical spheres, they are not easily excited by light and are thus known as dark eigenmodes. Bonding polaritons have net dipole moments and can thus be excited by light and are referred to as bright eigenmodes.

The polariton hybridization method was subsequently used to explain the extinction spectra of dimers of various morphologies including nanoshell dimers [3], nanorod dimers [4, 5], nanodisc dimers [6], nanocube dimers [7], nanoprism dimers [8, 9], nanoring dimers [10], and even heterodimers, i.e., two closely adjacent nanoparticles of different size, shape [11], or material composition [12].

In order to give a more in-depth explanation of polariton hybridization, we discuss heterodimers. In particular, we look at two Au nanospheres of different sizes [11]. This size asymmetry is the key for revealing the physical mechanisms for polariton hybridization in these structures: the dipolar ($l = 1$) polariton of the smaller

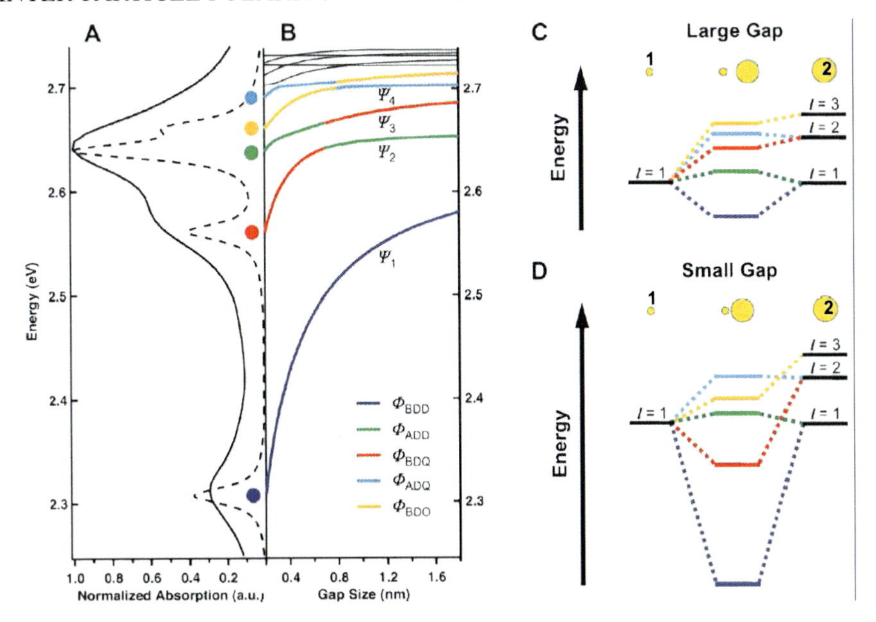

Figure 4.2: Heterodimers consisting of two Au nanospheres of different sizes as examples to elaborate polariton hybridization theory. (A) Absorption spectra calculated by FDTD (solid line) and by the polariton hybridization method (dashed line). The eigenmodes are color coded, and the codes are shown in (C) and (D). (B) Energy of the eigenmodes as a function of the inter-particle gap size. (C) and (D) are polariton hybridization diagrams for large and small gap sizes. In these diagrams, the energies of dipolar surface polaritons of two nanoparticles are the same because the particles are in the quasi-static limit. Reprinted (adapted) with permission from Reference [11]. © 2010 American Chemical Society.

nanoparticle can couple strongly to the dipolar ($l = 1$) and also to the higher multipolar ($l > 1$) polaritons of the larger nanoparticle. This coupling can be either symmetric (bonding) or anti-symmetric (anti-bonding), as shown in Fig. 4.2. These hybridized polaritons of the heterodimer change with a decrease of the distance between the two nanoparticles, as clearly seen in Fig. 4.2. The separation of the energy levels of the hybridized eigenmodes increases with decreasing gap distance. Loosely speaking, reduction of the gap increases the coupling strength, leading to "stronger hybridization". This is a general feature common to any materials, as the physical mechanism of hybridization - interaction of surface charges induced in individual nanoparticles - remains the same regardless of the nanoparticle materials.

4.1.2 Conductive coupling: charge-transfer polaritons

Consider a dimer, similar to those described in the previous section. Now imagine decrease of the inter-particle separation. When the separation is reduced so much

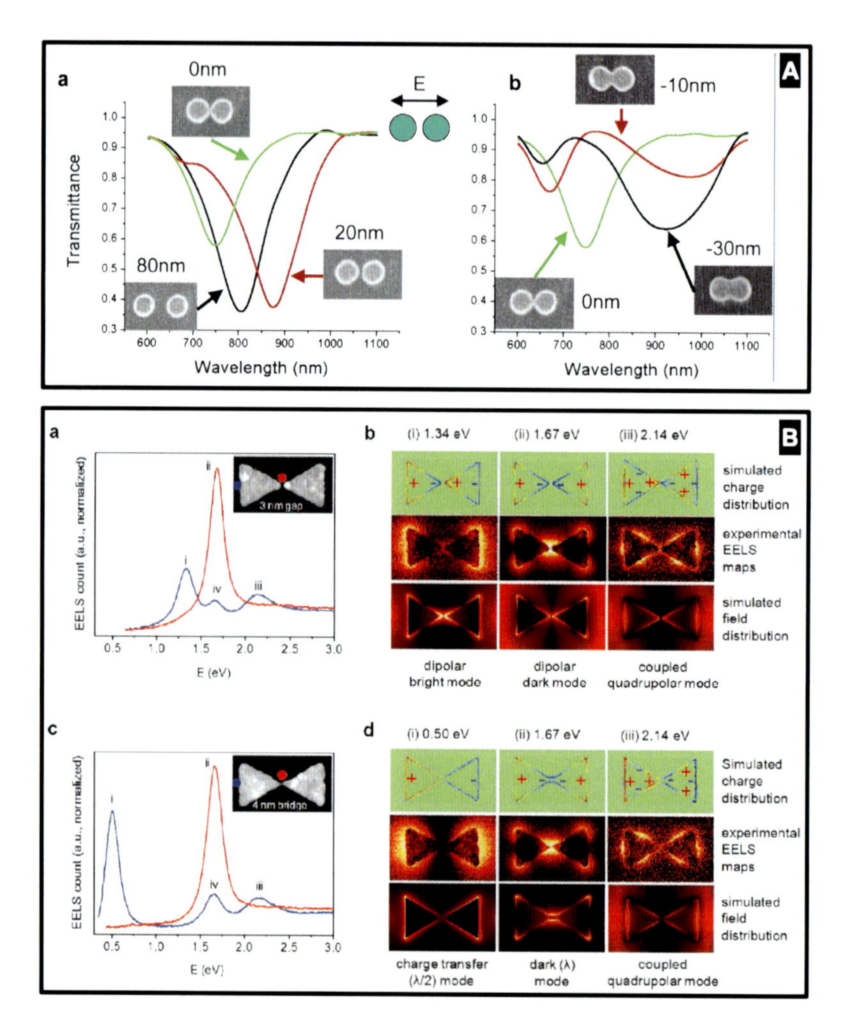

Figure 4.3: Charge-transfer polaritons as examples of conductive coupling. Panel A: Transmission spectra of gold nanoparticle dimers. Particles are not in contact (a) and are in contact with each other (b). Reprinted (adapted) with permission from Reference [13]. © 2004 American Chemical Society. Panel B: Polariton coupling of two nanoprisms separated by a gap [(a) and (b)] and connected by a bridge [(c) and (d)]. Experimental EELS spectra [(a) and (c)] and simulations [(b) and (d)]. Reprinted (adapted) with permission from Reference [9]. © 2012 American Chemical Society.

that the particles are in conductive contact, we enter a different coupling regime. This is known unsurprisingly as the *conductive coupling* regime. Work in this field was pioneered by Nurmikko et al. in 2004 [13], right after the development of the polariton hybridization method. They showed that the electromagnetic signature of a pair of metal nanoparticles will change dramatically as a conductive bridge is formed, resulting in a sharp and abrupt change of the polariton energy. This transition is illustrated in Fig. 4.3 (panel A).

This work inspired the study of eigenmodes in nearly touching or touching metallic nanoparticle dimers, both theoretically [14–18] and experimentally [9, 19–21]. In these works, the new long wavelength resonance observed in touching dimers was denoted the *charge-transfer polariton.*[1]

Figure 4.3 (panel B) shows a classic example of the nanoprism dimer [8, 9]. The electron energy loss spectroscopy (EELS) spectrum shows a good comparison of the eigenmodes of touching and nearly touching nanoprisms, clearly seen in the behavior of peak "i" in (a) and (c). The two separate parallel dipoles (b) become a single dipole (d). This doubles the polariton wavelength, hence the energy of the mode drops by approximately half, explaining the redshift of peak "i" observed in (a) and (c).

4.1.3 Link between capacitive and conductive coupling

In the preceding section, we discussed the charge-transfer polaritons (CTPs) between two metallic nanoparticles. The examples we saw all involved a metallic bridge, but a physical understanding leads us to the conclusion that similar polaritons also exist for bridges of other materials as long as they are conductive at optical frequencies. This direction was championed by P. Nordlander and J. Aizpurua [16]. Their theoretical work suggested that the optical properties of a nanoparticle dimer bridged by a conductive junction depend strongly on the junction's conductivity. More specifically, the hybridized dipolar mode arising from the capacitive coupling blueshifts with increasing junction conductivity (see Fig. 4.4, panel A). When the conductance is sufficiently large, a new polariton mode, the CTP, appears. Three types (semiconductor, tunneling, and molecular) of junction materials that support both capacitive and conductive coupling modes (theoretically predicted or experimentally verified) will be reviewed in this section.

Semiconducting junctions

By having a semiconductor as the gap material, the conductivity of the gap can theoretically be tuned by changing the free-carrier density of the semiconductor (Fig. 4.4, panel B) [22]. The range of tuning can be so large that the transition from the capacitive $N_{eh} = 0\ \text{cm}^{-3}$ to conductive $N_{eh} = 10^{22}\ \text{cm}^{-3}$ coupling regimes becomes possible. This enables a strong modification of the dimer's response, which can be observed

[1]In the field of plasmonics, charge-transfer polaritons are often called *charge-transfer plasmon-polaritons*. Another common term for them is charge-transfer plasmons, which is misleading, as it confuses two definitions of plasmons and polaritons (for details, see Section 1.3.3), and therefore is not used in this book.

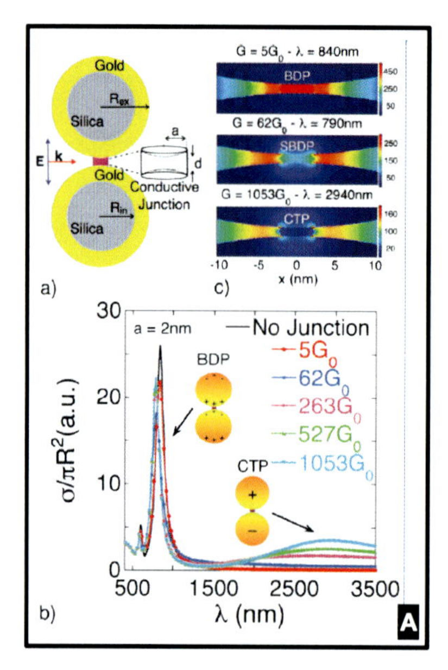

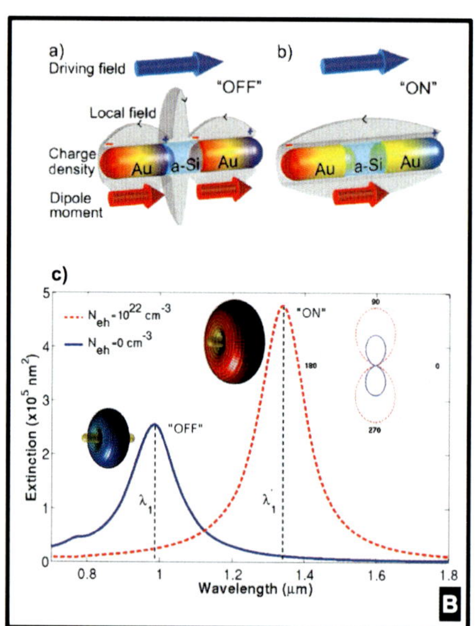

Figure 4.4: Link between the capacitive and conductive coupling. Panel A: The properties of a dimer depend strongly on the conductivity of the junction. Reprinted (adapted) with permission from Reference [16]. © 2010 American Chemical Society. Panel B: By changing the conductivity of the junction, the eigenmodes of a nanoantenna can be tuned accordingly. Reprinted (adapted) with permission from Reference [22]. © 2010 American Chemical Society.

both at far-field and local near-field intensities [Fig. 4.4, panel B: (c)]. The large modulation depth, low switching pumping energy threshold, and potentially ultrafast time responses of antenna switches hold promise for applications ranging from integrated nanophotonic circuits to quantum information devices.

Tunneling junctions

Classically, we expect the junction of a non-touching dimer to be insulating when it is dielectric and conducting when it is metallic due to the presence or lack of charge carriers, respectively. However, the limitations of the classical electromagnetic model become clear when nearly touching dimers separated by a sub-nanometer dielectric gap are considered [14]. At such small separations, it is theoretically expected that the charge carriers tunnel directly from one electrode to the other across the gap, forming the tunneling charge-transfer polariton (*t*CTP).

Following this theoretical prediction, Halas et al. investigated the polariton resonances of a dimer as their separation was reduced from the non-touching to the touching regimes [19]. This was followed by the first fully quantum mechanical investigation of the polariton resonances in a nanoparticle dimer as a function of the inter-particle separation [15]. The authors found that for separations of less than

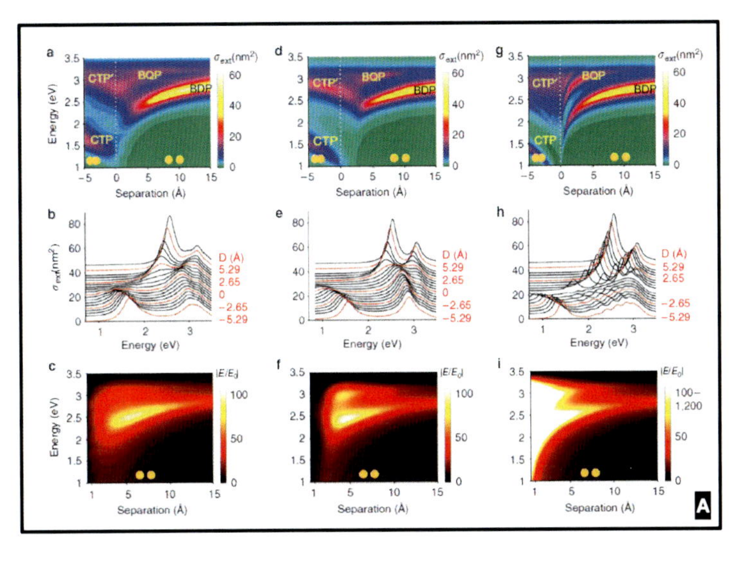

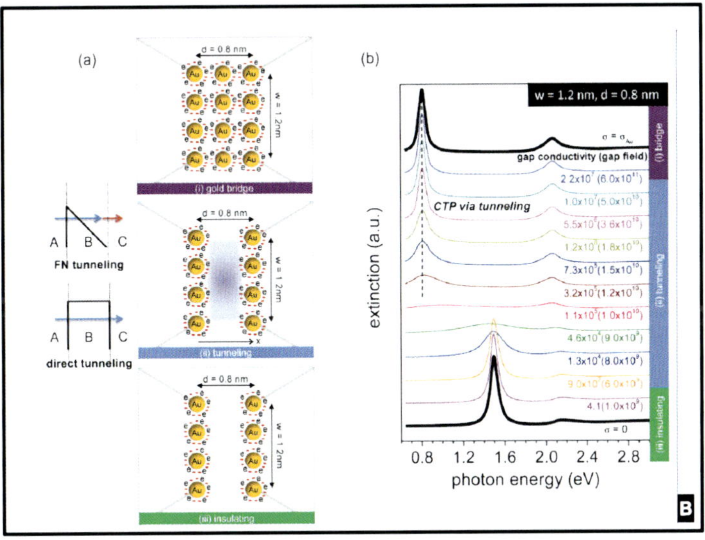

Figure 4.5: Polariton coupling in tunneling junctions. Panel A: Comparison of the optical properties of a metallic dimer as predicted by the full-QM [(a)–(c)], QCM [(d)–(f)], and classical electromagnetic models (CEM) [(g)–(i)]. Reprinted (adapted) with permission from Reference [17]. © 2012 Nature Publishing Group. Panel B: Three regimes of a nanogap system: conductive gap (i), tunneling gap (ii), and insulating gap (iii). The properties are mediated by the gap conductivity σ_{gap}, which is varied. Reprinted (adapted) with permission from Reference [18]. © 2013 American Chemical Society.

1 nm, quantum mechanical effects, namely tunneling, begin to significantly influence the response of the dimer. This results in significantly weaker hybridization and a large reduction of the electromagnetic field enhancements across the junction. For separations smaller than 0.5 nm, the *t*CTP appears. This polariton blueshifts with decreasing inter-particle separation. This article led to the new field of *quantum plasmonics*, triggering a lot of research that continues even up to the point of writing [17, 18, 20, 23, 24]. Two key instances of the recent theoretical efforts are discussed here to further explain the tunneling regime.

The first example is the famous quantum-corrected model (QCM), which incorporates tunneling into classical electromagnetic simulations of the optical properties of metallic nanostructures. The junction material is taken to have an effective conductivity, where it accounts for the tunneling current across the junction. The QCM successfully reproduces the optical spectra and near fields obtained using the full quantum model (QM) in all three separation regimes [Fig. 4.5, panel A: (a)–(c), and (d)–(f)]. Notably, the QCM accurately describes the mode transition between the contact and non-contact regimes and the sharp decrease of local field enhancement for $D = 4$ Å.

The second example deals with the predicted systems, where the tunneling can be observed experimentally. While direct tunneling is significant for small gaps, the Fowler-Nordheim (FN) [18] tunneling dominates when a high electric field is present in the gap. Unlike direct tunneling, in FN tunneling the electrons first tunnel to the middle of the barrier before leaving it completely on the other side [Fig. 4.5, panel B: (a)]. This makes the *t*CTP easier to observe in realistic dielectric junctions. Alternatively, the barrier height may be reduced by using Al_2O_3 in the gap instead of a vacuum [25].

We see that what was previously thought to be a discontinuous change between the touching and non-touching regimes can be explained as a continuous change when tunneling is considered. This intermediate tunneling regime lies between the contact and non-contact situations (1 Å $< D <$ 5 Å). In addition, it is predicted to appear in non-conductive dimers if the materials of two particles are properly chosen for tunneling junction. At the same time, the theoretical predictions of *t*CTP have yet to be demonstrated experimentally because it is still very challenging to fabricate sub-nanometer dielectric gaps.

Molecular junctions

The third type of junction we consider is a molecular junction. Typically, the dimer is an organic-inorganic hybrid [26, 27], where the tunneling barrier width is determined by the thickness of the molecular layer, while the barrier height is determined by the difference in energies of the Fermi levels of the electrodes and the molecular frontier orbitals. It can also be thought of as the energy gap between the highest occupied molecular orbital (HOMO) and lowest unoccupied molecular orbital (LUMO).

By using a self-assembled monolayer (SAM) of organic molecules as the junction material, we can satisfy the theoretical conditions to observe *t*CTP. First, a monolayer of molecules is clearly less than a nanometer. Second, the tunneling barrier height is reduced relative to the heights of vacuum gaps. Excitation of *t*CTP in

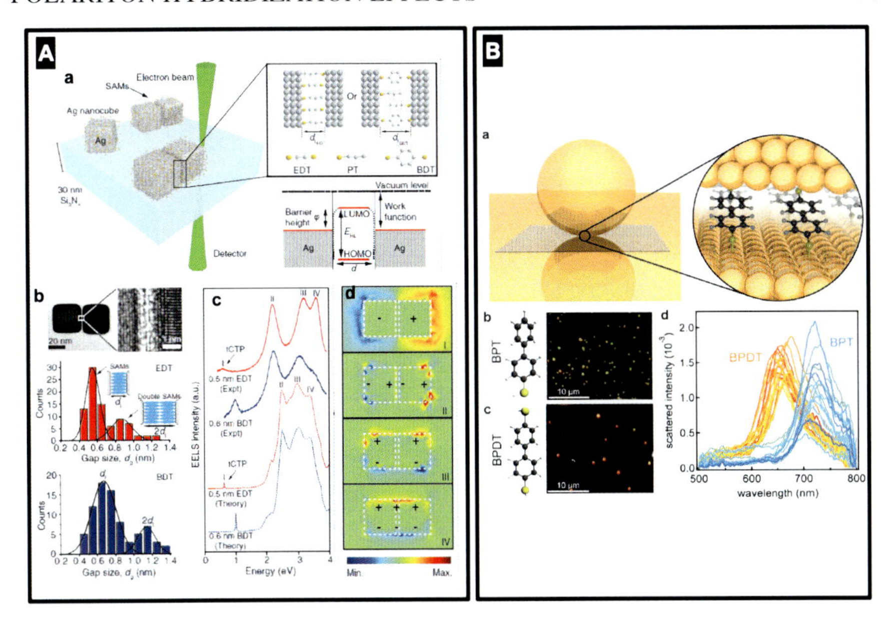

Figure 4.6: Polariton coupling in molecular junctions. Panel A: Direct observation of quantum tunneling in the molecular tunnel junctions made of two silver nanoparticles bridged by a SAM. Reprinted (adapted) with permission from Reference [26]. © 2014 American Association for the Advancement of Science. Panel B: Conductive and non-conductive self-assembled monolayers in nanoparticle on mirror geometry: a gold nanoparticle is placed on a gold film separated by a thin molecular spacer layer. Reprinted (adapted) with permission from Reference [27]. © 2015 American Chemical Society.

molecular junctions has been achieved experimentally (Fig. 4.6, panel A). The three higher energy peaks observed correspond to the bonding dipolar polaritons. Based on calculations of the charge transfer between the cuboids, the new low energy polariton is the *t*CTP. This *t*CTP is particularly sensitive to the types of molecules used in the junctions. It was subsequently noted that changing a single atom in the gap caused a significant blueshift of the bonding dipolar polariton mode (Fig. 4.6, panel B). These two works open the door for investigation of dimers with molecular junctions. For example, SAM-bridged dimers currently operate at frequencies larger than the optical. Using molecules from the field of molecular electronics [28, 29], it may be possible to shift the charge transport mechanisms into the optical regime.

4.2 Polariton hybridization effects

In the preceding section, we discussed inter-particle eigenmode hybridizations for dimers, the most widely studied two-particle systems, where the coupling occurs along the axis of two nanoparticles. In fact, there are many other two-particle sys-

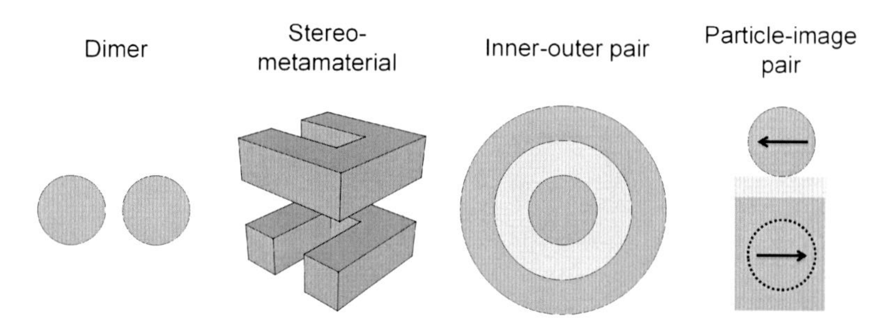

Figure 4.7: Different types of two-particle systems: laterally coupled dimer, stereo-metamaterial, radially coupled inner-outer nanoparticle pair, and nanoparticle image coupling of a nanoparticle above a substrate.

tems, for example, vertically coupled stereo-metamaterials, radially coupled inner-outer nanoparticle pairs, and nanoparticle-image coupled systems (see Fig. 4.7). Besides these two-particle systems, there exist other multi-particle systems, in which more than two particles are coupled. This includes 1D multiple-particle chains [30], 2D plasmonic oligomers [31], DNA-assembled 3D polyhedra such as pyramids and cubes [32], and 3D stacked split-ring resonator metamaterials [33], as shown in Fig. 4.8. These systems can have many different configurations, while their inter-particle coupling results in various hybridization effects. Consequently, this section is organized by hybridization effects that occur in multi-particle systems. In particular, we consider electric and magnetic responses (Section 4.2.1), directional radiation (Section 4.2.2), Fano resonances (Section 4.2.3), and chirality resonances (Section 4.2.4).

4.2.1 Electric and magnetic responses

In this section, we overview electric and magnetic responses in plasmonic and photonic multi-particle systems as a result of their eigenmode hybridization. Two systems will be presented: stereo-metamaterials and all-dielectric dimers.

Stereo-metamaterials

When the coupling between two particles is vertical or parallel to the direction of the incident light, the structures are known as stereo-metamaterials [34, 35]. A typical example is the metallic split-ring dimer. Each dimer consists of two identical split rings, one on top of the other, but the two split rings are rotated with respect to each other (see Fig. 4.9). In this configuration, the incident light is able to excite circulating currents in the structure, giving rise to magnetic dipole moments. For the metallic dimers discussed in earlier sections, the response was purely electric.

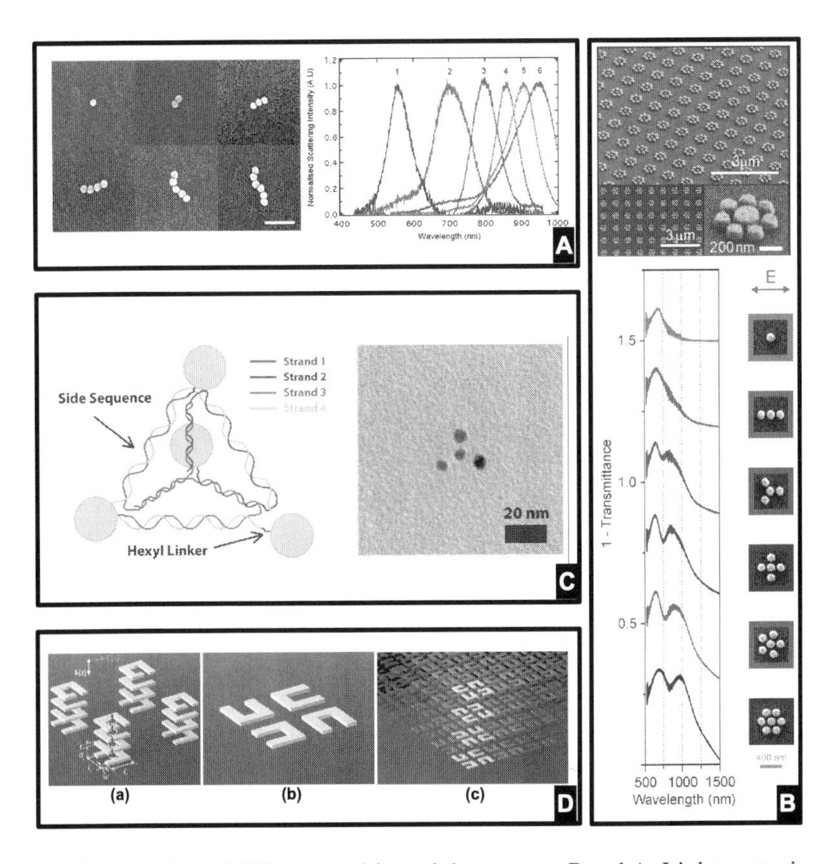

Figure 4.8: Examples of different multi-particle systems. Panel A: Light scattering of linear chains of gold nanoparticles. As the chain length increases, the surface polariton resonance redshifts. Reprinted (adapted) with permission from Reference [30]. © 2011 American Chemical Society. Panel B: The optical properties of oligomers. The experimental extinction spectra of the metallic oligomers are dependent on the number of particles in the outer ring. Reprinted (adapted) with permission from Reference [31]. © 2011 American Chemical Society. Panel C: DNA-nanocrystal pyramids. Reprinted (adapted) with permission from Reference [32]. © 2009 American Chemical Society. Panel D: Three-dimensional metamaterials: (a) Four-layer twisted split-ring resonator (SRR) metamaterial; (b) A unit cell of a two-dimensional magnetic metamaterial; (c) A hypothetical three-dimensional magnetic ("spin") metamaterial. Each layer consists of SRRs that are twisted by 90 degree to the layer directly below. Reprinted (adapted) with permission from Reference [33]. © 2008 Optical Society of America.

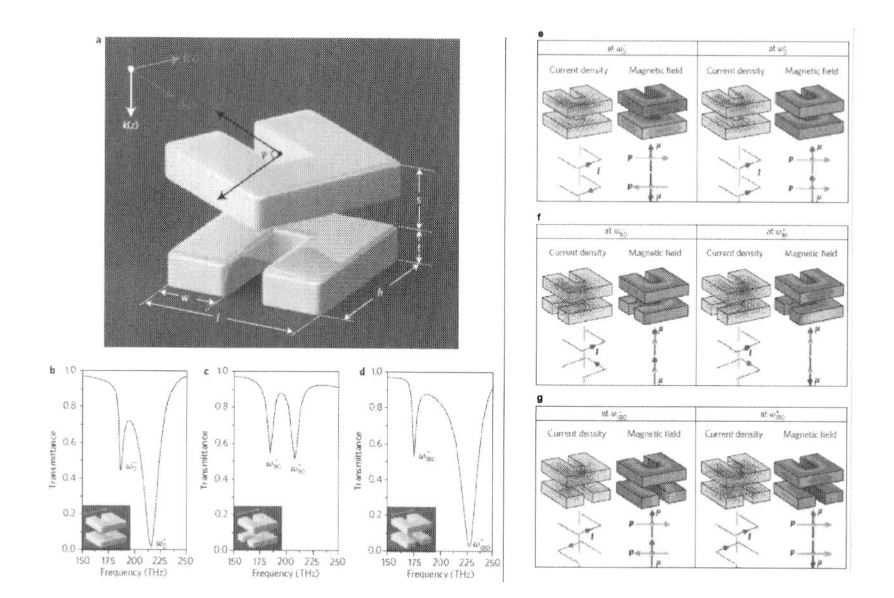

Figure 4.9: Electric and magnetic responses in stereo-metamaterials. Schematic (a) of the stereo-SRR (split ring resonator) dimer metamaterial. Simulated transmittance spectra for the 0 (b), 90 (c) and 180 degree (d) twisted SRR dimer metamaterials. Current and magnetic field distributions at respective resonances for the 0 (e), 90 (f) and 180 degree (g) twisted SRR dimer metamaterials. Lower left: schematics of currents (I) in two SRRs. Lower right: schematics of the alignments of the magnetic (m) and electric (p) dipoles. Reprinted from Reference [34]. © 2009 Nature Publishing Group.

The magnetic response of stereo-metamaterials allows us to control the interplay of electric and magnetic interactions, making them highly versatile structures.

All-dielectric dimers

Up to this point, we have discussed multi-particle systems made of metals. In fact, the hybridization theory is applicable to non-conductive systems as well. Such systems are fascinating because they have low losses and exhibit strong magnetic resonances (on the TE or quasi-TE polaritons) in addition to the electric resonances (on the TM or quasi-TM eigenmodes) common to metal structures. As shown in Chapter 3, single particles generally support both TM- and TE-polarized polaritons. However, for metal nanoparticles, only the TM absorption polaritons can feature $|\omega''_{nml}| \ll |\omega'_{nml}|$, while for dielectric particles, the principle holds for both the TM and TE absorption polaritons. As a result, dielectric particles feature a broader range of resonances on eigenmodes. In addition, we can manipulate them by simply changing the shape, size, or arrangement of the particles. In multi-particle systems, overlapping of the electric and magnetic resonances of different particles gives rise to exciting phenom-

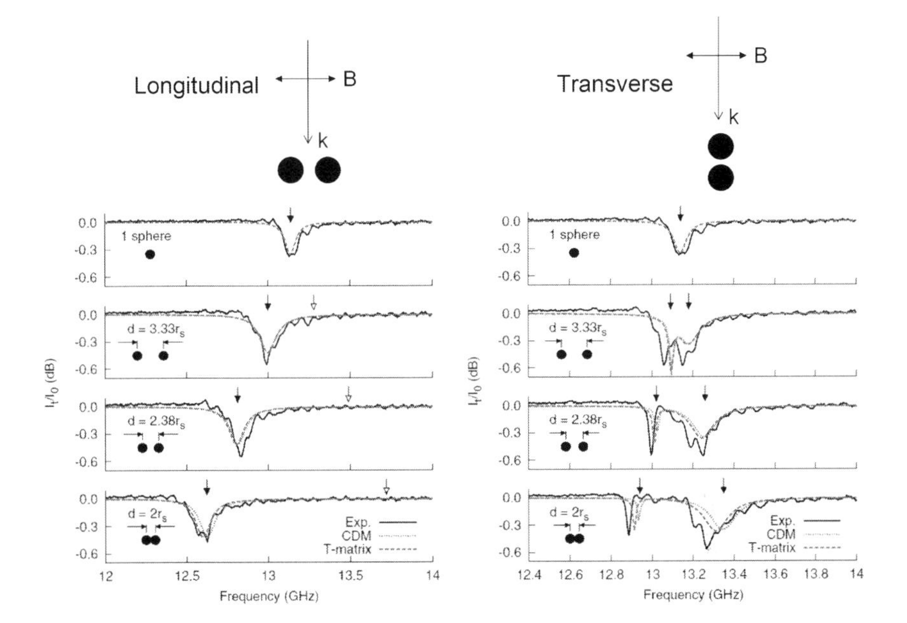

Figure 4.10: Magnetic responses of all-dielectric dimers. Transmission-frequency plots for a pair of identical MgO-CaO-TiO_2 spheres of radius $r_s = 1.07$ mm with various spacings for longitudinal (left) and transverse (right) orientations. The arrows show the resonant frequencies of the eigenmodes; empty arrowheads denote uncoupled polaritons. Reprinted from Reference [36]. © 2010 Optical Society of America.

ena. We consider below the simplest case, which is the magnetic response of a dimer of dielectric particles, since the excitation of electric responses of dielectric dimers is similar to the excitations of their metallic counterparts (see Section 4.1.1) and can be understood by using the same principles.

The dimer we discuss here consists of a pair of identical spheres made of a mixture of MgO, CaO, and TiO_2. Each sphere has a radius of $r_s = 1.07$ mm. The estimated magnetic dipole resonant frequency is 13.2 GHz for a single sphere. In Fig. 4.10, the interactions of two such spheres with various spacings are considered for two orientations: longitudinal (left) and transverse (right). For the longitudinal orientation, where the dimer is parallel to the incident magnetic field (Fig. 4.10, left), only the resonances on the parallel longitudinal dipolar polaritons are excited (indicated by arrows), whereas those on the anti-parallel longitudinal dipolar polaritons are not (indicated by empty arrowheads), as the incident wave propagating perpendicularly to the dimer axis induces the local fields equal for each sphere. On the other hand, for the transverse orientation, where the dimer is along the incidence direction (Fig. 4.10, right), there is a phase difference in the incident wave for two spheres and hence their local fields are not necessarily equal. In this case, both the parallel

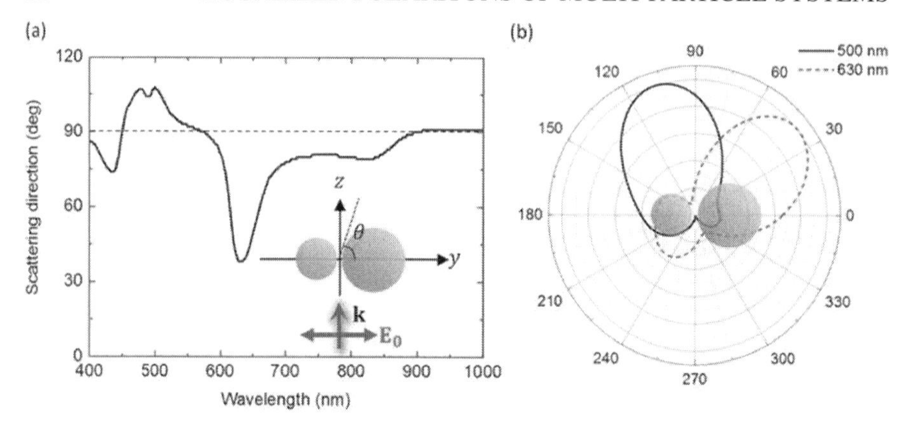

Figure 4.11: Directional radiation of asymmetric silicon dimers. By choosing the geometry of an asymmetric dimer appropriately, the scattering direction (which is a function of wavelength) can be tuned. Reprinted from Reference [39]. © 2015 Nature Publishing Group.

and anti-parallel transverse dipolar polaritons resonate with the incident field. For either longitudinal or transverse orientations, we can see that the single resonance for an isolated sphere splits into two resonances that shift further away when the two spheres get closer.

This study was conducted in the microwave range (with millimeter-sized dielectric particles), but the understanding of the phenomena is applicable to nanoparticles as long as the wavelength is much larger than the dimer length. This applicability is provided by scalability of the Maxwell's equations if the dielectric properties of the materials are counted properly [36]. Similar to electric resonances, there are higher order (quadrupolar, octupolar, and so on) magnetic modes [37], but they are harder to excite. This study was extended to the visible and near-infrared ranges [38].

4.2.2 *Directional radiation*

This section continues the discussions on hybridization effects. We now consider the directional scattering provided by polariton hybridization.

Asymmetric dielectric dimers

The coexistence of electric and magnetic resonances in a dielectric dimer enables the directional control of scattered light [39]. By overlapping or separating various electric and magnetic resonances in frequency space, we can manipulate the scattering direction [40]. Using the analytical dipole-dipole model [38], the authors showed that asymmetric silicon dipoles on a silica substrate can scatter light (incident perpendicularly to the dimer axis with a parallel electric field) from -7 to 27 degree (Fig. 4.11), where the scattering angle is wavelength-dependent [39].

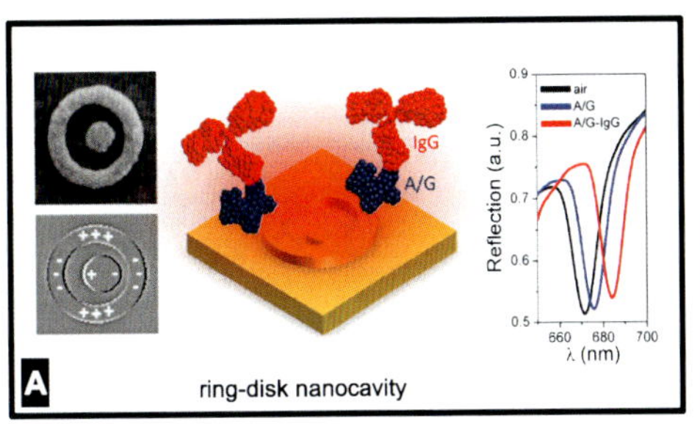

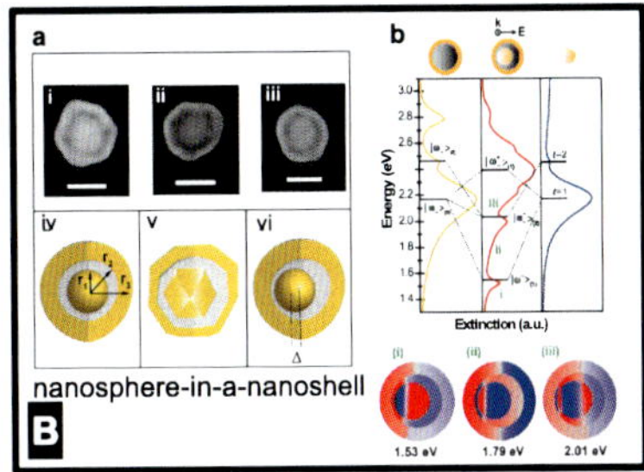

Figure 4.12: Fano resonances in two-particle systems. Panel A: Resonant ring-disk metallic nanocavities on a conducting substrate. Reprinted (adapted) with permission from Reference [44]. © 2012 American Chemical Society. Panel B: Au/SiO_2/Au core-shell particles. Reprinted (adapted) with permission from Reference [45]. © 2010 American Chemical Society.

1D dielectric nanoparticle chain

The directional radiation can be further tuned with one-dimensional (1D) dielectric nanoparticle chains. In certain configurations, they can act as Yagi-Uda antennae [41–43]. As compared to their metallic counterparts, such dielectric antennas are capable of operating in a much larger range of wavelengths [41], have better directivity [43] and lower losses, and, thus provide better radiation efficiency [42, 43]. Besides the support of directional radiation, such dielectric nanoparticle chain configurations can also guide waves with low losses. This will be addressed in more detail in Chapter 10.

4.2.3 Fano resonances

Fano resonances have been popular research topics in recent years, extensively discussed in several good review articles [46, 47]. Consequently, we limit our discussion here to three representative systems that support Fano resonances: two-particle metallic systems, multi-particle metallic systems, and dielectric oligomer systems. The key elements for all Fano systems are the same: (i) radiative interference and (ii) symmetry breaking. Radiative interference can be controlled by designing strongly coupled nanostructures featuring broad (super-radiant) and narrow (sub-radiant) resonances. By breaking the symmetry of a nanostructure, we can make the sub-radiant and super-radiant resonances coupled, leading to pronounced Fano effects.

Two-particle Fano systems

Here, we consider an interesting two-particle system called the inner-outer pair (see Fig. 4.12). It consists of two concentric particles that are not in conductive contact with each other: two-dimensional ring-disk nanocavities (panel A) [44, 50–53], three-dimensional nanosphere-in-a-nanoshell (panel B) [45, 54–56], and concen-

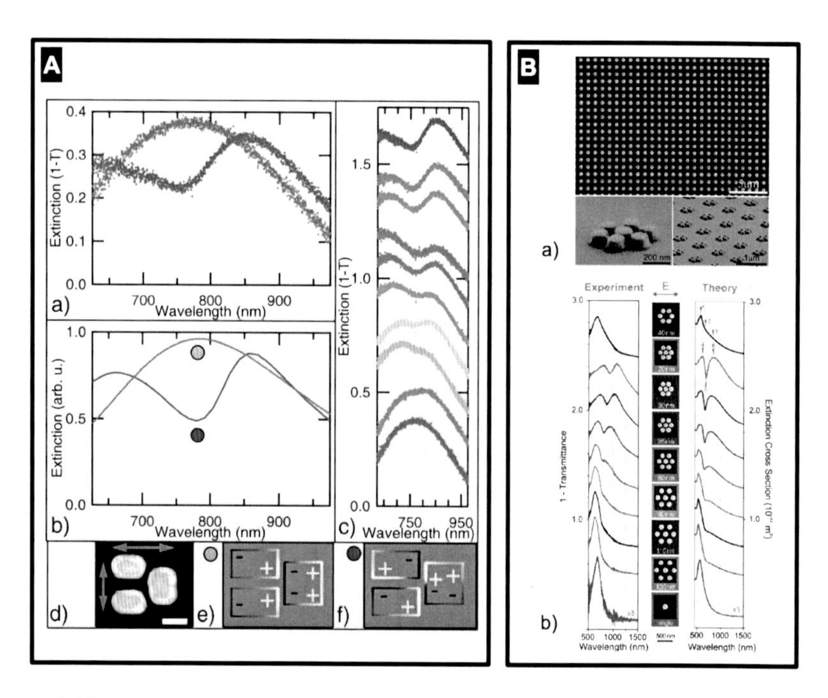

Figure 4.13: Fano resonances in multi-particle systems. Panel A: Fano resonance of a dolmen structure. (a) Experimental and (b) simulated extinction spectra. The colors of the trend lines represent different polarizations, as shown by the arrows in (d). (c) The change of the experimentally measured extinction as the polarization direction is changed in regular 10 degree steps. (e) Calculated surface charge distributions of the dipolar mode and (f) the Fano extinction dip, respectively. Reprinted (adapted) with permission from Reference [48]. © 2009 American Chemical Society. Panel B: Fano resonance of a metallic heptamer. (a) SEM images of a typical gold heptamer sample. (b) Extinction spectra of a gold monomer, a gold hexamer, and gold heptamers with different inter-particle gap separations. Left column: The experimental extinction spectra. Middle column: SEM images of the corresponding samples. Right column: Simulated extinction cross-section spectra. In the gold monomer and hexamer, dipolar polariton resonances are observed. The transition from isolated to collective modes is clearly visible in the different heptamers when decreasing the inter-particle gap distance. Specifically, a pronounced Fano resonance is formed as characterized by the distinct resonance dip when the inter-particle gap distance is below 60 nm. The presence or absence of the central nanoparticle can switch on or off the formation of the Fano resonance. Reprinted (adapted) with permission from Reference [49]. © 2010 American Chemical Society.

tric nanoshells [1, 57]. The presence of multiple overlapping resonances in this concentric-sphere geometry gives rise to coherent coupling effects such as Fano resonances [44, 45] and strongly enhanced fields that can be used for biosensing (panel A) [44]. Both super-radiant and sub-radiant eigenmodes are also observed in such structures [53]. Like dimers described in an earlier section, the behavior of inner-outer systems can also be understood with the polariton hybridization model [1].

Multi-particle Fano systems

We will describe two examples of multi-particle systems (Fig. 4.13) that exhibit Fano resonances. They are the dolmen structure (panel A) [48] and the metallic nanoparticle cluster (panel B) [49, 58, 59]. The dolmen structure possesses two resonances with different polarization charge distributions. When the incident electric field is perpendicular to the symmetry axis (blue curve), it excites counter-propagating currents in the two parallel slabs, resulting in the dark bonding quadrupolar eigenmode. This quadrupolar eigenmode is weakly coupled to the incident light. However, the presence of the third slab in this structure allows the dispersive coupling between the sharp bonding and broad dipolar eigenmodes. This explains the asymmetric Fano line shape (blue curve) with the central dip at about $\lambda = 780$ nm.

Fano resonances are also observed in metallic nanoparticle clusters (e.g., heptamers) [49] (Fig. 4.13, panel B). A heptamer consists of a central particle surrounded by six equivalent nanoparticles of the same diameter. Owing to the specific symmetry of this structure, the collective polaritons are the hybridized states formed by the interaction of the central particle eigenmodes with the collective eigenmodes of the surrounding six-particle ring structure. Here, we see how the coupling between the nanoparticles increases with decreasing inter-particle distance, resulting in a strong Fano resonance for the compact heptamer. When the central particle is removed, the counter-oscillating dipole is absent, and the destructive interference that causes the Fano interference is turned off.

All-dielectric Fano systems

The eigenmodes of symmetric dielectric quadrumers sorted by the resonance frequencies are shown in Fig. 4.14 [36]. Similar to dimers, the properties of incident radiation determine which eigenmodes resonate. A more detailed description of their behavior can be found in Reference [36]. Fano resonances have also been reported in anti-symmetric dielectric quadrumers [60]. They arise from the different environments of the particles in the oligomer. In the vicinity of the Fano resonance is a large overlap between the electric and magnetic modes; this leads to dichroism of chiral dielectric oligomers due to different loss mechanisms (absorption or scattering), depending on the handedness of the incoming light [61].

4.2.4 *Chirality resonances*

The chiral systems are desirable for two reasons. First, chirality is important in biology and chemistry, where the handedness defines physical and chemical properties

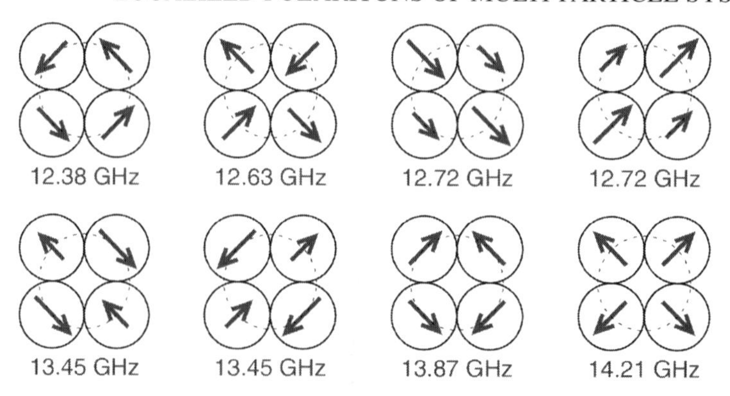

Figure 4.14: Eigenmodes of a symmetric dielectric quadrumer. The polaritons in the top row have a smaller frequency (redshifted) than a single particle, while those in the lower row have a higher frequency (blueshifted). Reprinted from Reference [36]. © 2010 Optical Society of America.

of chiral molecules. As a result, there is a need to detect and distinguish molecules responses to the right and left circularly polarized light. This is difficult for most non-chiral systems. Second, planar plasmonic structures which are "handed", despite not being truly chiral, interact very strongly with circularly polarized light. This demonstrates the great potential of structures that are truly chiral. Consequently, we see that chiral plasmonics and photonics play key roles in the detection of chiral molecules and design of optical devices providing chiral resonances.

Chiral structures can be fabricated by direct laser writing (Fig. 4.15, panel A) [62, 66, 67], multilayer e-beam lithography to make stereo-metamaterials (panel B) [34, 63, 68] and 3D oligomers (panel D) [65, 69, 70], ordinary self-assembly, [71, 72] and DNA-controlled self-assembly (panel C) [64, 73]. More recent fabrication techniques include hybrid techniques combining direct laser writing with electron-beam lithography [74] and large-scale low-cost manufacturing methods [75]. Operation and performance of chiral structures will be discussed in more detail in Chapter 7.

4.3 Polariton hybridization in complex multi-particle systems

This section will touch on polariton hybridization in complex multi-particle systems, namely image-coupled nanoparticle-on-mirror systems (Section 4.3.1), metal-dielectric hybrid systems (Section 4.3.2), 1D chains and 2D arrays (Section 4.3.3).

4.3.1 Image-coupled nanoparticle-on-mirror systems

One special type of two-particle system consists of a nanoparticle near a thick metal or high-index substrate (Fig. 4.7). The substrate acts as a mirror, enabling the nanoparticle to interact with its image charge [76, 77]. Consequently, such a system is similar to a dimer. The interaction with its image alone will redshift the nanoparticle polaritons. In addition to the particle-image coupling, there is a strong interac-

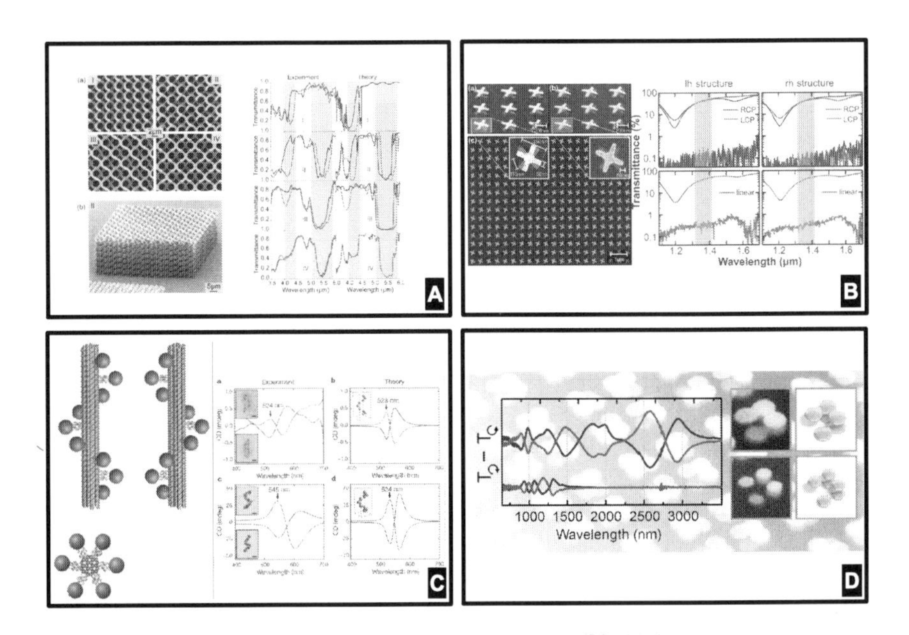

Figure 4.15: Examples of resonant chiral systems. Panel A: Three-dimensional photonic superlattices composed of polymeric helices in various spatial checkerboard-like arrangements. Depending on the relative phase shift and handedness of the chiral building blocks, different circular dichroism resonances appear or are suppressed. Samples corresponding to four different configurations are fabricated by direct laser writing. Reprinted (adapted) with permission from Reference [62]. © 2010 Optical Society of America. Panel B: Twisted gold crosses are circularly dichroic. The dichroism handedness is different at shorter and longer wavelengths. Reprinted (adapted) with permission from Reference [63]. © 2009 Optical Society of America. Panel C: Gold nanoparticles can self-assemble on DNA, leading to chiral structures. Reprinted (adapted) with permission from Reference [64]. © 2012 Nature Publishing Group. Panel D: 3D oligomers: the nanoparticles are touching (top) and isolated (bottom). The broadband response of the touching nanoparticles is a result of the charge transfer between the individual nanoparticles. The ohmic contact causes a strong redshift of the lower-order eigenmode, while the geometrical shape of the resulting oligomer still allows efficient excitation of the higher-order polaritons. Reprinted (adapted) with permission from Reference [65]. © 2012 American Chemical Society.

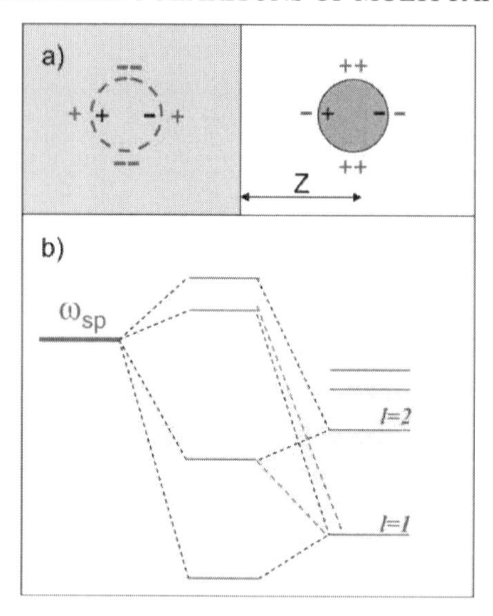

Figure 4.16: Image-coupled nanoparticle-on-mirror system. (a) Schematic illustrating the interaction of a nanosphere with a metallic surface. The nanoparticle with its polarization charge distribution (right) and the corresponding image (left). (b) Hybridization of the nanoparticle localized polaritons and the propagating surface interface polaritons. Reprinted (adapted) with permission from Reference [76]. © 2004 American Chemical Society.

tion between the localized eigenoscillations of the nanostructure and the propagating eigenwaves of the substrate. Hybridization in this system produces either a redshift or a blueshift in the nanoparticle polaritons, depending on the relative energies of the nanoparticle and substrate eigenmodes (Fig. 4.16). Using such a system, a small gap distance can be reliably achieved for a large number of particles [78].

The particle-substrate coupling is altered significantly when the substrate is sufficiently thin. However, this is out of the scope of the present chapter. A good review on plasmonic particles on a metal interface can be found in Reference [77], while the discussion on dielectric particles is given in Reference [79].

4.3.2 Metal-dielectric hybrid systems

We reviewed the main properties of metallic and dielectric particles individually. But what happens in a system that contains both of them? It has been shown that by using an array of alternating dielectric and metallic particles, we can achieve more control over the electric and magnetic responses [80–82]. These systems allow us to access fascinating properties such as "anti-ferromagnetic light" [83].

Another example of a metal-dielectric hybrid system is microsphere interference enhanced polariton resonance, in which metallic nanoparticles are encapsulated by a larger semiconductor microsphere. By studying a simpler system with only

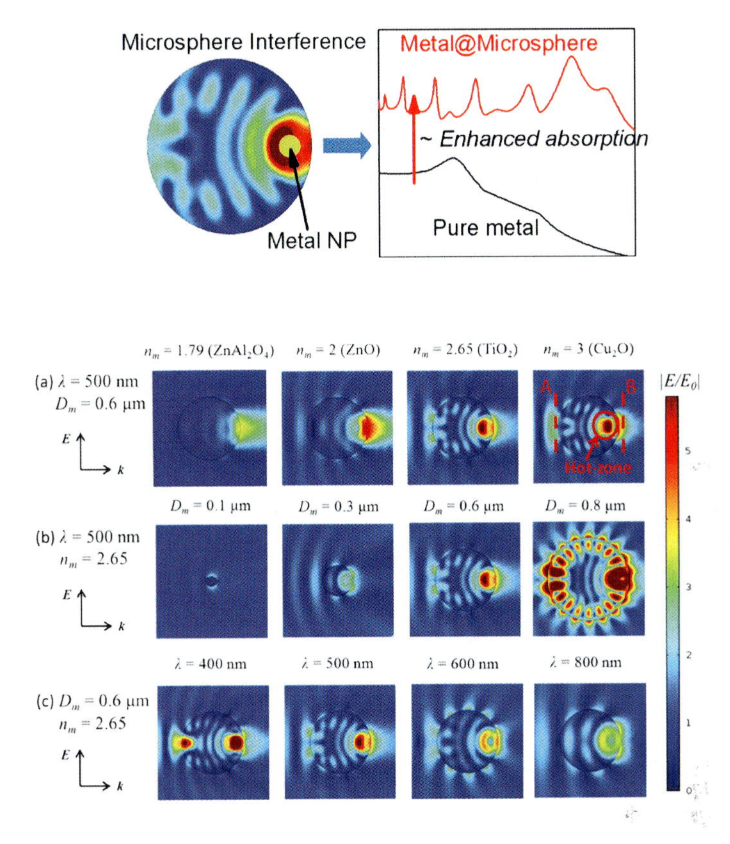

Figure 4.17: An example of a metal-dielectric hybrid system encapsulating metallic nanoparticles in a larger semiconductor microsphere. An interference pattern is formed inside the semiconductor microsphere due to the optical reflection and refraction at the interface between the microsphere and the catalytic medium. Reprinted (adapted) with permission from Reference [84]. © 2014 American Chemical Society.

one metallic nanoparticle in a larger semiconductor particle, we can understand the physics of such systems [84]. The study was extended to multiple metallic nanoparticles in a larger semiconductor particle [85]. This type of multiple-particle system has been successfully fabricated and used in photo-catalysis [86–92]. Such systems exploit interference to achieve broadband absorption over the entire visible range (Fig. 4.17).

In particular, it has been shown that an interference pattern is uniquely generated

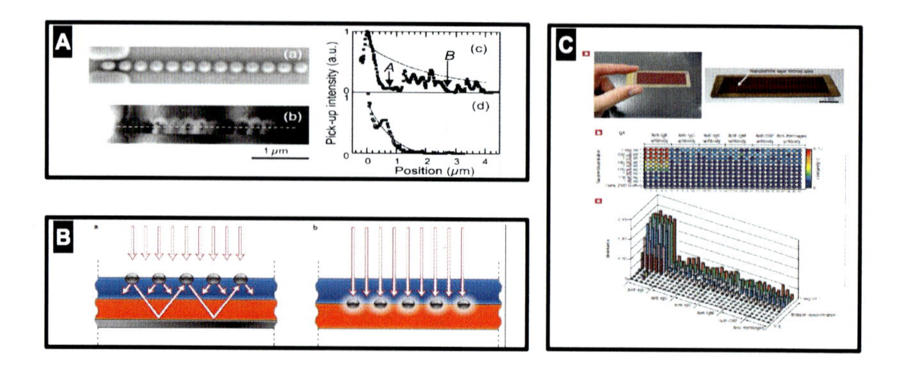

Figure 4.18: Examples of 1D chains and 2D arrays. Panel A: A linearly chained metallic nanodot coupler. Reprinted (adapted) with permission from Reference [93]. © 2005 American Institute of Physics. Panel B: Light trapping (a) by scattering from metal nanoparticles at the surface of the solar cell and (b) by the excitation of localized surface polaritons in metal nanoparticles embedded in the semiconductor. Reprinted (adapted) with permission from Reference [94]. © 2010 Nature Publishing Group. Panel C: (a) Biochip for multiplexed detection. (b) and (c) Absorbance measurements at individual locations on the array. Reprinted (adapted) with permission from Reference [95]. © 2006 American Chemical Society.

inside the semiconductor microsphere due to the optical reflection and refraction at the interface between the microsphere and the catalytic medium. Based on the properties of the interference, the broadband absorption enhancement can be obtained everywhere inside the microsphere and is particularly large at the microsphere hot zone. The studies also showed that a microsphere consisting of higher refractive index semiconductor can maximize the interference-induced broadband absorption enhancement. Besides, nanoparticles with different materials can be mixed to tune the overall absorption band for flexible energy harvesting and enhanced selectivity. At the same time, the evanescent nature of the near fields could be better exploited to enhance the catalytic rate if the nanoparticles are placed close to the microsphere surface.

4.3.3 Periodically ordered particles

We refer to periodically ordered nanoparticles as a particle array system. Each unit cell of the array may contain a single nanoparticle or a nanoparticle cluster. In this section, we will analyze such periodic systems by reviewing the interactions between unit cells.

One-dimensional (1D) chains

Near-field coupling in linear chains of resonant nanoparticles causes a number of interesting phenomena. The polaritons supported by such chains are no longer localized on individual nanoparticles, but can propagate along the chain, allowing electromagnetic energy transport. Consequently, these 1D chains act as waveguides with dimensions significantly smaller, than the free space wavelength of the guided light (see Fig. 4.18, panel A) [93, 96]. The main limitation of metallic nanoparticle chain waveguides is the significant loss factor. Consequently, the electromagnetic energy can be transmitted over short distances only. In particular, Reference [93] reports on the 4.0 μm transmission length of closely spaced metallic nanoparticles, which is three times longer than that of a metallic core waveguide owing to the efficient near-field coupling between the localized surface polaritons of neighboring nanoparticles. Current research efforts strive to seek alternatives to metals (for example, silicon) that enable longer transmission with reasonable energy confinement. This will be addressed in more detail in Chapter 10.

Two-dimensional (2D) arrays

Unlike 1D arrays of metallic nanoparticles, the 2D counterparts have been more popular due to their usefulness in a variety of applications. Here we briefly discuss three examples of such applications (Fig. 4.18, panels B and C).

In order to increase the efficiency of photovoltaic devices, metallic or dielectric nanoparticles are used to improve light trapping inside solar cells. Depending on where the nanoparticles are placed, the efficiency is increased by either increasing the optical path of the light inside the active layer [Fig. 4.18, panel B: (a)] or by resonating on localized polaritons that then create electron-hole pairs [Fig. 4.18, panel B: (b)]. More details can be found in the comprehensive review in Reference [94]. Metallic particle arrays can also be used for sensing [95, 97]. For example, the biochip (shown in Fig. 4.18, panel C) provides rapid, label-free detection of protein concentrations in small sample volumes. By functionalizing specific spots on the chip, various chemical and biological compounds may be detected. More examples of particle array based sensing will be illustrated in Chapter 8. Another creative application of such arrays is color printing [98] at the optical diffraction limit (100,000 DPI). This color printing application will be elaborated in more detail in Chapter 5.

4.4 Summary

In this chapter, we reviewed the fundamentals of polariton hybridization in multi-particle systems. We have shown that the polaritons of complex systems can be described as the hybridizations of the individual subsystems' eigenmodes. We briefly discussed the basic effects caused by such hybridizations in photonic, plasmonic, and hybrid structures. The next two parts of the book will cover the use of coupled polaritons in multi-particle systems with focus on specific optical applications of localized and propagating eigenmodes.

Bibliography

[1] E. Prodan, C. Radloff, N. J. Halas, and P. Nordlander, "A hybridization model for the plasmon response of complex nanostructures," *Science*, vol. 302, pp. 419–422, 2003.

[2] P. Nordlander, C. Oubre, E. Prodan, K. Li, and M. I. Stockman, "Plasmon hybridization in nanoparticle dimers," *Nano Lett.*, vol. 4, pp. 899–903, 2004.

[3] D. W. Brandl, C. Oubre, and P. Nordlander, "Plasmon hybridization in nanoshell dimers," *J. Chem. Phys*, vol. 123, p. 024701, 2005.

[4] P. K. Jain, S. Eustis, and M. A. El-Sayed, "Plasmon coupling in nanorod assemblies: optical absorption, discrete dipole approximation simulation, and exciton-coupling model," *J. Phys. Chem. B*, vol. 110, pp. 18243–18253, 2006.

[5] A. M. Funston, C. Novo, T. J. Davis, and P. Mulvaney, "Plasmon coupling of gold nanorods at short distances and in different geometries," *Nano Lett.*, vol. 9, pp. 1651–1658, 2009.

[6] P. K. Jain, W. Huang, and M. A. El-Sayed, "On the universal scaling behavior of the distance decay of plasmon coupling in metal nanoparticle pairs: a plasmon ruler equation," *Nano Lett.*, vol. 7, pp. 2080–2088, 2007.

[7] N. Grillet, D. Manchon, F. Bertorelle, C. Bonnet, M. Broyer, E. Cottancin, J. Lerme, M. Hillenkamp, and M. Pellarin, "Plasmon coupling in silver nanocube dimers: resonance splitting induced by edge rounding," *ACS Nano*, vol. 5, pp. 9450–9462, 2011.

[8] A. L. Koh, A. I. Fernndez-Domnguez, D. W. McComb, S. A. Maier, and J. K. W. Yang, "High-resolution mapping of electron-beam-excited plasmon modes in lithographically defined gold nanostructures," *Nano Lett.*, vol. 11, pp. 1323–1330, 2011.

[9] H. Duan, A. I. Fernndez-Domnguez, M. Bosman, and S. A. Maier, "Nanoplasmonics: classical down to the nanometer scale," *Nano Lett.*, vol. 12, pp. 1683–1689, 2012.

[10] C.-Y. Tsai, J.-W. Lin, C.-Y. Wu, P.-T. Lin, T.-W. Lu, and P.-T. Lee, "Plasmonic coupling in gold nanoring dimers: observation of coupled bonding mode," *Nano Lett.*, vol. 12, pp. 1648–1654, 2012.

[11] L. V. Brown, H. Sobhani, J. B. Lassiter, P. Nordlander, and N. J. Halas, "Heterodimers: plasmonic properties of mismatched nanoparticle pairs," *ACS Nano*, vol. 4, pp. 819–832, 2010.

[12] S. Sheikholeslami, Y. wook Jun, P. K. Jain, and A. P. Alivisatos, "Coupling of optical resonances in a compositionally asymmetric plasmonic nanoparticle dimer," *Nano Lett.*, vol. 10, pp. 2655–2660, 2010.

[13] T. Atay, J.-H. Song, and A. V. Nurmikko, "Strongly interacting plasmon nanoparticle pairs: from dipole-dipole interaction to conductively coupled regime," *Nano Lett.*, vol. 4, pp. 1627–1631, 2004.

[14] I. Romero, J. Aizpurua, G. W. Bryant, and F. J. G. de Abajo, "Plasmons in

nearly touching metallic nanoparticles: singular response in the limit of touching dimers," *Opt. Express*, vol. 14, pp. 9988–9999, 2006.

[15] J. Zuloaga, E. Prodan, and P. Nordlander, "Quantum description of the plasmon resonances of a nanoparticle dimer," *Nano Lett.*, vol. 9, pp. 887–891, 2009.

[16] O. Perez-Gonzalez, N. Zabala, A. G. Borisov, N. J. Halas, P. Nordlander, and J. Aizpurua, "Optical spectroscopy of conductive junctions in plasmonic cavities," *Nano Lett.*, vol. 10, pp. 3090–3095, 2010.

[17] R. Esteban, A. G. Borisov, P. Nordlander, and J. Aizpurua, "Bridging quantum and classical plasmonics with a quantum-corrected model," *Nat. Commun.*, vol. 3, p. 825, 2012.

[18] L. Wu, H. Duan, P. Bai, M. Bosman, J. K. W. Yang, and E. Li, "Fowler-Nordheim tunneling induced charge transfer plasmons between nearly touching nanoparticles," *ACS Nano*, vol. 7, pp. 707–716, 2013.

[19] J. B. Lassiter, J. Aizpurua, L. I. Hernandez, D. W. Brandl, I. Romero, S. Lal, J. H. Hafner, P. Nordlander, and N. J. Halas, "Close encounters between two nanoshells," *Nano Lett.*, vol. 8, pp. 1212–1218, 2008.

[20] K. J. Savage, M. M. Hawkeye, R. Esteban, A. G. Borisov, J. Aizpurua, and J. J. Baumberg, "Revealing the quantum regime in tunnelling plasmonics," *Nature*, vol. 491, pp. 574–577, 2012.

[21] F. Wen, Y. Zhang, S. Gottheim, N. S. K. andYu Zhang, P. Nordlander, and N. J. Halas, "Charge transfer plasmons: optical frequency conductances and tunable infrared resonances," *ACS Nano*, vol. 9, pp. 6428–6435, 2015.

[22] N. Large, M. Abb, J. Aizpurua, and O. L. Muskens, "Photoconductively loaded plasmonic nanoantenna as building block for ultracompact optical switches," *Nano Lett.*, vol. 10, pp. 1741–1746, 2010.

[23] D. Marinica, A. Kazansky, P. Nordlander, J. Aizpurua, and A. G. Borisov, "Quantum plasmonics: nonlinear effects in the field enhancement of a plasmonic nanoparticle dimer," *Nano Lett.*, vol. 12, pp. 1333–1339, 2012.

[24] R. Esteban, G. Aguirregabiria, A. G. Borisov, Y. M. Wang, P. Nordlander, G. W. Bryant, and A. J, "The morphology of narrow gaps modifies the plasmonic response," *ACS Photonics*, vol. 2, pp. 295–305, 2015.

[25] Y.-C. Yeo, T.-J. King, and C. Hu, "Metal-dielectric band alignment and its implications for metal gate complementary metal-oxide-semiconductor technology," *J. Appl. Phys.*, vol. 92, pp. 7266–7271, 2002.

[26] S. F. Tan, L. Wu, J. K. W. Yang, P. Bai, M. Bosman, and C. A. Nijhuis, "Quantum plasmon resonances controlled by molecular tunnel junctions," *Science*, vol. 343, pp. 1496–1499, 2014.

[27] F. Benz, C. Tserkezis, L. O. Herrmann, B. de Nijs, A. Sanders, D. O. Sigle, L. Pukenas, S. D. Evans, J. Aizpurua, and J. J. Baumberg, "Nanooptics of molecular-shunted plasmonic nanojunctions," *Nano Lett.*, vol. 15, pp. 669–674, 2015.

[28] A. Nitzan and M. A. Ratner, "Electron transport in molecular wire junctions," *Science*, vol. 300, pp. 1384–1389, 2003.

[29] C. Joachim and M. A. Ratner, "Molecular electronics: some views on transport junctions and beyond," *Proc. Natl. Acad. Sci.*, vol. 102, pp. 8801–8808, 2005.

[30] S. J. Barrow, A. M. Funston, D. E. Gmez, T. J. Davis, and P. Mulvaney, "Surface plasmon resonances in strongly coupled gold nanosphere chains from monomer to hexamer," *Nano Lett.*, vol. 11, pp. 4180–4187, 2011.

[31] M. Hentschel, D. Dregely, R. Vogelgesang, H. Giessen, and N. Liu, "Plasmonic oligomers: the role of individual particles in collective behavior," *ACS Nano*, vol. 5, pp. 2042–2050, 2011.

[32] A. J. Mastroianni, S. A. Claridge, and A. P. Alivisatos, "Pyramidal and chiral groupings of gold nanocrystals assembled using dna scaffolds," *J. Am. Chem. Soc.*, vol. 131, pp. 8455–8459, 2009.

[33] N. Liu and H. Giessen, "Three-dimensional optical metamaterials as model systems for longitudinal and transverse magnetic coupling," *Opt. Express*, vol. 16, pp. 21233–21238, 2008.

[34] N. Liu, H. Liu, S. Zhu, and H. Giessen, "Stereometamaterials," *Nat. Photonics*, pp. 157–162, 2009.

[35] N. Liu and H. Giessen, "Coupling effects in optical metamaterials," *Angew. Chem. Int. Ed.*, vol. 49, pp. 9838–9852, 2010.

[36] M. S. Wheeler, J. S. Aitchison, and M. Mojahedi, "Coupled magnetic dipole resonances in sub-wavelength dielectric particle clusters," *J. Opt. Soc. Am. B*, vol. 27, no. 5, pp. 1083–1091, 2010.

[37] A. Mirzaei and A. E. Miroshnichenko, "Electric and magnetic hotspots in dielectric nanowire dimers," *Nanoscale*, vol. 7, no. 14, pp. 5963–5968, 2015.

[38] P. Albella, M. A. Poyli, M. K. Schmidt, S. A. Maier, F. Moreno, J. J. Senz, and J. Aizpurua, "Low-loss electric and magnetic field-enhanced spectroscopy with subwavelength silicon dimers," *J. Phys. Chem. C*, vol. 117, no. 26, pp. 13573–13584, 2013.

[39] P. Albella, T. Shibanuma, and S. A. Maier, "Switchable directional scattering of electromagnetic radiation with subwavelength asymmetric silicon dimers," *Sci. Rep.*, vol. 5, p. 18322, 2015.

[40] I. Staude, A. E. Miroshnichenko, M. Decker, N. T. Fofang, S. Liu, E. Gonzales, J. Dominguez, T. S. Luk, D. N. Neshev, I. Brener, and Y. Kivshar, "Tailoring directional scattering through magnetic and electric resonances in subwavelength silicon nanodisks," *ACS Nano*, vol. 7, no. 9, pp. 7824–7832, 2013.

[41] A. E. Krasnok, A. E. Miroshnichenko, P. A. Belov, and Y. S. Kivshar, "Huygens optical elements and Yagi-Uda nanoantennas based on dielectric nanoparticles," *JETP Lett.*, vol. 94, no. 8, pp. 593–598, 2011.

[42] A. E. Krasnok, A. E. Miroshnichenko, P. A. Belov, and Y. S. Kivshar, "All-dielectric optical nanoantennas," *Opt. Express*, vol. 20, no. 18, pp. 20599–

20604, 2012.

[43] A. E. Krasnok, C. R. Simovski, P. A. Belov, and Y. S. Kivshar, "Superdirective dielectric nanoantennas," *Nanoscale*, vol. 6, no. 13, pp. 7354–7361, 2014.

[44] A. E. Cetin and H. Altug, "Fano resonant ring/disk plasmonic nanocavities on conducting substrates for advanced biosensing," *ACS Nano*, vol. 6, pp. 9989–9995, 2012.

[45] S. Mukherjee, H. Sobhani, J. B. Lassiter, R. Bardhan, P. Nordlander, and N. J. Halas, "Fanoshells: nanoparticles with built-in Fano resonances," *Nano Lett.*, vol. 10, pp. 2694–2701, 2010.

[46] B. Lukyanchuk, N. I. Zheludev, S. A. Maier, N. J. Halas, P. Nordlander, H. Giessen, and C. T. Chong, "The Fano resonance in plasmonic nanostructures and metamaterials," *Nat. Mater.*, vol. 9, pp. 707–715, 2010.

[47] A. E. Miroshnichenko, S. Flach, and Y. S. Kivshar, "Fano resonances in nanoscale structures," *Rev. Mod. Phys.*, vol. 82, pp. 2257–2297, 2010.

[48] N. Verellen, Y. Sonnefraud, H. Sobhani, F. Hao, V. V. Moshchalkov, P. V. Dorpe, P. Nordlander, and S. A. Maier, "Fano resonances in individual coherent plasmonic nanocavities," *Nano Lett.*, vol. 9, pp. 1663–1667, 2009.

[49] M. Hentschel, M. Saliba, R. Vogelgesang, H. Giessen, A. P. Alivisatos, and N. Liu, "Transition from isolated to collective modes in plasmonic oligomers," *Nano Lett.*, vol. 10, pp. 2721–2726, 2010.

[50] F. Hao, P. Nordlander, M. T. Burnett, and S. A. Maier, "Enhanced tunability and linewidth sharpening of plasmon resonances in hybridized metallic ring/disk nanocavities," *Phys. Rev. B*, vol. 76, p. 245417, 2007.

[51] F. Hao, Y. Sonnefraud, P. V. Dorpe, S. A. Maier, N. J. Halas, and P. Nordlander, "Symmetry breaking in plasmonic nanocavities: subradiant lspr sensing and a tunable fano resonance," *Nano Lett.*, vol. 8, pp. 3983–3988, 2008.

[52] F. Hao, P. Nordlander, Y. Sonnefraud, P. V. Dorpe, and S. A. Maier, "Tunability of subradiant dipolar and Fano-type plasmon resonances inmetallic ring/disk cavities: implications for nanoscale optical sensing," *ACS Nano*, vol. 3, pp. 643–652, 2009.

[53] Y. Sonnefraud, N. Verellen, H. Sobhani, G. A. E. Vandenbosch, V. V. Moshchalkov, P. V. Dorpe, P. Nordlander, and S. A. Maier, "Experimental realization of subradiant, superradiant, and Fano resonances in ring/disk plasmonic nanocavities," *ACS Nano*, vol. 4, pp. 1664–1670, 2010.

[54] D. Wu and X. Liu, "Tunable near-infrared optical properties of three-layered gold-silica-gold nanoparticles," *Appl. Phys. B*, vol. 97, pp. 193–197, 2009.

[55] Y. Hu, S. J. Noelck, and R. A. Drezek, "Symmetry breaking in gold-silica-gold multilayer nanoshells," *ACS Nano*, vol. 4, pp. 1521–1528, 2010.

[56] R. Bardhan, S. Mukherjee, N. A. Mirin, S. D. Levit, P. Nordlander, and N. J. Halas, "Nanosphere-in-a-nanoshell: a simple nanomatryushka," *J. Phys. Chem. C*, vol. 114, pp. 7378–7383, 2010.

[57] C. Radloff and N. J. Halas, "Plasmonic properties of concentric nanoshells," *Nano Lett.*, vol. 4, pp. 1323–1327, 2004.

[58] J. A. Fan, C. Wu, K. Bao, J. Bao, R. Bardhan, N. J. Halas, V. N. Manoharan, P. Nordlander, G. Shvets, and F. Capasso, "Self-assembled plasmonic nanoparticle clusters," *Science*, vol. 328, pp. 1135–1138, 2010.

[59] K. Bao, N. Mirin, and P. Nordlander, "Fano resonances in planar silver nanosphere clusters," *Appl. Phys. A*, vol. 100, pp. 333–339, 2010.

[60] A. Ahmadivand and N. Pala, "Multiple coil-type Fano resonances in all-dielectric antisymmetric quadrumers," *Opt. Quant. Electron.*, vol. 47, p. 2055, 2015.

[61] B. Hopkins, A. N. Poddubny, A. E. Miroshnichenko, and Y. S. Kivshar, "Circular dichroism induced by Fano resonances in planar chiral oligomers," *Laser Photon. Rev.*, vol. 10, no. 1, pp. 137–146, 2016.

[62] M. Thiel, H. Fischer, G. von Freymann, and M. Wegener, "Three-dimensional chiral photonic superlattices," *Optics Letters*, vol. 35, pp. 166–168, 2010.

[63] M. Decker, M. Ruther, C. E. Kriegler, J. Zhou, C. M. Soukoulis, S. Linden, and M. Wegener, "Strong optical activity from twisted-cross photonic metamaterials," *Opt. Lett.*, vol. 34, pp. 2501–2503, 2009.

[64] A. Kuzyk, R. Schreiber, Z. Fan, G. Pardatscher, E.-M. Roller, A. Hogele, F. C. Simmel, A. O. Govorov, and T. Liedl, "DNA-based self-assembly of chiral plasmonic nanostructures with tailored optical response," *Nature*, vol. 483, pp. 311–314, 2012.

[65] M. Hentschel, L. Wu, M. Schferling, P. Bai, E. P. Li, and H. Giessen, "Optical properties of chiral three-dimensional plasmonic oligomers at the onset of charge-transfer plasmons," *ACS Nano*, vol. 6, pp. 10355–10365, 2012.

[66] J. K. Gansel, M. Thiel, M. S. Rill, M. Decker, K. Bade, V. Saile, G. v. Freymann, S. Linden, and M. Wegener, "Gold helix photonic metamaterial as broadband circular polarizer," *Science*, vol. 325, pp. 1513–1515, 2009.

[67] A. Radke, T. Gissibl, T. Klotzbucher, P. V. Braun, and H. Giessen, "Three-dimensional bichiral plasmonic crystals fabricated by direct laser writing and electroless silver plating," *Adv. Mater.*, vol. 23, pp. 3018–3021, 2011.

[68] H. Liu, J. Cao, S. Zhu, N. Liu, R. Ameling, and H. Giessen, "Lagrange model for the chiral optical properties of stereometamaterials," *Phys. Rev. B*, vol. 81, p. 241403(R), 2010.

[69] M. Hentschel, M. Schferling, T. Weiss, N. Liu, and H. Giessen, "Three-dimensional chiral plasmonic oligomers," *Nano Lett.*, vol. 12, pp. 2542–2547, 2012.

[70] M. Hentschel, M. Schferling, B. Metzger, and H. Giessen, "Plasmonic diastereomers: adding up chiral centers," *Nano Lett.*, vol. 13, pp. 600–606, 2013.

[71] D. Zerrouki, J. Baudry, D. Pine, P. Chaikin, and J. Bibette, "Chiral colloidal clusters," *Nature*, vol. 455, pp. 380–382, 2008.

[72] M. A. Olson, A. Coskun, R. Klajn, L. Fang, S. K. Dey, K. P. Browne, B. A. Grzybowski, and J. F. Stoddart, "Assembly of polygonal nanoparticle clusters directed by reversible noncovalent bonding interactions," *Nano Lett.*, vol. 9, pp. 3185–3190, 2009.

[73] J. Sharma, R. Chhabra, A. Cheng, J. Brownell, Y. Liu, and H. Yan, "Control of self-assembly of DNA tubules through integration of gold nanoparticles," *Science*, vol. 323, pp. 112–116, 2009.

[74] I. Staude, M. Decker, M. J. Ventura, C. Jagadish, D. N. Neshev, M. Gu, and Y. S. Kivshar, "Hybrid high-resolution three-dimensional nanofabrication for metamaterials and nanoplasmonics," *Adv. Mater.*, vol. 25, pp. 1260–1264, 2012.

[75] B. Frank, X. Yin, M. Schaferling, J. Zhao, S. M. Hein, P. V. Braun, and H. Giessen, "Large-area 3D chiral plasmonic structures," *ACS Nano*, vol. 7, pp. 6321–6329, 2013.

[76] P. Nordlander and E. Prodan, "Plasmon hybridization in nanoparticles near metallic surfaces," *Nano Lett.*, vol. 4, pp. 2209–2213, 2004.

[77] N. J. Halas, S. Lal, W.-S. Chang, S. Link, and P. Nordlander, "Plasmons in strongly coupled metallic nanostructures," *Chem. Rev.*, vol. 111, pp. 3913–3961, 2011.

[78] J. Mertens, A. Eiden, D. Sigle, F. Huang, A. Lombardo, Z. Sun, R. Sundaram, A. Colli, C. Tserkezis, J. Aizpurua, and S. Milana, "Controlling subnanometer gaps in plasmonic dimers using graphene," *Nano lett.*, vol. 13, no. 11, pp. 5033–5038, 2013.

[79] E. Xifré-Pérez, L. Shi, U. Tuzer, R. Fenollosa, F. Ramiro-Manzano, R. Quidant, and F. Meseguer, "Mirror-image-induced magnetic modes," *ACS nano*, vol. 7, no. 1, pp. 664–668, 2012.

[80] B. Garc´ia-Cámara, F. Moreno, F. González, and O. J. F. Martin, "Light scattering by an array of electric and magnetic nanoparticles," *Opt. Express*, vol. 18, no. 10, pp. 10001–10015, 2010.

[81] Y. Hong, Y. Qiu, T. Chen, and B. M. Reinhard, "Rational assembly of optoplasmonic hetero-nanoparticle arrays with tunable photonic–plasmonic resonances," *Adv. Funct. Mater.*, vol. 24, no. 6, pp. 739–746, 2014.

[82] B. M. Reinhard, W. Ahn, Y. Hong, S. V. Boriskina, and X. Zhao, "Template-guided self-assembly of discrete optoplasmonic molecules and extended optoplasmonic arrays," *Nanophotonics*, vol. 4, no. 1, pp. 250–260, 2015.

[83] A. E. Miroshnichenko, B. Luk'yanchuk, S. A. Maier, and Y. S. Kivshar, "Optically induced interaction of magnetic moments in hybrid metamaterials," *ACS Nano*, vol. 6, no. 1, pp. 837–842, 2011.

[84] S. Sun, H. Liu, L. Wu, C. Png, and P. Bai, "Interference-induced broadband absorption enhancement for plasmonic-metal@semiconductor microsphere as visible light photocatalyst," *ACS Catal.*, vol. 4, pp. 4269–4276, 2014.

[85] S. Sun, L. Wu, E. C. Png, and P. Bai, "Nanoparticle loading effects on the broadband absorption for plasmonic-metal@semiconductor-microsphere pho-

tocatalyst," *Catalysis Today*, no. doi:10.1016/j.cattod.2016.01.023, 2016.

[86] Y. Deng, Y. Cai, Z. Sun, J. Liu, C. Liu, J. Wei, W. Li, C. Liu, Y. Wang, and D. Zhao, "Multifunctional mesoporous composite microspheres with well-designed nanostructure: a highly integrated catalyst system," *J. Am. Chem. Soc.*, vol. 132, pp. 8466–8473, 2010.

[87] E. C. Cho, S. W. Choi, P. H. C. Camargo, and Y. Xia, "Thiol-induced assembly of au nanoparticles into chainlike structures and their fixing by encapsulation in silica shells or gelatin microspheres," *Langmuir*, vol. 26, pp. 10005–10012, 2010.

[88] L. C. Kong, G. T. Duan, G. M. Zuo, W. P. Cai, and Z. X. Cheng, "Rattle-type Au@TiO2 hollow microspheres with multiple nanocores and porous shells and their structurally enhanced catalysis," *Mater. Chem. Phys.*, vol. 123, pp. 421–426, 2010.

[89] Q. Zhang, D. Lima, I. Lee, F. Zaera, M. Chi, and Y. Yin, "A highly active titanium dioxide based visible-light photocatalyst with nonmetal doping and plasmonic metal decoration," *Angew. Chem.*, vol. 123, pp. 7226–7230, 2011.

[90] G. N. Wang, X. F. Wang, J. F. Liu, and X. M. Sun, "Mesoporous Au/TiO2 nanocomposite microspheres for visible-light photocatalysis," *Chem. Eur. J*, vol. 18, pp. 5361–5366, 2012.

[91] F. Dong, Q. Li, Y. Zhou, Y. Sun, H. Zhang, and Z. Wu, "In situ decoration of plasmonic ag nanocrystals on the surface of (BiO)2CO3 hierarchical microspheres for enhanced visible light photocatalysis," *Dalton Trans.*, vol. 43, pp. 9468–9480, 2014.

[92] R. Wang, X. Li, W. Cui, Y. Zhang, and F. Dong, "In situ growth of Au nanoparticles on 3D Bi2O2CO3 for surface plasmon enhanced visible light photocatalysis," *New J. Chem.*, vol. 39, pp. 8446–8453, 2015.

[93] W. Nomura, M. Ohtsu, and T. Yatsui, "Nanodot coupler with a surface plasmon polariton condenser for optical far/near-field conversion," *Appl. Phys. Lett.*, vol. 86, p. 181108, 2005.

[94] H. A. Atwater and A. Polman, "Plasmonics for improved photovoltaic devices," *Nat. Mater.*, vol. 9, pp. 205–213, 2010.

[95] T. Endo, K. Kerman, N. Nagatani, H. M. Hiepa, D.-K. Kim, and Y. Yonezawa, "Multiple label-free detection of antigen-antibody reaction using localized surface plasmon resonance-based core-shell structured nanoparticle layer nanochip.," *Anal. Chem.*, vol. 78, pp. 6465–6475, 2006.

[96] S. A. Maier, P. G. Kik, H. A. Atwater, S. Meltzer, E. Harel, B. E. Koel, and A. A. Requicha, "Local detection of electromagnetic energy transport below the diffraction limit in metal nanoparticle plasmon waveguides," *Nat. Mater.*, vol. 2, pp. 229–232, 2003.

[97] J. N. Anker, W. P. Hall, O. Lyandres, N. C. Shah, J. Zhao, and R. P. V. Duyne, "Biosensing with plasmonic nanosensors," *Nat. Mater.*, vol. 7, pp. 442–453, 2008.

[98] K. Kumar, H. Duan, R. S. Hegde, S. C. W. Koh, J. N. Wei, and J. K. W. Yang, "Printing colour at the optical diffraction limit," *Nat. Nanotechnol.*, vol. 7, pp. 557–561, 2012.

Part II

Applications of localized eigenmodes

Chapter 5

Nanostructural coloration

Ravi S. Hegde
Indian Institute of Technology Gandhinagar, India

This chapter will discuss the use of resonances on eigenmodes in coupled multi-particle systems for structural coloration applications. Section 5.1 will give a brief introduction to Colorimetry. Section 5.2 will discuss the principles of structural coloration. Section 5.3 will discuss emerging material platforms for coloration. The chapter will conclude with a summary in Section 5.4.

5.1 Introduction and background

Color is fundamental to the human experience. The human visual system has three different kinds of photoreceptors with peak sensitivities at three different wavelengths. The relative degree to which each of these individual receptors responds to a particular spectral distribution evokes various categories of color. The perceived color of an object by humans is often complex and subjective; however, it does depend upon the physical properties of the object such as its surface morphology, transmission, and emission spectra and also upon the ambient illumination, arrangement and properties of nearby objects. The generalized notion of spectral discrimination is also important for the functioning of non-human organisms, although the specifics of the spectral discrimination are unique to a species.

Coloration can be achieved via pigments, surface structuring, and a combination of pigments and surface structuring. In pigmentation-based coloration, some spectral bands of the incident illumination are absorbed by the pigment (dye, metallic compound, or other substance) thereby imparting a color. In structure-based coloration [1], interference, diffraction, and scattering of light are involved (in addition to optional absorption) to generate the observed colors.

Structural coloration is very prevalent in the biological world where organisms utilize it to achieve coloring related signaling [3]. The Morpho butterfly, a member of the lepidopteran species, shown in Fig. 5.1 is often cited as one of the finest examples of structural coloration. It relies on complex nanostructures [2, 4] and a variety of optical mechanisms including interference and diffraction to achieve its wing coloration. Humans used noble metal nanoparticle suspensions to impart rich colors to objects, particularly to glass. The Lycurgus cup, a Roman era cage glass

Figure 5.1: Structural coloration found in butterfly wings. (a) Splendid blue coloration on the wings of the Morpho Anaxibia butterfly photographed in the butterfly house at Kuala Lumpur. Reprinted under the GNU free documentation license. © *Erin Silversmith*. (b) Transmission electron micrographs of sections of the structurally colored scales of various lepidopteran species. Scale bars: A, D, F, G: 500 nm. B, C, E, J, L: 200 nm. H: 2 μm. K: 1 μm. Reprinted with permission from Reference [2]. © 2016 Society for Experimental Biology.

cup, uses dichroic glass that imparts a color changing effect to the cup; under frontal illumination, it appears green and under backlit light it appears red (see Fig. 5.2). The dichroicity is achieved by the small amounts of gold and silver nanoparticle colloidal suspensions that strongly scatter light in the green to blue end of the spectrum. The red color in transmission results from the subtraction of green and blue colors from the white illumination; whereas in front lit configuration the scattered green and blue hues dominate.

Surface structuring in the forms of diffractive elements and holograms is ubiquitous in modern times. Current state-of-the-art nanofabrication capabilities enable the creation of well-controlled nanostructures and arrays of nanostructures. Nanoscale resonators [5] made of certain materials (metals exhibiting strong resonant response in the visible and near infrared (IR) regions such as gold, silver, and aluminum; and, high refractive index [6] semiconductors such as silicon and gallium nitride) were recently reported to exhibit strong optical interactions despite their subwavelength sizes. Precisely controlled nanostructures made of certain materials with ultra-small form factors are thus expected to enable unprecedented manipulation of the local spectral content of an incident wavefront at subwavelength scales. Nanophotonics- and plasmonics-based structural coloration is thus expected to be an enabling platform for several exciting applications: ultra-high resolution color image sensors [7–9], multispectral sensors [10–12], low-power, low-cost, wide-area, and high-performance displays [13–15], colorimetric chemical and biochemical sens-

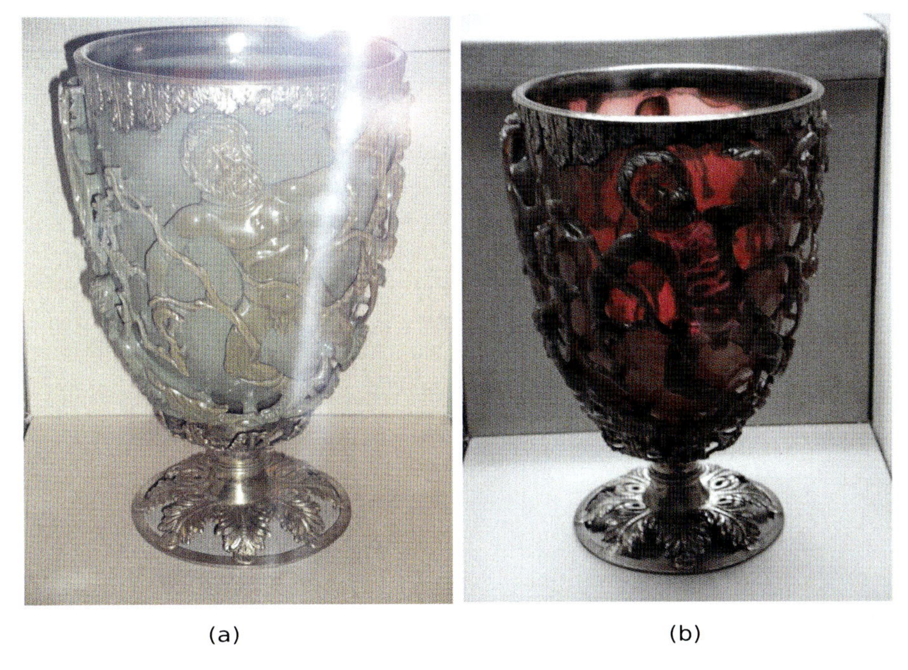

(a) (b)

Figure 5.2: The Lycurgus cup, one of the earliest known examples of nanostructure-based coloration, photographed at the British museum as seen with (a) frontal illumination and (b) backlit illumination. Reprinted under the Creative Commons license.

ing [16, 17], building integrated photovoltaics [14, 18], optical information security [19, 20] devices and high-density data storage [20, 21].

5.1.1 *Quantification of color*

The human visual system is sensitive to electromagnetic radiation in the approximate wavelength range of 380 nm to 780 nm. When an object generates, reflects, or scatters electromagnetic radiation in this wavelength range, we perceive color. An object is perceived to have a certain color based on the exact spectral power distribution of electromagnetic energy intercepted by the eye. *Colorimetry* is the science that deals with quantification of the color perception by human beings. In 1931, the International Commission on Illumination (CIE) recommended a system for the specification of color stimuli. This system is in widespread use even today. Since human color vision under medium and high illumination conditions is primarily derived from three kinds of photoreceptive cells (which have peak sensitivities at different wavelengths), any color stimulus can be described by specifying three numerical values (the so called *tristimulus values*). A human observer cannot notice the difference between excitation at a single wavelength and excitation by the three primary colors with properly chosen tristimulus values; this is called the *additive color mixing principle* [22].

The development of the color matching functions was a very important advancement in colorimetry. These functions describe the amounts of the three primaries that in combination will provide a similar color perception as a unit-intensity monochromatic source at a given wavelength. The color matching functions depend on what primaries are chosen and the 1931 CIE publication provided a standard set of color matching functions. The CIE also defined tables of spectral power distribution for many standard light sources. By knowing the spectral reflectance of any object, we can thus quantitatively determine the color [22]. The tristimulus values X, Y, and Z for an object whose surface exhibits a spectral reflectivity $R(\lambda)$ for an illumination with spectral power distribution $I(\lambda)$ are given by:

$$X = \frac{\sum_{360}^{830} I(\lambda)R(\lambda)\bar{r}(\lambda)}{100/\sum_{360}^{830} I(\lambda)\bar{g}(\lambda)}, \quad Y = \frac{\sum_{360}^{830} I(\lambda)R(\lambda)\bar{g}(\lambda)}{100/\sum_{360}^{830} I(\lambda)\bar{g}(\lambda)}, \quad Z = \frac{\sum_{360}^{830} I(\lambda)R(\lambda)\bar{b}(\lambda)}{100/\sum_{360}^{830} I(\lambda)\bar{g}(\lambda)}, \quad (5.1)$$

where $\bar{r}$, $\bar{g}$, and $\bar{b}$ denote the CIE color matching functions and the summation is carried over the wavelength range of 360 nm to 830 nm in steps of 1 nm. This summation is a useful approximation in practice considering that analytical expressions are not available for the color matching functions (these are usually given as a table at 1nm wavelength intervals) and also due to the discrete form in which experimental data for spectral power distribution is obtained [22].

The tristimulus values can be more conveniently converted to chromaticity coordinates x, y, z using:

$$x = \frac{X}{X+Y+Z}, \quad y = \frac{Y}{X+Y+Z}, \quad z = \frac{Z}{X+Y+Z}. \quad (5.2)$$

Since $z = 1 - x - y$, we can use a two-dimensional plot called the *chromaticity diagram* to represent all the possible colors with the caveat that colors with different luminance values are collapsed into a single point. This simple system is used to assess the colors of different nanostructures. Reflection, transmission, or extinction spectra can then be used to determine a color by using these equations.

5.2 Coloration by nanostructures

Metallic nanoparticles have been used as colorants since antiquity; well known examples are colloidal suspensions of gold and other noble metal nanoparticles used in stained glass windows. The coloration strongly depends on the size and shape of the nanoparticles, its immediate environment and the nanoparticle material. The recently renewed interest in metals stems from the ability to precisely control the shape and arrangement of nanoparticles afforded by modern nanofabrication techniques.

Structural coloration with metallic and other high-index dielectric nanostructures has several advantages:

- Wide variation in resulting colors by slight tweaking of geometrical parameters.

- Long-term chemical stability and reduced susceptibility to radiation enables colors to retain their quality for a long duration in contrast to dyes and other pigments.
- Extremely compact size gives an advantage in comparison to thin film, multilayer or photonic crystal based structural coloration.
- Ability to juxtapose spectral filtering elements with pixel sizes far smaller than the current state-of-the-art allows very high resolution image sensors and full color printing/display, and other related applications.

Various classes of nanostructures have been explored in connection with structural coloration: regular and irregular arrays of nanoparticles [23, 24]; regular and irregular arrays of nanoapertures [9, 25–28]; and, various coupled multilayer geometries [29, 30]. The coupled multilayer configuration has two or more layers; each individual layer is composed of either a nanoparticle array, or a nanoaperture array, or an unstructured layer. The coupled multilayer geometries can be further distinguished based on the predominant mechanism of coupling [31] between the constituent layers: Bragg like coupling [32] or near-field coupling [20]. In this chapter, we will restrict the scope to arrays of nanoparticles, nanoapertures, and to split-complementary structures. The split-complementary structure is composed of a nanoparticle layer and a second layer of nanoapertures shifted so that upon overlaying these two layers a uniform unpatterned thin film results.

5.2.1 *Structural color in nanoparticle arrays*

The starting point for understanding the latest developments in structural coloration is the electromagnetic behavior (scattering, absorption) of nanoscale resonators made of metallic or high refractive index dielectrics. Such nanostructures exhibit scattering cross-sections larger than their physical cross-sections near eigenmode resonance and are interesting from a structural coloration viewpoint[1] because their optical behavior is highly wavelength-dependent. Furthermore, their spectral properties (and, hence, their apparent color) are highly tunable via alteration of the size, shape or permittivity of the particles [33], surrounding medium, substrate, and particle arrangement. An understanding of the influence of these factors on the resulting color is thus essential for the rational design of structural colorants.

First, let us consider the simplest case of the spectral response of an isolated spherical nanoparticle fully surrounded by a homogeneous medium. Figure 5.3(a) shows the colors observable in the reflection regime for nanoparticles of various sizes and for the three common metals (gold, silver, and aluminum). In the case of gold, for instance, the localized surface polariton resonance (LSPR) causes a strong absorption of the blue-green parts of the visible spectrum yielding reddish colors; for larger sizes, the resonant absorption shifts to produce greenish tones. Figure 5.3(b) shows the colors observable by light scattering from subwavelength silicon nanospheres.

Classical electrodynamics can explain the optical properties of nanoparticles even

[1] Subwavelength resonators made of low refractive index materials exhibit significantly reduced levels of interaction with incident light in comparison to those made of metals or high-index dielectrics and are of little interest.

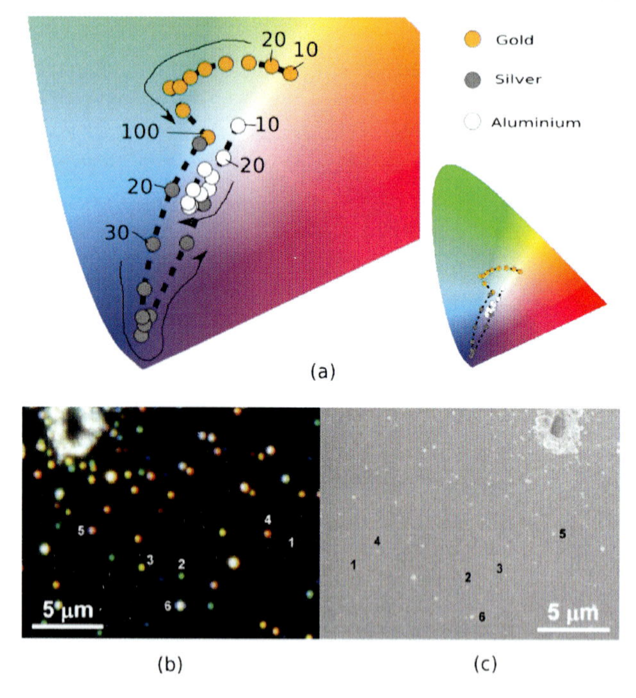

Figure 5.3: (a) Evolution of colors observable in reflection from a dilute solution of gold, silver, and aluminum nanospheres in water, over a wide range of particle sizes. The colors are obtained by converting the numerically calculated spectra of the scattering cross-sections into chromaticity coordinates as described in Section 5.1.1. (b) Dark-field microscope images of silicon nanoparticles produced by the laser ablation method and (c) scanning electron micrograph of the same scene show the strong size-dependent light scattering by subwavelength silicon nanospheres. The particles are labeled according to their diameter: 1 = 100 nm, 2 = 140 nm, 3 = 150 nm, 4 = 182 nm, 5 = 220 nm, 6 = 270 nm. Reprinted from Reference [34]. © 2012 Macmillan Publishers Limited.

down to the size of a few nanometers [35]. Gustav Mie provided the first mathematical description of the spectral dependence of the scattering by a spherical nanoparticle [36]. Following Mie, the mathematical description of the scattering properties of non-spherical particles progressed slowly [37, 38]; only recently, with the burgeoning interest in the field of plasmonics has there been a renewed interest. With the advent of colloidal chemistry and advanced nanofabrication techniques, it is possible to create a variety of nanostructures. However, only the simplest shapes remain analytically tractable; numerical tools [39] are essential to analyze and design nanostructures.

Non-retarded limit

Instead of solving the problem of nanoparticle scattering with the fully general Maxwell's equations, we can instead use the quasi-static approximation. This simplification improves analytical tractability although the quasi-static approximation is valid only for particles up to a few tens of nanometers [38]. In the quasi-static approximation, the electric and magnetic fields become decoupled. The electric field becomes irrotational, allowing us to define a scalar potential field and solving for the potential. In this manner, the multi-polar contributions to the polarizability α can be obtained.

For the simplest case of a spherical object of radius R and permittivity ε enclosed by a surrounding medium of permittivity ε_m, the dipolar polarizability can simply be expressed as:

$$\alpha(\lambda) = 3\varepsilon_m V \frac{\varepsilon - \varepsilon_m}{\varepsilon + 2\varepsilon_m}, \tag{5.3}$$

where V is the volume of the spherical nanoparticle and ε and ε_m are the wavelength-dependent permittivities of the nanoparticle and surrounding medium. The poles of the polarizability,

$$\varepsilon + 2\varepsilon_m = 0, \tag{5.4}$$

determine the wavelength at which resonant scattering occurs (the Frohlich condition). Note that this resonance is on the surface absorption polaritons (for details, see Section 3.4), as it requires negative Re(ε). As a result, only metallic nanoparticles can exhibit strong scattering and absorption in the quasi-static limit. To observe similar effects in dielectric particles, sizes should be on a wavelength-scale when they exhibit resonances on volume absorption polaritons. Note that the resonant scattering condition predicted by the quasi-static approximation is independent of the size of the nanoparticle and does not explain the commonly observed redshifts associated with the size increase although the size-dependent scattering is included.

For structural coloration purposes, the far-field response of the nanoparticles is more important than the near-field response. The far-field spectral properties can be described by the wavelength-dependent absorption and scattering cross-sections σ_{abs} and σ_{sca}. For periodic structures, the spectra of reflectance and transmittance can be used instead. The dipolar polarizability can be used to determine the various cross-sections as follows:

$$\sigma_{ext} = \sigma_{abs} + \sigma_{sca} = \frac{2\pi}{\lambda \varepsilon_m} \mathrm{Im}(\alpha), \tag{5.5}$$

$$\sigma_{sca} = \frac{8\pi^3}{3\lambda^4} |\alpha|^4. \tag{5.6}$$

The simplification introduced by the quasi-static approximation also enables the determination of the polarizability of non-spherical particles. For a small ellipsoidal nanoparticle [38] with half-axis lengths R_1, R_2, R_3, respectively, and permittivity ε, surrounded by a medium with permittivity ε_m, the dipolar polarizability along the principal axis j is given by:

$$\alpha(\lambda) = \varepsilon_m V \frac{\varepsilon - \varepsilon_m}{\varepsilon_m + L_j(\varepsilon - \varepsilon_m)}, \tag{5.7}$$

where V is the volume of the nanoparticle and L_j are the depolarization factors given by:

$$L_j = \frac{R_1 R_2 R_3}{2} \int_0^{\infty} \frac{ds}{(s+R_j^2)\sqrt{(s+R_1^2)(s+R_2^2)(s+R_3^2)}}. \tag{5.8}$$

The depolarization factors sum to unity, $L_1 + L_2 + L_3 = 1$, and the ellipsoidal particle has a strong anisotropic response. The resonant wavelengths for an electric field polarized along any particular axis is then given by the poles of Eq. (5.7),

$$\varepsilon_m + L_j(\varepsilon - \varepsilon_m) = 0. \tag{5.9}$$

The simplest case to consider is when two of the semi-axis lengths are equal $R_1 = R_2 = R$; when $R_3 = R$, we recover the sphere results. Consider the electric field polarized along either of the axes with half-axis length R. When $R_3/R < 1$, the resonance wavelength redshifts. We can thus expect that the Localized Surface Plasmon Resonance (LSPR) resonance of a metallic cylinder with electric field perpendicular to the cylinder axis is redshifted in comparison to a sphere which has the same radius R. In addition to the sphere and ellipsoid cases, the quasi-static approximation has also been used for other shapes like regular polyhedra and cylinders using simple numerical solution methods [40]. The quasi-static theory is a qualitative tool for explaining the shape dependent effects.

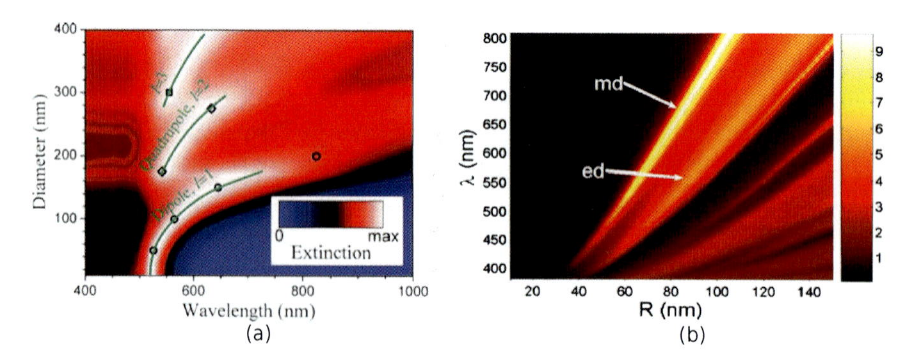

Figure 5.4: (a)The extinction spectra of gold nanospheres in water for wide range of particle sizes. The extinction values have been normalized so as to track the peak locations. Reprinted with permission from Reference [38]. © 2008 Royal Society of Chemistry. (b) The scattering efficiency in air of silicon nanospheres given as a function of particle radius R and wavelength λ. Reprinted with permission from Reference [41] © 2010 American Physical Society.

Influence of retardation effects

The predictions of the quasi-static approximation do not hold for nanoparticle sizes above few tens of nanometers although they provide a qualitative understanding of the influence of shape and other effects. Retardation effects arising from the finite speed of transmission of electromagnetic waves must be taken into consideration for a more accurate prediction of the scattering by nanoparticles. A quantitative understanding of the influence of retardation can be obtained by considering the Mie theory solutions for the scattering and can be seen in the spectral dependence of nanosphere extinction (Fig. 5.4). In the case of silicon, the absorption is small and the extinction is basically dominated by scattering.

Retardation related effects become noticeable when the largest length scale $L_{\max}$ associated with the nanoparticle begins to approach the wavelength of the light in the medium surrounding the nanoparticle (i.e. for $L_{\max} > 0.1\lambda$). The effects associated with retardation manifest most strongly for dipolar scattering modes and successively less strongly for quadrupole and higher order multipolar modes. A phase delay of $4\pi R/\lambda$ is associated with the local fields at one end of the nanoparticle to changes in the charge distribution occurring at the opposite end in the dipolar mode of a spherical nanoparticle of radius R; the oscillation period will thus increase in proportion to this delay. The effective interaction distance is reduced due to the multiple nodes occurring in higher order scattering modes that reduce the influence of the retardation effect on the oscillation period. The result of retardation effects can be clearly seen in Fig. 5.4. In the case of silicon, the scattering arises from the induced magnetic and electric dipoles as seen in Fig. 5.4 (b) and a detailed mathematical treatment can be found in [41].

For aspherical nanoparticles, the resonant wavelengths are influenced by both the retardation effects arising from the increased length scale and by the depolarization effects associated with the aspect ratio. Figure 5.5 shows the influence of both these effects for ellipsoidal particles. The size and aspect ratio of the ellipsoids are varied; the resonant wavelength of the dipolar mode is plotted for the incident wave polarized such that it excites the dipolar mode along the longest major axis of the ellipsoid (the other two semi-axes are equal in length). Keeping a fixed aspect ratio, an increase in the length L results in redshifting of the resonance wavelengths as a result of retardation effects. For a given length L, increasing the aspect ratio also leads to redshifted resonance as a result of the depolarization effects discussed in Section 5.2.1.

Influence of underlying substrate

Most structural coloration applications involve an underlying substrate (usually a dielectric one) adjacent to the nanoparticle. The presence of the substrate breaks the spatial symmetry and also alters the coupling strength between the nanoparticle eigenmodes and the free-space propagating electromagnetic waves [42].

The influence of an underlying substrate can be analyzed with the quasi-static approximation by including the influence of a mirror dipole. When an ellipsoid is placed at a height h above a substrate with permittivity ε_s so that two axes are parallel to the interface and the third one is perpendicular to it, the effective polarizability is given by:

$$\alpha_j^{eff} = \alpha_j(1-\beta)\left[1-\frac{\alpha_j\beta}{32\pi\varepsilon_m(R_3+d)^3}\right]^{-1}, \tag{5.10}$$

where R_1, R_2, and R_3 are the ellipsoid's half-axis lengths along the two parallel and

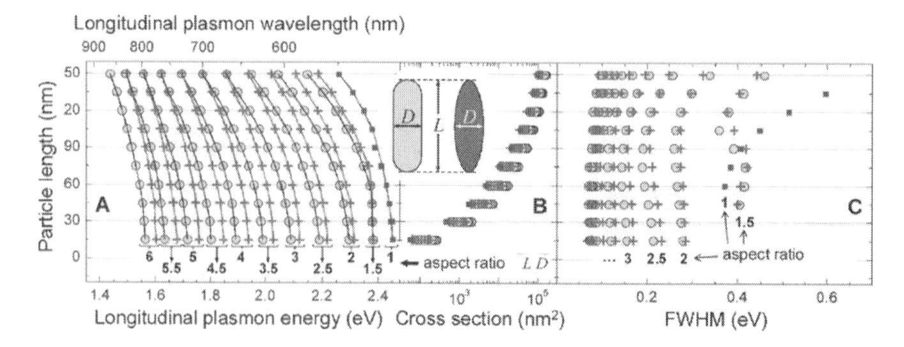

Figure 5.5: Influence of gold nanorod shape (ellipsoidal or hemispherical capped cylinders) and aspect ratio on the resonance wavelengths using numerical calculations. (A) Longitudinal resonance (with electric field polarized along the long edge of the ellipsoid) wavelength in gold ellipsoids (plus symbols) and rods with hemispherical caps (circles) as a function of particle length (vertical axis) and aspect ratio (labels at the bottom). Dipolar polaritons in spheres (unit aspect ratio) are shown as reference (squares). (B) Extinction cross-section at the resonant wavelength for illumination with incident electric field along the particle's axis of symmetry. (C) Full width at half maximum of the longitudinal eigenmode. Reprinted with permission from Reference [38]. © 2008 Royal Society of Chemistry.

one perpendicular axes, with $\beta = (\varepsilon_s - \varepsilon_m)/(\varepsilon_s + \varepsilon_m)$. The modified polarizability changes the resonance conditions. For an ellipsoid lying on the substrate ($d = 0$), the resonance occurs when the permittivity ε satisfies the condition [38]:

$$\varepsilon = \frac{1 - L_1 + \beta\gamma/24}{L_1 - \beta\gamma/24}\varepsilon_m. \tag{5.11}$$

The most noticeable effect of an adjacent substrate is to redshift the resonance wavelength of the nanoparticle in close proximity to the interface in comparison to the resonance of the same particle far away from it. The influence of the mirror dipole is expected to be more pronounced for larger particles. Retardation effects become important in the substrate-nanoparticle interaction-induced changes in the optical spectra of nanoparticles. The breaking of spatial symmetry results in eigenmode hybridization [43] that may appear as an anomalous line-width broadening [42] under unpolarized illumination conditions. In general, the presence of the substrate can induce interactions between the various multipoles of the nanoparticles and the corresponding mirror multipoles causing alterations in the resulting spectral response. Nordlander and coworkers applied the polariton hybridization concept to the analysis of a silver cube lying adjacent to a dielectric substrate [44] and highlight the Fano interference related spectral modifications in this system. To summarize, accurate numerical simulations must be performed to understand and exploit the spectral modifications of a metallic nanoparticle in close proximity to a supporting substrate.

It is suggested to perform two separate simulations: one accurately incorporating the interface and a second one using a homogenized surrounding medium to clearly isolate the substrate effects. Consider the numerical simulations for the reflectance spectra of nanodisks shown in Fig. 5.6 in vacuum and in the presence of a substrate. In vacuum, the peak position of the reflectance redshifts from 430 nm to 580 nm with increasing diameters of the nanodisk. The reflectance intensity also increases with increasing diameter of the nanodisks. Thus, reddish tones should appear brighter in comparison to bluish tones. Next, we consider the influence of the substrate on the reflectance spectrum. When the disks are placed directly on a reflective surface, all visible wavelengths appear to be reflected almost equally, which would result in a dull gray color. As soon as the nanodisks are allowed to hover above the surface even at a distance of 20 nm, the spectrum shifts drastically for both the larger and smaller disks. At a distance of 180 nm, the lower wavelengths are reflected, resulting in a bluish tinge to the colors observed. The behavior of nanodisks in the vicinity of a back-reflector can be explained via the existence of a screening dipole, which is a mirror dipole that cancels the effect of the original dipole; the cancellation is most effective, when the dipoles are closest to each other. Thus, when resting right on top, there is almost full cancellation and as we start hovering the disk, the cancellation becomes less effective.

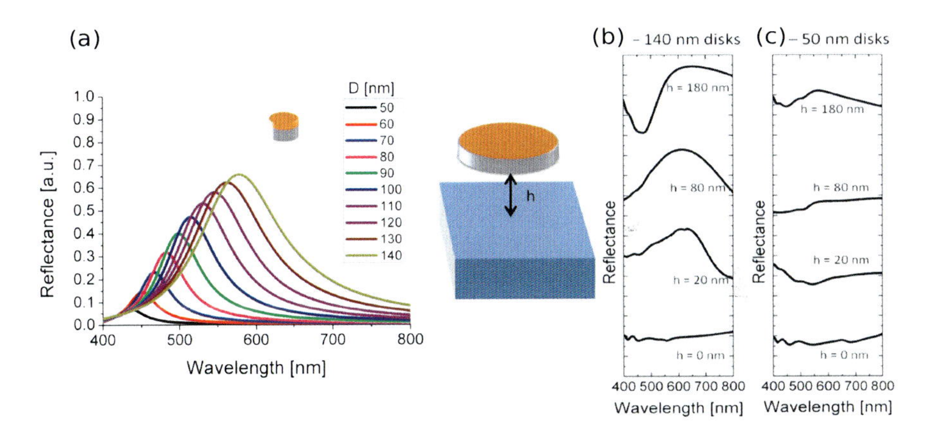

Figure 5.6: Numerically simulated reflection spectra of nanodisks in vacuum and on a substrate. (a) Simulation of the reflectances of an array of silver disks with various diameters and a separation of 120 nm in vacuum. (b) and (c) Simulated spectra of a silver nanodisk array with a gap of 30 nm hovering above a reflective silicon surface and diameters of (b) 140 nm and (c) 50 nm. The simulations also account for a thin protective gold coating. Reprinted with permission from Reference [20]. © 2016 Nature Publishing Group.

5.2.2 *Structural color in nanoaperture arrays*

The first class of nanostructured substrates explored for spectral filtering with spatial control were arrays of subwavelength holes in thin metallic layers. A rather surprising optical behavior of light interaction with a regular array of subwavelength holes was reported in 1998 by Ebbesen and coworkers [45]. An unexpectedly large value of light transmission termed *extraordinary optical transmission* (EOT) for certain wavelengths was observed. The physical origin of EOT is the wave tunneling mediated by resonances on the surface polaritons that exist on the interfaces of a structured thin film [46, 47]. From the spectral filtering viewpoint, the design of the peak transmission wavelength is critical. A subwavelength hole array can be characterized by the lattice structure, constants of the array, size and shape of the holes, and thickness of the aperture. In the long-wavelength approximation where the hole dimensions and lattice periodicity are small compared to the excitation wavelength, the peak wavelength does not depend on the hole shape.

For a rectangular lattice, the peak wavelength $\lambda_{\max}$ is given by [48]:

$$\lambda_{\max} = \frac{P}{\sqrt{i^2 + j^2}} \sqrt{\frac{\varepsilon_m \varepsilon_d}{\varepsilon_m + \varepsilon_d}}, \tag{5.12}$$

where P is the lattice periodicity; the indices i and j refer to the order of the resonance with $i = j = 0$ denoting the first order peak with the highest amount of transmission; ε_m and ε_d are the permittivities of the metal and surrounding medium. For a triangular lattice, the peak wavelength is given by [49]:

$$\lambda_{\max} = \frac{P}{\sqrt{4.3(i^2 + ij + j^2)}} \sqrt{\frac{\varepsilon_m \varepsilon_d}{\varepsilon_m + \varepsilon_d}}. \tag{5.13}$$

The peak wavelength obtained from the simple theory is shown to be dependent on the periodicity P and there can be several distinct peak wavelengths depending on the indices i and j. The hole array can thus serve as a spectral transmission filter (or a color filter), that can filter different colors depending on the value of the periodicity P. Figure 5.7 shows the performances of various spectral filters based on subwavelength hole arrays in aluminum. Contrary to the simple theory, it shows that the hole shape and array type influence the characteristics of the filter. It reveals that while the transmission values are lowered for a triangular hole, the linewidths are lowered as well, leading to sharper colors. Furthermore, a hexagonal arrangement also can improve the color quality (or purity) due to the increased separation of various resonance orders. By shaping the hole, highly angle-insensitive color filtering action has been demonstrated [50].

Digital cameras are now ubiquitous consumer electronics products; the image sensor is at the heart of digital imaging. Color filter arrays (CFAs) are integral parts of the image sensor, and advancements in color filtering technologies are essential in the further scaling down of pixel size beyond 2 μm. Scaling down the pixel size is difficult with conventional color filtering techniques [7, 8] due to the increased spatial cross-talk and fabrication-related complexities. Additionally, the conventionally absorptive filters also suffer from aging- and radiation-related degradation. Subwavelength hole array based color filtering is thus an attractive option. Figure 5.8 shows images of a full 360 by 320 pixel arrayed image sensor integrated with an aluminum hole array filter array [9]. Improvements in the design of hole array filters [25, 26, 28]

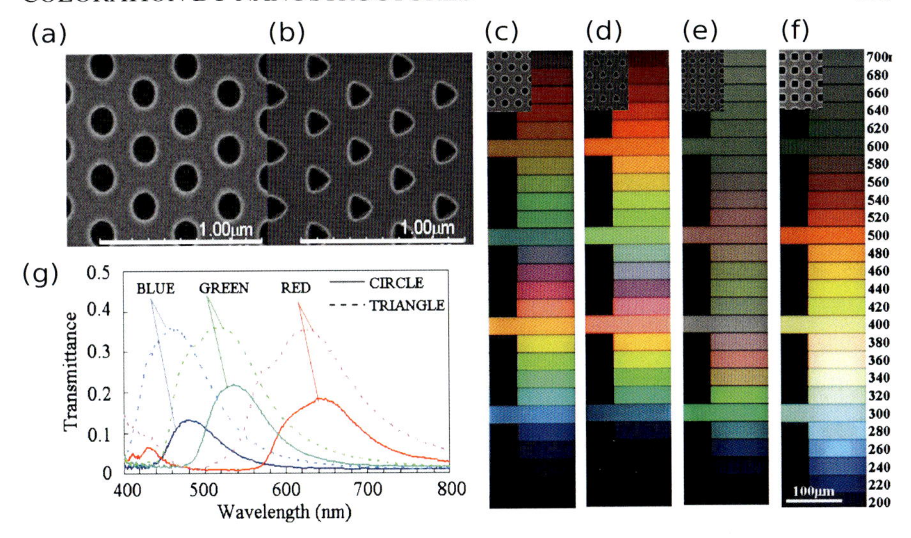

Figure 5.7: Structural color in aluminum subwavelength hole arrays. (a) and (b) SEM micrographs of the subwavelength hole arrays in aluminum with (a) circular and (b) triangular-shaped holes located in a triangular lattice. (c) through (f) Optical microscope images of the fabricated color filters showing the influence of hole shape and lattice array type. The insets show that (c) and (d) are hexagonal lattices while (e) and (f) are rectangular lattices. The numbers at the side denote the periodicity in nanometers. (g) Transmission spectra of RGB filters with circular and triangular hole shapes. Reprinted with permission from Reference [51]. © 2011 American Physical Society.

are targeted to further reduce spatial cross-talk and improve efficiency and color purity.

5.2.3 Split-complementary nanostructured reflective color filters

While the color filters discussed in the previous section in the context of image sensors operate in the transmission mode, the color filters operating in the reflective mode are of particular interest for applications [23] such as displays, identification, security tags, and colorimetric sensing [16, 17]. For these applications, the reflective filters should be bright enough to be seen under ambient illumination conditions; at the same time, they should be of higher resolution (or smaller pixel size). The split-complementary nanoantenna structure was the first reported nanostructure to achieve wide range of colors observable under ambient illumination conditions and large pixel density [20]. The split-complementary structure is a dual layered using a layer of nanoparticles and their complementary shaped nanoaperture layer (by addition, we can get a uniform layer). This structure can be fabricated in a single step in comparison to other multilayer geometries. The motivation for use of the split-complementary structure arises from the limitations of the nanoparticle and nanoaperture array geometries: (i) it is difficult to achieve a broad range of colors,

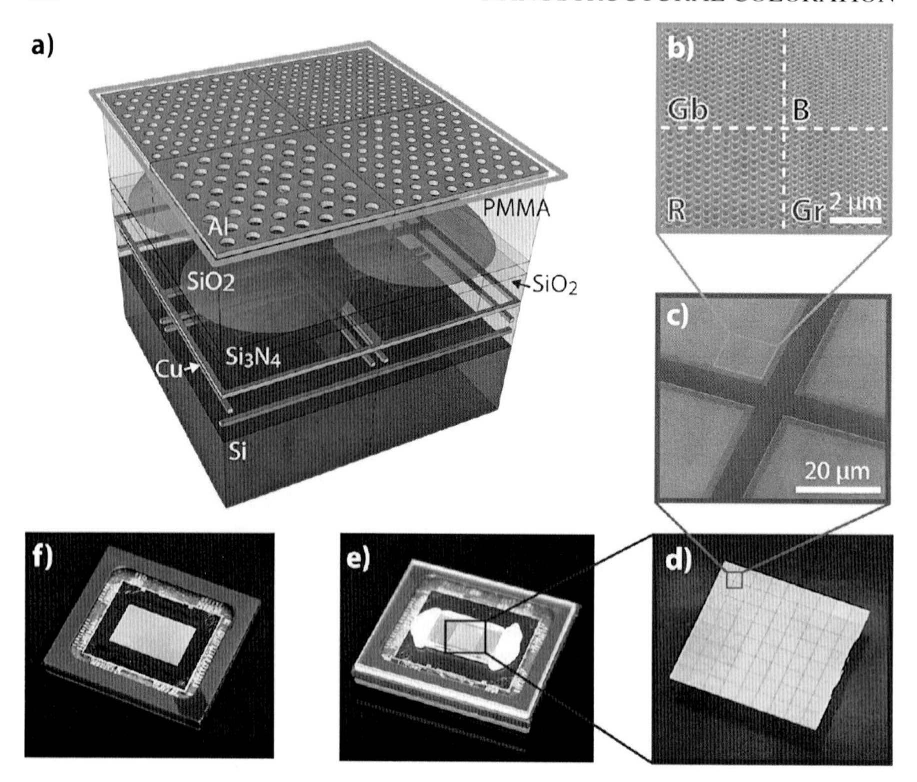

Figure 5.8: Metallic subwavelength hole array filter and integration with CMOS photodetector layer. (a) Schematic of CMOS image sensor pixel with the integrated metallic color filter in the Bayer mosaic layout (Red Green Blue Green). (b) and (c) SEM micrographs of the hole array filter showing the 40 by 40 element filter block. (d) View of the 360 by 320 pixel array filter. Pictures of the CMOS image sensor before (f) and after (e) integration with the metallic filter. Reprinted with permission from Reference [9]. © 2013 American Chemical Society.

when the nanoresonators are located on a substrate [23] due to the variable scattering from differently sized nanoresonators and the substrate; (ii) Nanoaperture array geometry, while producing vivid coloration, relies on periodicity and thus requires variations in periodicity for achieving filter mosaics. Additionally, the pixel sizes achievable through nanoapertures are much larger.

The schematic of the split-complementary nanostructure and its fabrication process [20] is shown in Fig. 5.9; small groups can be clustered together to form a pixel. The nanostructured surface consists of metallic nanodisks hovering in close proximity to a nanohole. The back reflector (made of the same metal) containing the nanoaperture, in addition to improving the nanodisk scattering, can also add its own resonances, allowing a broader range of colors to be achieved. The nanostructure can be designed to achieve arbitrary spatially dependent spectral filtering or the printing and display of arbitrary color images. The spatial variation can be achieved by changing the geometrical parameters like the diameter D of the nanodisks or the spacing g between them. The simple fabrication process is amenable to high-volume replication techniques such as nanoimprint lithography.

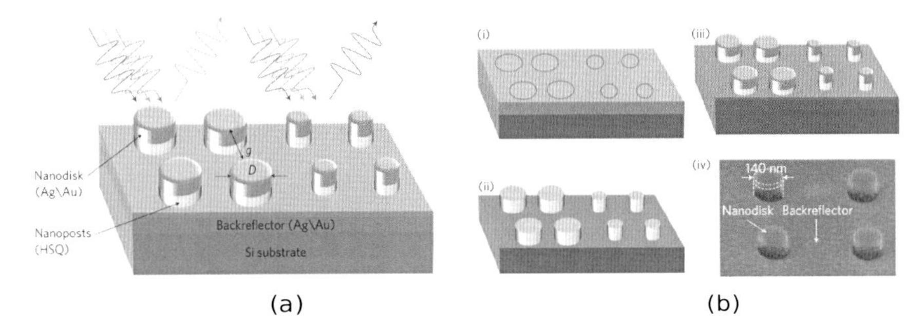

Figure 5.9: Schematic and fabrication process of split complementary nanoantenna array color filters. A small group of nanoantennas (four) forms a pixel, and the adjacent pixels can have different colors, as shown in (a), as a result of different diameters D and separations g of the nanodisks within each pixel. (b) Nanostructure fabrication procedure. Step (i): a 95-nm thick layer of hydrogen silsesquioxane (HSQ) resist is spin-coated onto a silicon wafer and patterned using e-beam lithography. Step (ii): the unexposed portions of the HSQ are developed away, leaving HSQ nanoposts. Step (iii): the nanoposts and back reflector are coated using metal evaporation. (iv) Side-angle SEM micrograph of nanostructures after metal deposition. Reprinted with permission from Reference [20]. © 2012 Nature Publishing Group.

The split-complementary nanoantennas, when arranged in a heterogeneous fashion, can create surfaces with spatial variation in spectral properties; the printing of arbitrary microscopic color images with a full degree of control over the color and tone has been reported by several groups. The printing of a photo-realistic image with dimensions 50 μm by 50 μm (the standard "Lena" test image) with pixel sizes 250 nm by 250 nm was first reported by Kumar and coworkers using this technique [20], as shown in Fig. 5.10. By variation of the structural features locally, a gray-scale image is first seen in Fig. 5.10(a). Color information from bitmap images can be encoded pixel by pixel by varying the geometrical parameters (here by changing the disk diameter D). The gray-scale image becomes colored subsequent to metalization. Rapid tonal variations in the image are seen to be reproduced with high fidelity in the printed color images that are observable in bright field. The pixel size of 250 nm by 250 nm is the theoretical resolution limit of the optical microscope at the mid-spectrum wavelength of 500 nm. Figures 5.10(d) and (e) show that the resolution is indeed at the diffraction limit corresponding to the mid-spectrum wavelength. The SEM micrographs demonstrate a set of checkerboard resolution test structures with alternating colors. In (d), each square in the checkerboard consists of a 3 by 3 array with pixel size of 325 nm; in (e), it consists of a 2 by 2 array of disks per pixel with a pixel size of 250 nm. While the colors are clearly resolved in (d), they are just barely resolved in (e), indicating that the diffraction limit has been approached.

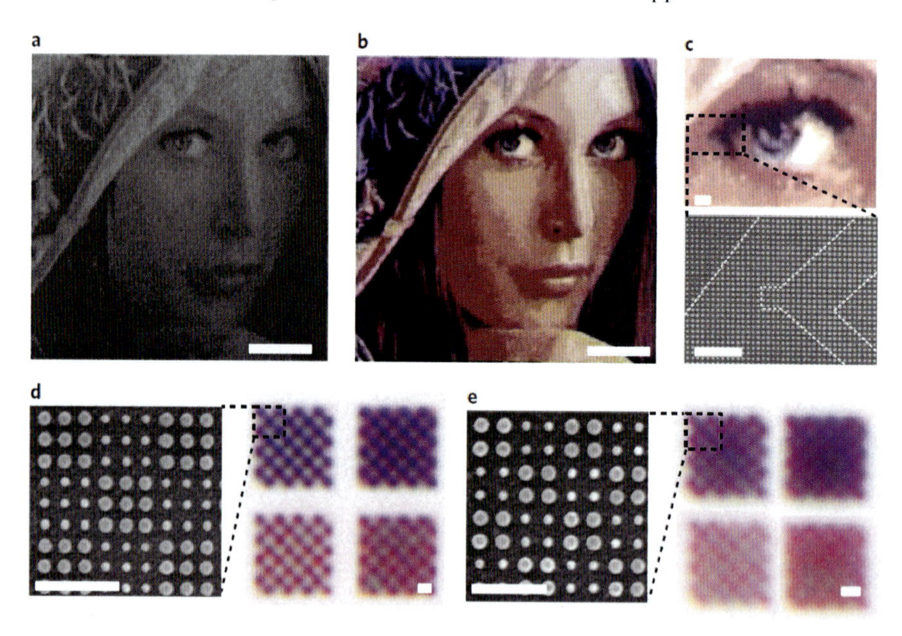

Figure 5.10: Full-color image printing with pixel sizes at the mid-visible-spectrum diffraction limit. The optical micrographs of the standard "Lena" test image are shown (a) in greyscale (before metalization step) and (b) in full color (after metalization step). (c) Optical micrograph of the enlarged region of the same image that shows high spatial frequency related detail and color rendition with rapid transitions. (d) and (e) Checkerboard-like patterns of two different colors are used to demonstrate that pixel sizes are indeed equal to the mid-spectrum diffraction limit. In the pattern (d), each pixel is 375 nm in size and is a 3 by 3 array of similarly sized disks; in (e), each pixel is 375 nm in size and is a 2 by 2 array. The checkerboard pattern (e) is barely discernible at the diffraction limit of red light, whereas it is clearly discernible in (d). Scale bars: 10 mm (a, b), 1 mm (c), 500 nm (d, e). Reprinted with permission from Reference [20]. © 2012 Nature Publishing Group.

Spectral response and tuning

To achieve a full palette of colors that spans the visible range, the diameter D and the separation g between the disks should be tuned. In Fig. 5.11(a), the spectra of pixels having a fixed g and varying diameter D are plotted. The role of localized eigenmode resonances in the color formation is evident: the metalization and structures with the same periodicity give rise to different colors. The spectra exhibit peaks and dips that may be tuned across the visible spectrum by varying D and thus the periodicity. Full wave simulations demonstrate qualitative agreement with the corresponding experimental results shown in Fig. 5.11(b). The experimental results were reported

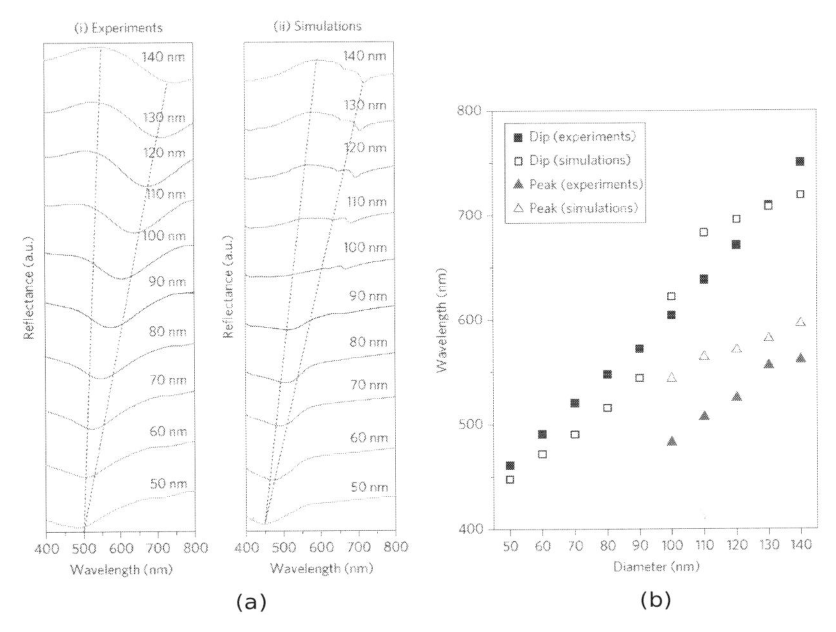

Figure 5.11: (a) Spectral responses of arrays of pixels, where each pixel consists of a square lattice of nanoposts with diameter D, gap g, and period $D+g$. Experimental (i) and simulation (ii) spectra of nanodisks with g of 120 nm and varying D as indicated. The trendlines approximate the movement of the peaks and dips with varying sizes of the nanostructures. (b) Correlation between dips and peaks observed in the experimental and simulation data. Reprinted with permission from Reference [20]. © 2012 Nature Publishing Group.

for the silver-based split-complementary nanostructure by Kumar et al. [20], with the disk diameter D varying from 50 nm to 140 nm and the gaps g changing from 30 nm to 120 nm. In the reported results, the peaks of reflectance occurred for λ between 430 nm and 580 nm, while the spectral dips could be tuned over the whole visible region.

The spectral signature consists of clearly seen dips and peaks that shift in response to a change in size and spacing of the nanoposts. The peaks correspond to the polariton resonances of the disks that intensify for larger disks because of their increased scattering strengths. The spectral dips, however, appear to originate from different effects depending on the diameter of the disk. The spectral dips observed for smaller disks are due to power absorption by the disks and, to a lesser extent, by the back reflector. The disk, post, and back reflector together effectively act as an anti-reflection stack at the dip wavelength. The dips for larger disks are due to the Fano resonances that arise from the interference between the broad resonance of the nanoholes and nanodisks with the sharp resonance on the surface polaritons (This resonance can be treated as the antenna-enhanced EOT where the optical power flows around the nanodisks through the nanoholes and is absorbed by the back reflector or silicon substrate).

The back layer is observed to exert little influence on the location of the eigenmode peaks; however it does lead to an apparent narrowing of the linewidth. Addition of metalization on the back layer results in a subwavelength aperture. In the case of the split-complementary structure, the light transmitted through the aperture gets absorbed in the substrate. The presence of the disk enables the coupling to surface polaritons very similar to what occurs in EOT discussed in the previous section. At the EOT-related resonance, the coupling of the broad resonance (namely the plane wave or disk dipole resonance) and the narrow linewidth traveling surface polaritons, a characteristic Fano-like signature can be distinctly seen in the reflectance spectrum. For the structure in Fig. 5.11, the characteristic S-shaped Fano signature appears beyond the visible spectrum (not shown in figure) at 900 nm for the periodicity of 120 nm and at 1200 nm for the periodicity of 240 nm (independent of the disk diameter). This is consistent with the observation that EOT resonance occurs as a function of structure periodicity. Although the EOT-related resonance occurs beyond the visible window, it has the effect of lowering the structure's reflection in the longer wavelengths for larger disk sizes.

The split-complementary structure achieves a wide range of colors without the need to vary the periodicity by combining various eigenmode resonances occurring in the nanodisk, nanoaperture, and interaction between them. Figure 5.12 presents the simulation results for structures with a constant periodicity of 120 nm and D varying between 50 nm and 90 nm, which exhibit multiple colors as indicated in the insets. In Fig. 5.12(a), the spectral responses of the split-complementary nanostructure is compared with those of an isolated nanodisk and an isolated nanoaperture; the silicon substrate and the low-index column supporting the nanodisk are presented in all three cases.

We can see that structures comprising only nanodisks or only the back reflector plane cannot produce the colors observed. The metallic back reflector acting alone (dotted lines) displays a fairly flat spectrum across arrays with the same periodicity, with a point of inflexion at 900 nm, the signature of a Fano resonance profile, and a dip at 450 nm attributed to the anti-reflection stack at this wavelength, as described earlier. Further evidence for the dip at 450 nm having a different physical origin from the feature at 900 nm can be seen in its invariance to changing periodicity. For structures with just disks (dashed lines), a single peak is observed corresponding to the nanodisk eigenmode resonance that blueshifts and intensifies with increasing diameter D. A broader span of colors is achieved only in the combined structure; as the scattering strength of the disks increases, the spectrum peak shifts in favor of the nanodisk resonance and away from the Fano resonance. The structures consisting of disks raised above a back reflector film without nanoholes would display similar colors but without the Fano resonance. However, the presence of the nanoaperture appears to narrow the eigenmode peaks of the nanodisk, resulting in purer colors.

By examining the near-field distribution of electric field and power flow, we can understand the origin and relative contributions of different resonances and their roles in tuning the color response. Figure 5.12(b) demonstrates the electric field enhance-

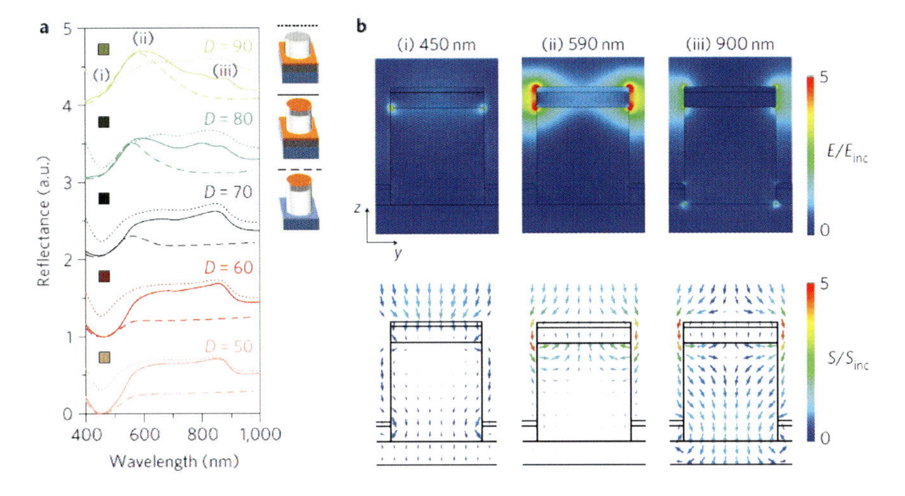

Figure 5.12: Numerical simulation for structures of periodicity 120 nm. (a) Simulated reflectance spectra with variation in the disk diameter D. Solid lines show reflectances for the combined structure (with disks and back reflector); dotted lines show where metal nanodisks are removed; dashed lines depict where the back reflector is removed. Note that the feature corresponding to the Fano resonance occurs at the constant wavelength of 900 nm for all values of D. Color variation at constant periodicity can be achieved only for the combined structure of nanodisks and back reflector. (b) Electric field enhancement plots (top) and time-averaged power flow plots (bottom) for a structure with diameter D of 90 nm are shown at three wavelengths: (i) 450 nm, (ii) 590 nm, and (iii) 590 nm. Plane wave illumination is incident from top in the z direction and polarized along the y axis. Reprinted with permission from Reference [20]. © 2012 Nature Publishing Group.

ment and shows the power flow (Poynting vector) plots for the combined structure extracted at three wavelengths: (i) 450 nm, (ii) 590 nm and (iii) 900 nm. The nanodisk appears to play different roles at each of these wavelengths: it is absorbing in the anti-reflection stack at (i), scattering at (ii) and enhancing absorption around the nanohole at (iii). These effects can also be seen in the channeling of the Poynting vectors into the nanodisk at (i), the strong fields signifying nanodisk's polariton resonance at (ii), and the directing of power flow around the nanodisk and into the base of the nanohole at (iii).

The spectral dip at 450 nm corresponds to light absorption by both the metal structures and silicon substrate. In the Poynting vector flow, we see a gradual drop in magnitude from top of the disk and in the substrate. This decay shows that the structure is acting like an absorber, while the field enhancement plots reveal the excitation of surface polaritons between the silver disk and the HSQ post. For smaller disks, this enhancement is even stronger. This may explain the dip appearing in smaller sized disks at low wavelengths

The peak at 590 nm is due to the eigenmode resonance of the disk acting as a dipole antenna that re-radiates light back to the observer. For the y-polarized incident E field, the dipole results in maximum field enhancements at the tip of the disks as can be seen in the field enhancement plot. The maximum joule heating takes place at the center of the disk at the polariton resonance. It is seen in the Poynting vector plots, where the power flows toward the disk center.

The inflexion point at 900 nm signifies a resonance where the power flows around the disk through the nanohole and is absorbed by the bottom rim of the nanohole array and substrate. At the 900 nm point, we see that the disk eigenmode resonance is still active, but away from resonance, with reduced field enhancements. A strong field enhancement is also observed toward the rim of the subwavelength aperture. A clearer picture emerges in the Poynting vector flow plot, where the power is channeled into the subwavelength aperture. At this point, EOT occurs through the aperture. Note that the power flow in the case of 590 nm is highly concentrated at the surface of the disk-air interface, and the stronger scattering prevents power from reaching the aperture. At 900 nm, the disk and aperture act in concert to channel the power into the back substrate. The Poynting vector shows the possibility that the disk and aperture can couple even strongly when the height is reduced.

5.3 Emerging materials for structural coloration

Initial reports on structural coloration used silver as the main material; however, aluminum has several advantages over silver for structural coloration applications [53]. Aluminum is an abundant material and thus less expensive than silver. It is also fully compatible with the CMOS fabrication process. Aluminum [54] has been studied for polariton resonances occurring in the ultraviolet region [55]. Aluminum-based nanoantennas can thus be expected to yield vivid blue colors; but a significant linewidth broadening effect is expected for resonance wavelengths corresponding to green and red colors that results in color purity degradation. The inter-band transition that occurs around 1.5 eV [53] has been hypothesized to be the cause of the increased spectral broadening resulting in dull coloration in the green and red tones reported in aluminum nanostructures [52, 56]. The purity of aluminum, i.e. the influence of its oxidation layer on polariton resonances, is also of concern for structural coloration applications [53].

Yang and coworkers proposed a superpixel approach (combining differently sized nanoantennas in each pixel in close proximity) to achieve a broad range of colors [52]. Figure 5.13 shows the strategy used to broaden the range of color tones. With the supercell configuration, each pixel is composed of two differently sized nanodisks (two of each size). By fixing the overall size of the superpixel it is possible to have free variations of the disks' sizes and spacing. The authors describe this in terms of the mixing and spacing palettes. Shrestra et al. proposed a subtractive color mixing principle [57] for expanding the coverage in the chromaticity diagram. They demonstrated cyan, magenta, and yellow colors; however, they did not demonstrate that subtractive color mixing is possible without adding multilayers. Clausen et al. also reported coloration with split-complementary aluminum-based nanostructures [56] and demonstrated the feasibility of high volume replication of coloration

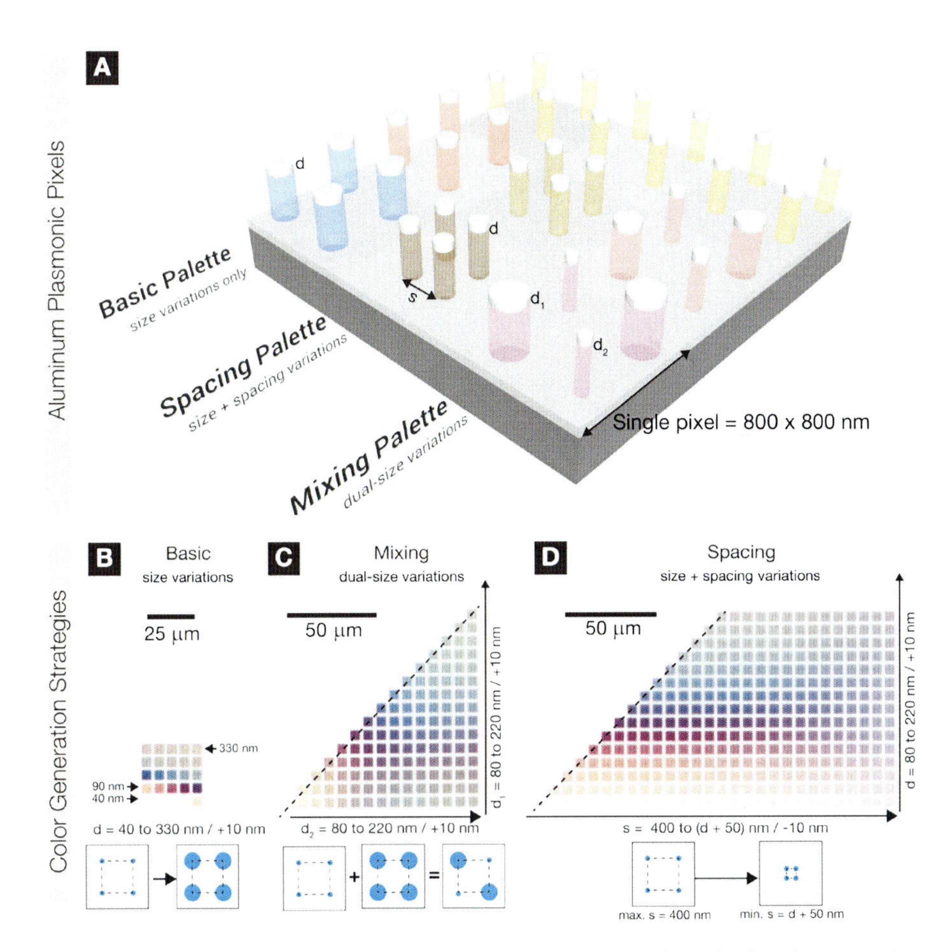

Figure 5.13: Aluminum split-complementary geometry and method to increase the tonal range. (A) Schematic illustrating the architecture of aluminum pixels. Actual experimental images of fabricated color palettes with different layout strategies: (B) basic color palette with only size variations (d = 40 to 330 nm) at the fixed pitch of 400 nm between nanodisks; (C) mixing color palette with two size variations (d_1, d_2 = 80 to 220 nm) among four nanodisks within an 800 x 800 nm pixel, at the fixed spacing of 400 nm between nanodisks; (D) spacing color palette with both size variations (d = 80 to 220 nm) and spacing variations ($s = d$ + 50 nm to 400 nm) among four nanodisks within an 800 x 800 nm pixel. Reproduced with permission from Reference [52] © 2014 American Chemical Society.

for consumer products. Additionally, the obtained colors are nearly independent of the viewing angle, as shown in Fig. 5.14.

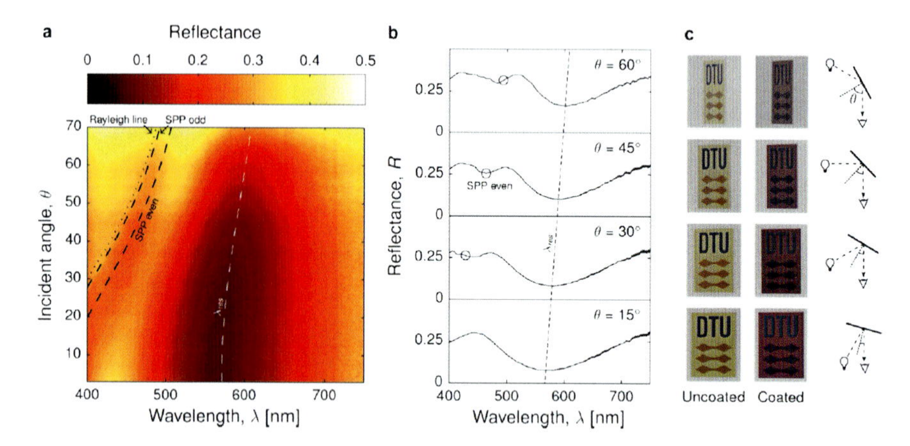

Figure 5.14: Reflected colors observable over wide viewing angles for the aluminum-based split-complementary design. (a) Measured angle-resolved spectra of coated sample with periodicity of 200 nm and disk diameter $D = 86 \pm 2$ nm. The reflectance minima tracked by the white dashed line is nearly invariant over a broad range of angles. The theoretical odd and even surface plasmon-polaritons (SPPs) for a 20 nm thick flat aluminum film are tracked by the dashed lines. The Rayleigh wavelength is tracked by the black dotted lines. (b) Single spectra at four different angles of the same sample as in panel (a). (c) Optical micrographs of macroscopic patterns viewed at four different angles. Reprinted with permission from Reference [56]. © 2014 American Chemical Society.

By utilizing structural anisotropy (for instance, by using elliptical disks instead of circular disks), simultaneous local spectral and polarization control becomes possible. Yang and coworkers utilized elliptical nanodisks to encode two color images simultaneously, where each image is given by one of the two possible orthogonal polarization states [58]. These reported results are promising and point towards aluminum structural coloration techniques competing with diffractive optical elements [59] for optical security applications [59] including anti-counterfeiting labels.

For high resolution and low power full-color display applications, however, the limited coverage of the chromaticity diagrams shown in Reference [52, 56] is problematic. It is apparent that the mixing and spacing palettes [52] do not expand the coverage in the chromaticity plane, but merely introduce fine tonal variations instead. Olson et al. recently reported a technique to address the broad linewidth of polariton resonances in aluminum nanostructures that relies on diffractive coupling between neighboring nanoantennas [53]. Red, blue, and green colors of high purity were demonstrated with reduced resonance linewidths. However, the reported structures show purer colors at an oblique incidence and the beneficial diffractive coupling appears to be suppressed at normal incidence. By improving the color purity obtainable in aluminum nanostructures (albeit at oblique incidence angles), metallic colorimetric sensors were reported by Halas and coworkers [17]. The reliance on oblique incidence (and consequently, a strong dependence on viewing angle) is a limitation for display and imaging applications [13].

Thus far, metallic nanostructures have been predominantly investigated for structural coloration and sensing applications. While metal nanostructures provide a number of advantages, they suffer from large absorption losses. Additionally, gold and silver — the conductive materials most suitable for visible frequency applications — are incompatible with the silicon CMOS process. Recently, nanostructures made of silicon (which has a high refractive index in the visible range) were of high interest [6, 60, 61]. All-dielectric optical nanoantennas exhibit drastically lowered absorption losses and enhanced magnetic response in comparison to metal counterparts.

Exploration of structural coloration effects is potentially interesting in the realization of monolithic low-powered active displays [62]. By using silicon nanowires and nanoparticles of different sizes, Brongersma and coworkers [62] reported a wide range of vivid colors. The resonant light scattering effect does not rely on long-range order, as required by photonic-crystal-based geometries [63], or on quantum-confinement-related effects, making it quite robust. These geometries [64] are especially interesting for large display areas provided economic fabrication techniques can be realized. The reported pixel sizes [65], however, are far larger [66] in comparison to the near diffraction limit sizes reported in metal split-complementary nanostructures [20].

The silicon nanowire geometry relies on the nanowire polariton resonances that lie in the visible range. While these resonances can be easily tuned, the vertical lengths required are often on the order of micrometers [67]. Recently, vivid color generation from silicon nanostructures with a comparatively lower aspect ratio was demonstrated [67]. Hegde and coworkers reported a numerical study utilizing a cross-shaped meta-atom [68] and wide gamut structural colorations observable in reflection mode (see figure 5.15). The inter-element spacing can be much smaller than a micron and can lead to very high resolution of spatial alteration in color. In contrast to metals, a design utilizing silicon is fully compatible with CMOS process flow and may make possible the fabrication of displays and image sensors integrated with VLSI electronic circuits. Further developments in small form-factor all-dielectric nanostructures will be attractive for structural coloration applications.

5.4 Summary

This chapter presented an overview of recent developments in photonic and plasmonic nanostructures for structural coloration applications. While the earlier reports demonstrated the feasibility and showcase the potential for designing structural coloration applications with nanostructures, further work is needed for translation to products. Several avenues exist for further research and development: improvement of color purity and overall coverage of the chromaticity range achievable in reflection and transmission modes, active tunability [69] of the spectral response, and investigation of high-volume fabrication and replication techniques [70].

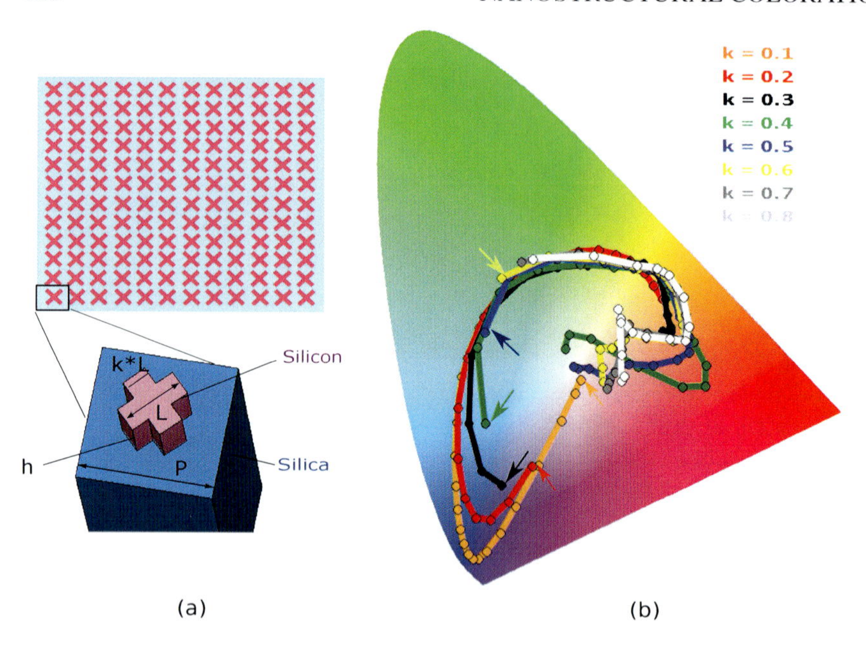

Figure 5.15: Numerical simulations of the obtainable colors with a nanocross element design. (a) Schematic of the all-dielectric ultra-thin spectral filter showing an arrangement of dielectric nanocrosses with the zoom-in showing the geometrical parameters of the individual element of this array. (b) Filter color variation with different geometrical parameters. For the silicon nanocross array with the periodicity $P = 333$ nm, the colors obtained in the reflection mode are shown on the CIE chromaticity diagram. Each set of geometrical parameters is represented by a point and grouped so that a line connects those with a fixed value of k (the width to length ratio). Along any line with a given k, the length of the nanocross changes from 100 nm to 280 nm in steps of 10 nm. The little arrowheads represent the 100 nm long nanocross. Reprinted with permission from Reference [68]. © 2016 Society of Photo-instrumentation Engineers.

Bibliography

[1] S. Kinoshita, S. Yoshioka, and J. Miyazaki, "Physics of structural colors," *Reports on Progress in Physics*, vol. 71, no. 7, p. 076401, 2008.

[2] R. O. Prum, T. Quinn, and R. H. Torres, "Anatomically diverse butterfly scales all produce structural colours by coherent scattering.," *The Journal of Experimental Biology*, vol. 209, no. 4, pp. 748–765, 2006.

[3] M. Srinivasarao, "Nano-optics in the biological world: beetles, butterflies, birds, and moths.," *Chemical Reviews*, vol. 99, no. 7, pp. 1935–1962, 1999.

[4] P. Vukusic, J. R. Sambles, C. R. Lawrence, and R. J. Wootton, "Quantified interference and diffraction in single Morpho butterfly scales," *Proceedings of*

the *Royal Society B: Biological Sciences*, vol. 266, no. 1427, p. 1403, 1999.

[5] Q-Han Park, "Optical antennas and plasmonics," *Contemporary Physics*, vol. 50, no. 2, pp. 407–423, 2009.

[6] A. E. Krasnok, P. A. Belov, A. E. Miroshnichenko, A. I. Kuznetsov, B. S. Luk'yanchuk, and Y. S. Kivshar, "All-dielectric optical nanoantennas (Review)," in *Progress in Compact Antennas, InTech, Chapter 6* (Dr. Laure Huitema, ed.), 2014.

[7] Q. Chen, D. Chitnis, K. Walls, T. D. Drysdale, S. Collins, and D. R. S. Cumming, "CMOS photodetectors integrated with plasmonic color filters," *IEEE Photonics Technology Letters*, vol. 24, no. 3, pp. 197–199, 2012.

[8] S. Yokogawa, S. P. Burgos, and H. A. Atwater, "Plasmonic color filters for CMOS image sensor applications," *Nano Letters*, vol. 12, no. 8, pp. 4349–4354, 2012.

[9] S. P. Burgos, S. Yokogawa, and H. A. Atwater, "Color imaging via nearest neighbor hole coupling in plasmonic color filters integrated onto a complementary metal-oxide semiconductor image sensor," *ACS Nano*, vol. 7, no. 11, pp. 10038–10047, 2013.

[10] H. Park and K. B. Crozier, "Multispectral imaging with vertical silicon nanowires," *Scientific Reports*, vol. 3, pp. 1–6, 2013.

[11] M. Najiminaini, F. Vasefi, B. Kaminska, and J. J. L. Carson, "Nanohole array based device for 2D snapshot multispectral imaging.," *Scientific Reports*, vol. 3, p. 2589, 2013.

[12] H. Aouani, M. Rahmani, H. Š´ipová, V. Torres, K. Hegnerová, M. Beruete, J. Homola, M. Hong, M. Navarro-C´ia, and S. A. Maier, "Plasmonic nanoantennas for multispectral surface-enhanced spectroscopies," *Journal of Physical Chemistry C*, vol. 117, no. 36, pp. 18620–18626, 2013.

[13] J. Olson, A. Manjavacas, T. Basu, D. Huang, A. E. Schlather, B. Zheng, N. J. Halas, P. Nordlander, and S. Link, "High chromaticity aluminum plasmonic pixels for active liquid crystal displays," *ACS Nano*, vol. 10, pp. 1108–1117, 2016.

[14] L. Wen, Q. Chen, S. Song, Y. Yu, L. Jin, and X. Hu, "Photon harvesting, coloring and polarizing in photovoltaic cell integrated color filters: efficient energy routing strategies for power-saving displays," *Nanotechnology*, vol. 26, no. 265203, pp. 1–10, 2015.

[15] J. A. Gordon and R. W. Ziolkowski, "Colors generated by tunable plasmon resonances and their potential application to ambiently illuminated color displays," *Solid State Commun.*, vol. 146, pp. 228–238, 2008.

[16] M. Khorasaninejad, S. Mohsen Raeis-Zadeh, H. Amarloo, N. Abedzadeh, S. Safavi-Naeini, and S. S. Saini, "Colorimetric sensors using nano-patch surface plasmon resonators," *Nanotechnology*, vol. 24, no. 35, p. 355501, 2013.

[17] N. S. King, L. Liu, X. Yang, B. Cerjan, H. O. Everitt, P. Nordlander, and N. J. Halas, "Fano resonant aluminum nanoclusters for plasmonic colorimetric sens-

ing," *ACS Nano*, vol. 9, no. 10, pp. 10628–10636, 2015.

[18] K.-T. Lee, J. Y. Lee, S. Seo, and L. J. Guo, "Colored ultrathin hybrid photovoltaics with high quantum efficiency," *Light: Science & Applications*, vol. 3, no. 10, pp. 1–7, 2014.

[19] Y. Cui, R. S. Hegde, I. Y. Phang, H. K. Lee, and X. Y. Ling, "Encoding molecular information in plasmonic nanostructures for anti-counterfeiting applications," *Nanoscale*, vol. 6, no. 1, pp. 282–288, 2014.

[20] K. Kumar, H. Duan, R. S. Hegde, S. C. W. Koh, J. N. Wei, and J. K. W. Yang, "Printing colour at the optical diffraction limit," *Nature Nanotechnology*, vol. 7, no. 9, pp. 557–561, 2012.

[21] Y. Cui, I. Y. Phang, R. S. Hegde, Y. H. Lee, and X. Y. Ling, "Plasmonic silver nanowire structures for two-dimensional multiple-digit molecular data storage application," *ACS Photonics*, vol. 1, no. 7, pp. 631–637, 2014.

[22] S. Westland and C. Ripamonti, *Computational Color Science*. John Wiley & Sons Ltd., 2004.

[23] S. Y. Lee, C. Forestiere, A. J. Pasquale, J. Trevino, G. Walsh, P. Galli, M. Romagnoli, and L. Dal Negro, "Plasmon-enhanced structural coloration of metal films with isotropic pinwheel nanoparticle arrays," *Optics Express*, vol. 19, no. 24, p. 23818, 2011.

[24] G. Si, Y. Zhao, J. Lv, M. Lu, F. Wang, H. Liu, N. Xiang, T. J. Huang, A. J. Danner, J. Teng, and Y. J. Liu, "Reflective plasmonic color filters based on lithographically patterned silver nanorod arrays," *Nanoscale*, vol. 5, no. 14, pp. 6243–8, 2013.

[25] R. Rajasekharan, E. Balaur, A. Minovich, S. Collins, T. D. James, A. Djalalian-Assl, K. Ganesan, S. Tomljenovic-Hanic, S. Kandasamy, E. Skafidas, D. N. Neshev, P. Mulvaney, A. Roberts, and S. Prawer, "Filling schemes at submicron scale: development of submicron sized plasmonic colour filters," *Scientific Reports*, vol. 4, 2014.

[26] Y. S. Do and K. C. Choi, "Matching surface plasmon modes in symmetry-broken structures for nanohole-based color filter," *IEEE Photonics Technology Letters*, vol. 25, no. 24, pp. 2454 – 2457, 2013.

[27] B. Zeng, Y. Gao, and F. J. Bartoli, "Ultrathin nanostructured metals for highly transmissive plasmonic subtractive color filters," *Scientific Reports*, vol. 3, p. 2840, 2013.

[28] Y. Yu, Q. Chen, L. Wen, X. Hu, and H.-F. Zhang, "Spatial optical crosstalk in CMOS image sensors integrated with plasmonic color filters," *Optics Express*, vol. 23, no. 17, p. 21994, 2015.

[29] K. Diest, J. A. Dionne, M. Spain, and H. A. Atwater, "Tunable color filters based on metal-insulator-metal resonators," *Nano Letters*, vol. 9, no. 7, pp. 2579–2583, 2009.

[30] Y. Yu, L. Wen, S. Song, and Q. Chen, "Transmissive/reflective structural color filters: theory and applications," *Journal of Nanomaterials*, vol. 2014,

no. 212637, pp. 1–17, 2014.

[31] D. Chanda, K. Shigeta, T. Truong, E. Lui, A. Mihi, M. Schulmerich, P. V. Braun, R. Bhargava, and J. A. Rogers, "Coupling of plasmonic and optical cavity modes in quasi-three-dimensional plasmonic crystals," *Nature Communications*, vol. 2, p. 479, 2011.

[32] V. R. Shrestha, S.-S. Lee, E.-S. Kim, and D.-Y. Choi, "Non-iridescent transmissive structural color filter featuring highly efficient transmission and high excitation purity," *Scientific Reports*, vol. 4, p. 4921, 2014.

[33] A. L. González, C. Noguez, J. Beránek, and A. S. Barnard, "Size, shape, stability, and color of plasmonic silver nanoparticles," *Journal of Physical Chemistry C*, vol. 118, pp. 9128–9136, 2014.

[34] A. I. Kuznetsov, A. E. Miroshnichenko, Y. H. Fu, J. Zhang, and B. Luk'yanchuk, "Magnetic light," *Scientific Reports*, vol. 2, p. 492, 2012.

[35] H. Duan, A. I. Fernández-Dom´inguez, M. Bosman, S. A. Maier, and J. K. W. Yang, "Nanoplasmonics: classical down to the nanometer scale," *Nano Letters*, vol. 12, no. 3, pp. 1683–1689, 2012.

[36] G. Mie, "Beiträge zur Optik trüber Medien, speziell kolloidaler Metallösungen," *Annalen der Physik*, vol. 330, no. 3, pp. 377–445, 1908.

[37] M. Kolwas, "Scattering of Light on Droplets and Spherical Objects: 100 Years of Mie Scattering," *Computational Methods in Science and Technology*, no. 2, pp. 107–113, 2010.

[38] V. Myroshnychenko, J. Rodr´iguez-Fernández, I. Pastoriza-Santos, A. M. Funston, C. Novo, P. Mulvaney, L. M. Liz-Marzán, and F. J. Garc´ia de Abajo, "Modelling the optical response of gold nanoparticles," *Chemical Society Reviews*, vol. 37, no. 9, p. 1792, 2008.

[39] B. Gallinet, J. Butet, and O. J. F. Martin, "Numerical methods for nanophotonics: standard problems and future challenges," *Laser and Photonics Reviews*, vol. 603, no. 6, pp. 577–603, 2015.

[40] A. Sihvola, "Dielectric polarization and particle shape effects," *Journal of Nanomaterials*, vol. 2007, no. 45090, pp. 1–9, 2007.

[41] A. B. Evlyukhin, C. Reinhardt, A. Seidel, B. S. Luk'Yanchuk, and B. N. Chichkov, "Optical response features of Si-nanoparticle arrays," *Physical Review B: Condensed Matter and Materials Physics*, vol. 82, no. 4, pp. 1–12, 2010.

[42] M. W. Knight, Y. Wu, J. B. Lassiter, P. Nordlander, and N. J. Halas, "Substrates matter: influence of an adjacent dielectric on an individual plasmonic nanoparticle," *Nano Letters*, vol. 9, no. 5, pp. 2188–2192, 2009.

[43] H. Chen, T. Ming, S. Zhang, Z. Jin, B. Yang, and J. Wang, "Effect of the dielectric properties of substrates on the scattering patterns of gold nanorods," *ACS Nano*, vol. 5, no. 6, pp. 4865–4877, 2011.

[44] S. Zhang, K. Bao, N. J. Halas, H. Xu, and P. Nordlander, "Substrate-induced

Fano resonances of a plasmonic nanocube: a route to increased-sensitivity localized surface plasmon resonance sensors revealed," *Nano Letters*, vol. 11, no. 4, pp. 1657–1663, 2011.

[45] T. Ebbesen, H. J. Lezec, H. F. Ghaemi, T. Thio, P. A.Wolff, T. Thio, and P. A.Wolff, "Extraordinary optical transmission through sub-wavelength hole arrays," *Nature*, vol. 86, no. 6, pp. 1114–1117, 1998.

[46] L. Mart´in-Moreno, F. J. Garc´ia-Vidal, H. J. Lezec, K. M. Pellerin, T. Thio, J. B. Pendry, and T. W. Ebbesen, "Theory of extraordinary optical transmission through subwavelength hole arrays," *Physical Review Letters*, vol. 86, no. 6, pp. 1114–1117, 2001.

[47] H. Liu and P. Lalanne, "Microscopic theory of the extraordinary optical transmission.," *Nature*, vol. 452, no. 7188, pp. 728–731, 2008.

[48] T. Xu, H. Shi, Y. K. Wu, A. F. Kaplan, J. G. Ok, and L. J. Guo, "Structural colors: From plasmonic to carbon nanostructures," *Small*, vol. 7, no. 22, pp. 3128–3136, 2011.

[49] I. J. H. McCrindle, J. Grant, T. D. Drysdale, and D. R. S. Cumming, "Hybridization of optical plasmonics with terahertz metamaterials to create multi-spectral filters," *Optics Express*, vol. 21, no. 16, pp. 19142 – 19152, 2013.

[50] L. Lin and A. Roberts, "Angle-robust resonances in cross-shaped aperture arrays," *Applied Physics Letters*, vol. 97, no. 6, pp. 2008–2011, 2010.

[51] D. Inoue, A. Miura, T. Nomura, H. Fujikawa, K. Sato, N. Ikeda, D. Tsuya, Y. Sugimoto, and Y. Koide, "Polarization independent visible color filter comprising an aluminum film with surface-plasmon enhanced transmission through a subwavelength array of holes," *Applied Physics Letters*, vol. 98, no. 9, pp. 2009–2012, 2011.

[52] J. K. W. Tan, S. J., Zhang, L., Zhu, D., Goh, X. M., Qiu, C-W., Yang, "Plasmonic color palettes for photorealistic color printing with aluminum nanostructures," *Nano letters*, vol. 14, pp. 4023–4029, 2014.

[53] J. Olson, A. Manjavacas, L. Liu, W.-S. Chang, B. Foerster, N. S. King, M. W. Knight, P. Nordlander, N. J. Halas, and S. Link, "Vivid, full-color aluminum plasmonic pixels," *Proceedings of the National Academy of Sciences of the United States of America*, vol. 111, no. 40, pp. 14348–14353, 2014.

[54] M. W. Knight, N. S. King, L. Liu, H. O. Everitt, P. Nordlander, and N. J. Halas, "Aluminum for plasmonics," *ACS Nano*, vol. 8, no. 1, pp. 834–840, 2014.

[55] G. Maidecchi, G. Gonella, R. Proietti Zaccaria, R. Moroni, L. Anghinolfi, A. Giglia, S. Nannarone, L. Mattera, H. L. Dai, M. Canepa, and F. Bisio, "Deep ultraviolet plasmon resonance in aluminum nanoparticle arrays," *ACS Nano*, vol. 7, no. 7, pp. 5834–5841, 2013.

[56] J. S. Clausen, E. Hjlund-Nielsen, A. B. Christiansen, S. Yazdi, M. Grajower, H. Taha, U. Levy, A. Kristensen, and N. A. Mortensen, "Plasmonic metasurfaces for coloration of plastic consumer products," *Nano Letters*, vol. 14, no. 8, pp. 4499–4504, 2014.

[57] V. R. Shrestha, S. S. Lee, E. S. Kim, and D. Y. Choi, "Aluminum plasmonics based highly transmissive polarization-independent subtractive color filters exploiting a nanopatch array," *Nano Letters*, vol. 14, pp. 6672–6678, 2014.

[58] X. M. Goh, Y. Zheng, S. J. Tan, L. Zhang, K. Kumar, C.-W. Qiu, and J. K. W. Yang, "Three-dimensional plasmonic stereoscopic prints in full colour," *Nature Communications*, vol. 5, no. November 2015, pp. 5361–6, 2014.

[59] O. Matoba, T. Nomura, E. Perez-Cabre, M. S. Millan, and B. Javidi, "Optical techniques for information security," *Proceedings of the IEEE*, vol. 97, no. 6, pp. 1128–1148, 2009.

[60] S. Liu, M. Sinclair, and T. Mahony, "Optical magnetic mirrors without metals," *Optica*, vol. 1, no. 4, pp. 250–256, 2014.

[61] I. Staude, A. E. Miroshnichenko, M. Decker, N. T. Fofang, S. Liu, E. Gonzales, J. Dominguez, T. S. Luk, D. N. Neshev, I. Brener, and Y. Kivshar, "Tailoring directional scattering through magnetic and electric resonances in subwavelength silicon nanodisks," *ACS Nano*, vol. 7, no. 9, pp. 7824–7832, 2013.

[62] L. Cao, P. Fan, E. S. Barnard, A. M. Brown, and M. L. Brongersma, "Tuning the color of silicon nanostructures," *Nano Letters*, vol. 10, no. 7, pp. 2649–2654, 2010.

[63] A. C. Arsenault, D. P. Puzzo, I. Manners, and G. A. Ozin, "Photonic-crystal full-colour displays," *Nature Photonics*, vol. 1, no. 8, pp. 468–472, 2007.

[64] K. Seo, M. Wober, P. Steinvurzel, E. Schonbrun, Y. Dan, T. Ellenbogen, and K. B. Crozier, "Multicolored vertical silicon nanowires," *Nano Letters*, vol. 11, no. 4, pp. 1851–1856, 2011.

[65] J. Walia, N. Dhindsa, M. Khorasaninejad, and S. S. Saini, "Color generation and refractive index sensing using diffraction from 2D silicon nanowire arrays," *Small*, vol. 10, no. 1, pp. 144–151, 2014.

[66] H. S. Ee, J. H. Kang, M. L. Brongersma, and M. K. Seo, "Shape-dependent light scattering properties of subwavelength silicon nanoblocks," *Nano Letters*, vol. 15, no. 3, pp. 1759–1765, 2015.

[67] S. C. Yang, K. Richter, and W. J. Fischer, "Multicolor generation using silicon nanodisk absorber," *Applied Physics Letters*, vol. 106, no. 081112, pp. 23–27, 2015.

[68] R. S. Hegde and K. S. Panse, "Design and optimization of ultra-thin spectral filters based on silicon nanocross antenna arrays," *SPIE Journal of Nanophotonics*, vol. 10, no. 2, p. 026030, 2016.

[69] D. Franklin, Y. Chen, A. Vazquez-Guardado, S. Modak, J. Boroumand, D. Xu, S.-T. Wu, and D. Chanda, "Polarization-independent actively tunable colour generation on imprinted plasmonic surfaces," *Nature Communications*, vol. 6, p. 7337, 2015.

[70] V. E. Johansen, L. H. Thamdrup, K. Smistrup, T. Nielsen, O. Sigmund, and P. Vukusic, "Designing visual appearance using a structured surface," *Optica*, vol. 2, no. 3, p. 239, 2015.

Chapter 6

Nanostructure-enhanced fluorescence emission

Song Sun
Institute of High Performance Computing, Singapore
Lin Wu
Institute of High Performance Computing, Singapore
Ping Bai
Institute of High Performance Computing, Singapore

In this chapter, we consider fluorescence emission enhanced with resonant nanostructures. Section 6.1 introduces the basic concept of fluorescence emission. Section 6.2 discusses the mechanisms of fluorescence emission including both linear and nonlinear types. Sections 6.3 and 6.4 briefly review the state-of-the-art technologies based on plasmonic and photonics materials to enhance fluorescence emission. Summary and future perspective are covered in Section 6.5.

6.1 Introduction and background

6.1.1 Application of fluorescent emitters

Fluorescence labels are powerful and inexpensive techniques for detection and imaging of biomolecules including proteins, DNA, RNA, and even viruses that offers the visual representation, characterization, and quantification of biological processes at both cellular and sub-cellular levels. By bonding a specific fluorophore to targeted molecules, we can access the cell culture and monitor molecular bio-processes such as gene expression and progression and regression of cancer, as shown in Fig. 6.1 [1, 2]. All these information ultimately contributes to the development of targeted drug and gene therapy. Moreover, fluorescence labels possess incomparable advantages in terms of their stability and ability to overcome the biological delivery barrier and access the molecular activities in previously inaccessible regions such as brain [3].

Various techniques have been developed to improve the applicability, quality, and sensitivity of fluorescence imaging. For instance, fusion of green fluorescent protein (GFP) with other proteins can monitor specific cell compartments and dynamic processes. Utilizing near-infrared fluorescence labels can maximize the penetration depth, while minimizing the autofluorescence from non-target tissue. Designing "smart" probes with protease activation functionality can significantly enhance the sensitivity and avoid nonspecific bonding. Combining with advanced reconstruction algorithms allows us to develop fluorescence-mediated tomography. By correlating the fluorescence and electron microscopies, the biological roles of molecules and the cellular ultrastructures can be revealed to provide more profound understandings. These techniques can be found in varous reviews [3–6].

6.1.2 *Types of fluorescent emitters*

To obtain high-quality imaging or high-accuracy detection, it is important to choose a proper fluorophore label for a particular target or event. Many factors have to be considered when choosing the fluorophore label to achieve optimal performance, including the excitation and emission features (e.g., Stokes shift), quantum yield, solubility, chemical and thermal stability, functional group for site-specific labeling, compatibility with the experimental setup for emission signal enhancement and collection. Especially for biomedical applications, additional caution has to been taken regarding the potential toxicity (cytotoxicity and nanotoxicity) of the label. These properties of the fluorophores not only affect the detection limit and dynamic range, but also define the reliability of the measurement and the potential for multiplexing (e.g., parallel detection of different targets).

There are two types of fluorophore labels commonly used in the industry: organic dyes and inorganic quantum dots (QDs). Organic dyes are molecular systems with well-defined structures, whose optical properties are determined by electronic transitions (e.g., delocalized optical transition and intra-molecular charge-transfer

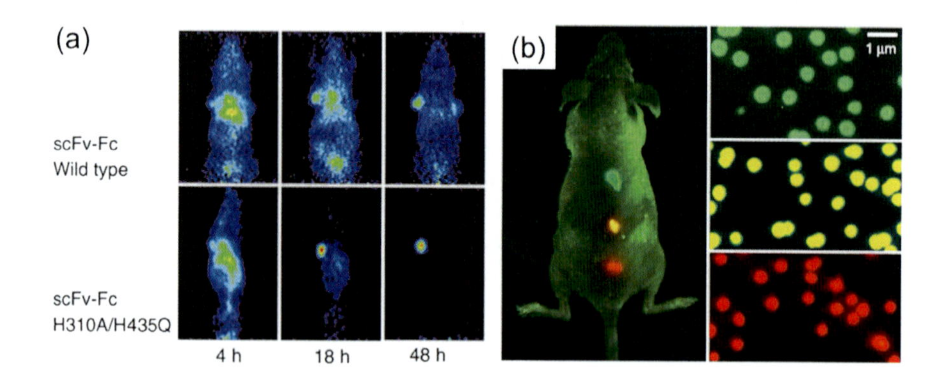

Figure 6.1: (a) Imaging of the expression of carcinoembryonic antigen (CEA) with fluorescence-labeled antibody fragments in mice. (b) Multiplexed optical imaging using quantum dots. Reprinted with permission. (a) Reference [1]. © 2005 Nature Publishing Group. (b) Reference [2]. © 2004 Nature Publishing Group.

transition) inside the molecule. Most dyes have narrow and mirrored absorption and emission bands, where the Stokes shift is insensitive to the solvent medium. On the other hand, QDs are inorganic nanocrystal chromophores with size-dependent optical, physical, and chemical properties, which can be synthesized from II/VI, III/V semiconductors, carbon or Si nanoparticles. As compared to the organic dyes, QDs have gradually increasing absorption toward shorter wavelengths with larger absorption coefficients, whereby the absorption and emission bands can be easily controlled with the particle size. Meanwhile, QDs possess high quantum yields from visible to near-infrared (NIR) regions, while dyes have high quantum yields only in the visible region. Furthermore, QDs have much better stability owing to the inorganic layer to protect the core material, whereas organic dyes are more sensitive to the microenvironment and have poor stability. Nevertheless, there are well-established protocols for organic dyes to attach to biomolecules and deliver dyes into cells, depending on their types. For QDs, however, it is difficult to develop general protocols as the surfaces of QDs are unique to a large extent, depending on the synthesis process. Fig. 6.2 summarizes the differences between organic dyes and QDs in various aspects [7].

6.1.3 Enhancement of fluorescence with nanostructures

Despite the applications and all the advantages provided by fluorescence emitters, effective detection of fluorescence signals is difficult using conventional optical microscopy. The image quality, in particular the resolution, is the core determinant of the performance of a microscope and is constrained by the well-known diffraction limit. In 1877, the famous equation defining the diffraction limit was derived by von Helmholtz and Stephenson. It states that the diffraction limit is proportional to the wavelength of the light and inversely proportional to the numerical aperture of the lens. Under this principle, the resolution in the visible spectrum (i.e., at the wavelength range from 400 nm to 800 nm) can only achieve 80 nm $\sim$ 100 nm using a conventional microscope, which is not enough for the detection of fluorescence-labeled molecular or cell activities, whereby a super resolution below 10 nm is required. As a result, finding new detection methods able to break the diffraction limit keeps driving modern research [8].

Besides the detection methods, another fundamental problem associated with fluorescence emission is a weak optical intensity of the light emitted by a single fluorophore that suffers from significant photobleaching after several measurements. Such a weak signal is very difficult to detect with conventional devices, such as CCD cameras, because of a low signal-to-noise ratio limited sensitivity. This problem becomes critical in some emerging applications requiring single-molecule fluorescence detection, such as DNA sequencing and detection of low levels of analytes in small volumes for early diagnosis. As a result, fluorescence signal enhancement techniques are consistently pursued to improve sensitivity.

To resolve these two issues, nanoscale optics were proposed. Nano-optics utilize interactions between light and nanometer-scale structures to squeeze the light into the subwavelength region, allowing us to overcome the diffraction limit that restricts the conventional optical microscopes. This is the key in achieving super resolution

Property	Organic dye	QD[a]
Absorption spectra	Discrete bands, FWHM[b] 35 nm[c] to 80–100 nm[d]	Steady increase toward UV wavelengths starting from absorption onset; enables free selection of excitation wavelength
Molar absorption coefficient	2.5×10^4–2.5×10^5 M^{-1} cm^{-1} (at long-wavelength absorption maximum)	10^5–10^6 M^{-1} cm^{-1} at first exitonic absorption peak, increasing toward UV wavelengths; larger (longer wavelength) QDs generally have higher absorption
Emission spectra	Asymmetric, often tailing to long-wavelength side; FWHM, 35 nm[c] to 70–100 nm[d]	Symmetric, Gaussian profile; FWHM, 30–90 nm
Stokes shift	Normally <50 nm[c], up to >150 nm[d]	Typically <50 nm for visible wavelength-emitting QDs
Quantum yield	0.5–1.0 (visible[e]), 0.05–0.25 (NIR[e])	0.1–0.8 (visible), 0.2–0.7 (NIR)
Fluorescence lifetimes	1–10 ns, mono-exponential decay	10–100 ns, typically multi-exponential decay
Two-photon action cross-section	1×10^{-52}–5×10^{-48} cm^4 s $photon^{-1}$ (typically about 1×10^{-49} cm^4 s $photon^{-1}$)	2×10^{-47}–4.7×10^{-46} cm^4 s $photon^{-1}$
Solubility or dispersibility	Control by substitution pattern	Control via surface chemistry (ligands)
Binding to biomolecules	Via functional groups following established protocols Often several dyes bind to a single biomolecule Labeling-induced effects on spectroscopic properties of reporter studied for many common dyes	Via ligand chemistry; few protocols available Several biomolecules bind to a single QD Very little information available on labeling-induced effects
Size	~0.5 nm; molecule	6–60 nm (hydrodynamic diameter); colloid
Thermal stability	Dependent on dye class; can be critical for NIR-wavelength dyes	High; depends on shell or ligands
Photochemical stability	Sufficient for many applications (visible wavelength), but can be insufficient for high-light flux applications; often problematic for NIR-wavelength dyes	High (visible and NIR wavelengths); orders of magnitude higher than that of organic dyes; can reveal photobrightening
Toxicity	From very low to high; dependent on dye	Little known yet (heavy metal leakage must be prevented, potential nanotoxicity)
Reproducibility of labels (optical, chemical properties)	Good, owing to defined molecular structure and established methods of characterization; available from commercial sources	Limited by complex structure and surface chemistry; limited data available; few commercial systems available
Applicability to single-molecule analysis	Moderate; limited by photobleaching	Good; limited by blinking
FRET	Well-described FRET pairs; mostly single-donor- single-acceptor configurations; enables optimization of reporter properties	Few examples; single-donor–multiple-acceptor configurations possible; limitation of FRET efficiency due to nanometer size of QD coating
Spectral multiplexing	Possible, 3 colors (MegaStokes dyes), 4 colors (energy-transfer cassettes)	Ideal for multi-color experiments; up to 5 colors demonstrated
Lifetime multiplexing	Possible	Lifetime discrimination between QDs not yet shown; possible between QDs and organic dyes
Signal amplification	Established techniques	Unsuitable for many enzyme-based techniques, other techniques remain to be adapted and/or established

Properties of organic dyes are dependent on dye class and are tunable via substitution pattern. Properties of QDs are dependent on material, size, size distribution and surface chemistry.

[a]Emission wavelength regions for QD materials (approximate): CdSe, 470–660 nm; CdTe, 520–750 nm; InP, 620–720 nm; PbS, >900 nm; and PbSe, >1,000 nm.

[b]FWHM, full width at half height of the maximum.

[c]Dyes with resonant emission such as fluoresceins, rhodamines and cyanines.

[d]CT dyes.

[e]Definition of spectral regions used here: visible, 400–700 nm; and NIR, > 700 nm.

Unless stated otherwise, all values were determined in water for organic dyes and in organic solvents for QDs, and refer to the free dye or QD.

Figure 6.2: Comparison of properties of organic dyes and quantum dots (QDs). Reprinted with permission. Reference [7]. © 2008 Nature Publishing Group.

imaging. Such an optical confinement also creates a strong field enhancement region (so-called hot-spot), which in turn can significantly enhance the intensity of the fluorescence located at that region [6]. Moreover, as compared to other fluorescence signal amplification strategies such as enzymatic amplification, avidin-biotin secondary detection technique, and nucleic acid amplification, among others. Nano-optic devices rely on manipulation of the electromagnetic field in the vicinity of nanostructures without disturbing the chemical environment of the fluorophore labels, thereby resulting in enhanced photostability. Generally, plasmonic and photonic nanostructures are made of noble metals such as Au and Ag or high-index dielectric materials such as Si. We will look at these systems in the latter sections of this chapter.

6.2 Fluorescence emission mechanism

6.2.1 *Classical description*

An excited molecule placed near a nanostructure can be considered a point transmitter. Meanwhile, a molecule in the ground state excited by the localized field near the optical antenna acts as a point receiver [9, 10]. Single fluorophores, organic dyes, and quantum dots are commonly treated as a two-level system (not accounting for nonlinear optical effects), represented with an electric dipole. The fluorescence emission rate γ_{em} can be expressed as the product of excitation rate γ_{exc} and quantum yield q, whereby q is defined as the fraction of the radiative decay rate over the total decay rate. It is based on the fact that the excitation and emission of a fluorophore are two incoherent processes that can be described separately. The fluorescence enhancement can thus be expressed as

$$\frac{\gamma_{em}}{\gamma_{em}^0} = \frac{\gamma_{exc}}{\gamma_{exc}^0} \cdot \frac{q}{q^0}, \tag{6.1}$$

where the superscript 0 represents the nanostructure-free quantity.

Below the fluorophore absorption saturation, the excitation rate is proportional to $\mathbf{E} \cdot \mathbf{p}$, where $\mathbf{E}$ is the electrical field at the location of the dipole and $\mathbf{p}$ is the transition dipole momentum. Enhancement of the local electric field $\mathbf{E}$ naturally results in an enhanced fluorescence excitation rate, which is the fundamental principle of metal and high-index dielectric nanostructures. The quantum yield in principle should be expressed by the weighted sum of all possible decay channels, as described by Fermi's golden rule. If the nanostructure is viewed as a non-radiative acceptor, its interaction with a fluorescence emitter actually mimics the fluorescence-resonant energy transfer shown in Fig. 6.3. Since the fluorophore is treated as an electric point dipole, the expression for the quantum yield can be simplified as

$$q = \frac{\Gamma_r/\Gamma_r^0}{\Gamma_r/\Gamma_r^0 + \Gamma_{abs}/\Gamma_r^0 + (1-q^0)/q^0}, \tag{6.2}$$

where Γ_r is the radiative decay in the presence of nanophotonic structure, Γ_{abs} is the additional non-radiative decay induced by the energy dissipation of the nanostructure, and q^0 is the intrinsic quantum yield of the fluorophore. Moreover, the radiative and nonradioactive decay rates can be correlated to the power of the unit harmonic

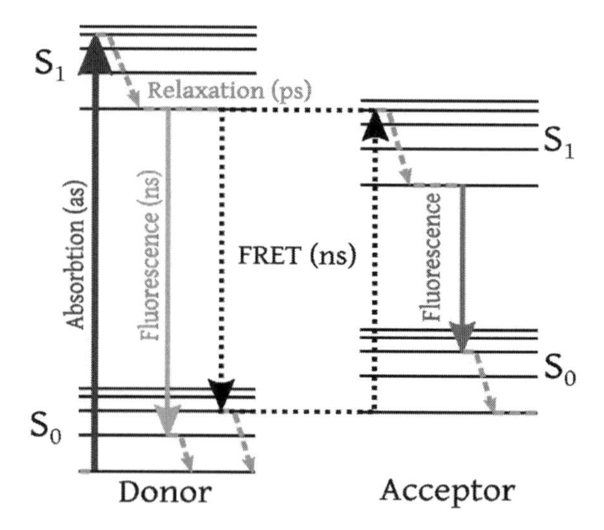

Figure 6.3: Schematic of fluorescence-resonant energy transfer mechanism. Wikipedia online (Source: https://en.wikipedia.org/wiki/Förster_resonance_energy_transfer).

dipole: $\Gamma_r/\Gamma_r^0 = P_r/P^0$ and $\Gamma_{abs}/\Gamma_r^0 = P_{abs}/P^0$, where P_r is the power radiated to the far field, P_{abs} is the power absorbed by the system, and P^0 is the power radiated by a free-standing unit harmonic dipole in a homogeneous environment. From Eqs. (6.1) and (6.2), the presence of nanostructure can improve the excitation rate by enhancing the local electric field and also manipulate the emission behavior of the fluorophore.

6.2.2 Two-photon absorption mechanism

Two-photon absorption is the process involving two photons used to excite a molecule from one state (usually the ground state) to a higher energy state, where the two photons may have identical or different frequencies. It is a nonlinear process that differs from the single-photon absorption in the atomic transition rate that depends on the square of the light intensity. As compared to the theory in Section 6.2.1 for a single-photon absorption, where the fluorescence is described with a two-level system, the two-photon absorption mechanism involves multi-state electronic transition and should be viewed as a system of three or more levels. The detail formulation can be found in Refs. [11, 12]. Two-photon excited fluorescence can significantly reduce the out-of-focus photobleaching and has a smaller autofluorescence signal, which eventually leads to higher image resolution [11]. For certain fluorescent dyes with good two-photon absorption, it is possible to realize excitation at wavelengths approximately double that of single-photon excitation (usually in the visible range).

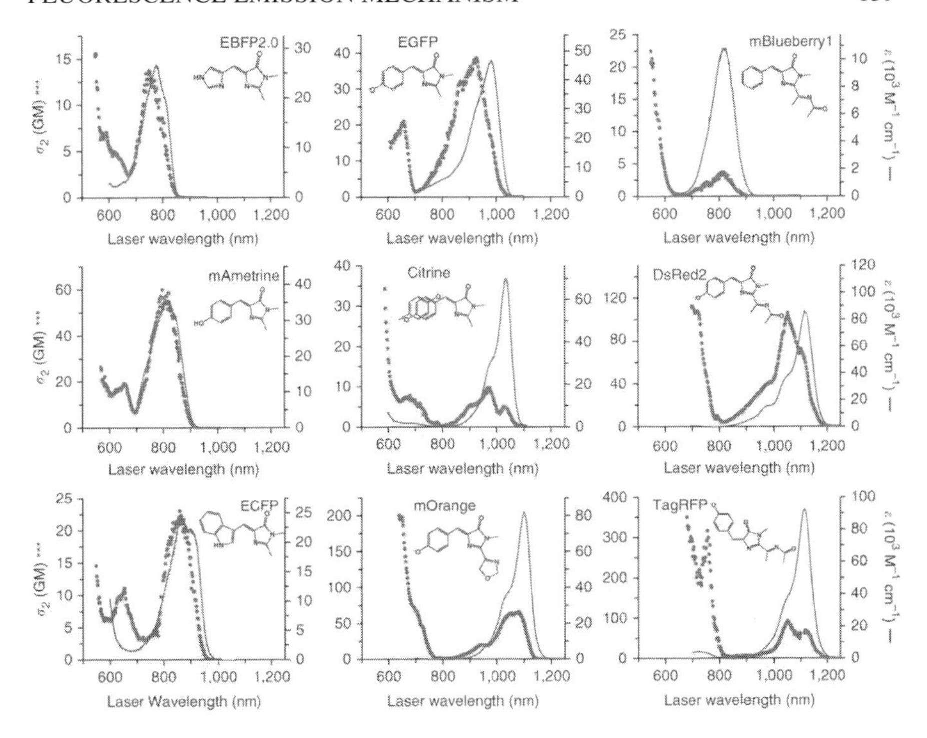

Figure 6.4: One- and two-photon absorption spectra of fluorescence with different chromophores. Two-photon absorption spectra are plotted against laser wavelength used for excitation. For the purpose of comparison, in one-photon absorption spectra the actual excitation wavelength is multiplied by a factor of two. Reprinted with permission. Reference [11]. © 2011 Nature American, Inc.

As a result, we can use excitation in the infrared range that offers deeper tissue penetration.

Because the chromophore in fluorophore is not centro-symmetric, the parity selection rules for one- and two-photon absorption are relaxed. Therefore, the same bands should appear in both one- and two-photon absorption spectra with different intensities. Generally, there should be an apparent overlap between one- and two-photon absorption spectra at the longest wavelength corresponding to the first electronic transition from the ground state to the first excited state. However, due to the resonant-enhancement effect, there is a distinct blueshift of the two-photon absorption band relative to the one-photon absorption band [12]. Fig. 6.4 illustrates the difference between two- and one-photon absorption bands of nine fluorophores, whereby the two-photon absorption band generally occurs at shorter wavelength, as compared to the one-photon absorption band.

6.3 Fluorescence enhancement with metal nanostructures

A distinct feature of metal nanostructures is their pronounced resonance on surface polaritons (see Chapters 2 and 3). These resonances exhibit strong fields near metal surfaces, which can tremendously enhance the excitation rate of fluorophores and modify the quantum yields. Eventually, it can enhance the fluorescence emission following the theoretical considerations noted in Section 6.2.1. In this section, we briefly review various fluorescence enhancement techniques relying on the use of metal nanostructures.

6.3.1 Metal nanoparticles

Metal nanoparticles (NPs) are probably the simplest structures that can be synthesized on large scales. Gold and silver are the most commonly used metals because NPs made of these materials exhibit stronger resonances on localized surface polaritons and feature lower optical losses compared to other metals. The resonances of subwavelength spherical gold and silver nanoparticles naturally occur around the free-space wavelengths of 520 nm and 400 nm, respectively. These can be easily tuned by varying the morphology of NPs or by changing the external medium to cover the entire visible and near-infrared regions. In addition, NPs can be coated with multiple ligands or probes to significantly increase the binding rates of emitting molecules and the resultant fluorescence intensity, essential in bioimaging and sensing applications [8].

Fluorescence emission modified by metal NPs has been studied extensively. First, researchers tried to understand the mechanism of fluorescence enhancement with a single NP, as shown in Fig. 6.5(a), and found that the strongest fluorescence enhancement occurs at moderate distances (around 10 nm) from the surface of the metal NP [9, 10]. Shorter distance results in quenching of fluorescence due to high non-radiative loss of the metal NP [see Fig. 6.5(b)]. Later, it was discovered that a much stronger fluorescence emission can occur when the emitter is placed in a nanometer-scale gap between two metal NPs [see Fig. 6.5(c)]. The Purcell factor (see Section 6.4.1) in this case can be as high as $10,000$, and the fluorescence enhancement can reach values above $1,000$ depending on the intrinsic quantum yield of the fluorescence emitter [13]. Such a strong fluorescence enhancement results from the hybridization of the near fields of two close NPs that creates a hot-spot inside the gap and tremendously boosts the excitation rate of the emitter. Similar dimer structures have been realized with metal nanowires, bowties, nanorods, and nanocubes. More details can be found in Ref. [8].

6.3.2 Metal thin films

Metal thin film is another commonly used configuration to enhance the fluorescence signal. Metal thin films rely on the excitation of transmission-polariton resonances. When an incident beam of p-polarized light strikes an electrically conducting metal layer sandwiched between the high- and low-index materials, it can resonate on the transmission polaritons, resulting in a reduced intensity of the reflected light under the conditions of total internal reflection. Such excitation provides a much longer penetration depth of the near fields, producing a much larger detection volume. On the other hand, it features weaker optical confinement than metal nanoparticles, which consequently leads to a smaller fluorescence enhancement.

The earliest studies [14, 15] used the Kretschmann or Otto configuration shown in Fig. 6.6(a) and (b), respectively. For the Kretschmann configuration, a thin metal layer is deposited on the substrate. With the correct film thickness (around 50 nm), the incident fields can excite the transmission polaritons that feature growing evanescent fields at the opposite surface of the metal film. On the other hand, for the Otto configuration, the metal film is separated from the substrate by a low-index dielectric layer (e.g., an air slit). In this configuration, the excited transmission polaritons provide strong evanescent fields at the metal-dielectric interface, but require a low-index layer $\sim 1\ \mu$m thick that limits its capability in loading samples. For this reason, the Otto configuration is not as widely used as the Kretschmann one.

More advanced models were developed based on two basic configurations shown in Figs. 6.6(c) and (d). The long-range structure [16] can extend the evanescent fields to micrometer range, which significantly increases the detection volume and consequently the sensitivity of the device. Such a structure can also be integrated with an optical waveguide to enhance the binding efficiency of the targeted analytes or be combined with diffractive elements to further enhance the fluorescence intensity. Moreover, embedding periodic patterns (e.g., holes, pillars, slits, etc.) in the metal film can also significantly enhance the fluorescence signal, as shown in Fig. 6.6(d), achieving fluorescence enhancements as high as a few thousands [17].

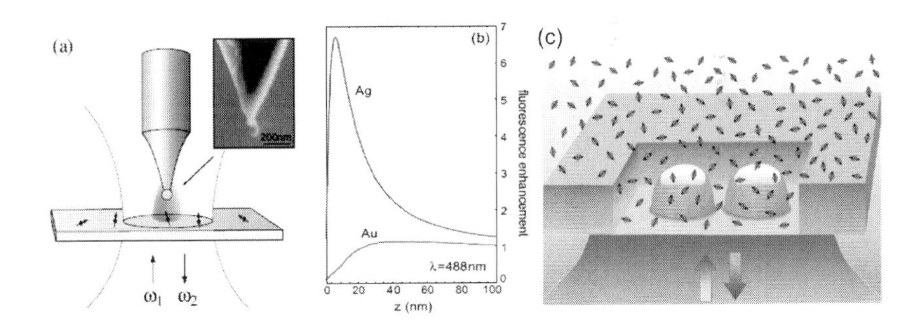

Figure 6.5: Fluorescence enhancement with metal nanoparticles. (a) A single metal nanoparticle is attached to a single glass tip to interact with a single fluorescence molecule. (b) Fluorescence enhancement of Ag or Au nanoparticles with 80 nm diameter. (c) A metal ‘antenna-in-box’ dimer structure. Reprinted with permission. (a) and (b) Reference [9]. © 2007 Optical Society of America. (c) Reference [13]. © 2013 Macmillan Publishers Limited.

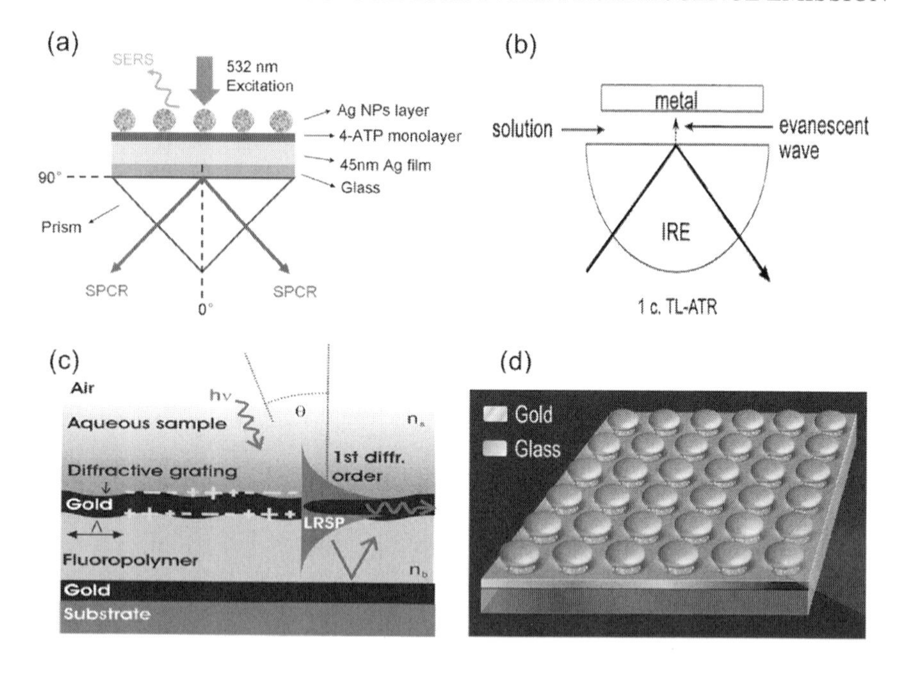

Figure 6.6: Fluorescence enhancement with metal thin films. Schematics of (a) Kretschmann configuration, (b) Otto configuration, (c) long-range configuration, and (d) three-dimensional nano-antenna array. Reprinted with permission. (a) Reference [14]. © 2015 American Chemical Society. (b) Reference [15]. © 2001 American Chemical Society. (c) Reference [16]. © 2012 American Chemical Society. (d) Reference [17]. © 2012 American Chemical Society.

6.3.3 *Modified emission directivity*

Both the magnitude and the directivity of the fluorescence emission can be modified by properly designing the metal nanostructure. The directivity is the ability of the nanostructure to concentrate the radiated power of the fluorescence into a certain direction, which is crucial in determining the amount of power to be collected with the detector, e.g., a CCD camera with a finite numerical aperture. Lokowicz et al. first demonstrated that the directivity of the fluorescence emission can be enhanced by coupling with the transmission polaritons of a thin metal film, as shown in Fig. 6.7(a) [18]. At least 50% of the total emission can be collected accompanied by simultaneous reduction of the unwanted background noise. Later, Wenger et al. constructed a nanoaperture surrounded by periodic corrugation (also known as a bull's eye aperture) [see Fig. 6.7(b)] to constrain the fluorophore emission within ±15 degrees, while enhancing the fluorescence emission 120 times [19]. Hulst et al. recently demonstrated the directional control with a half-wavelength metal nanorod, as shown in Fig. 6.7(c), whereby the directivity of fluorescence emission can be eas-

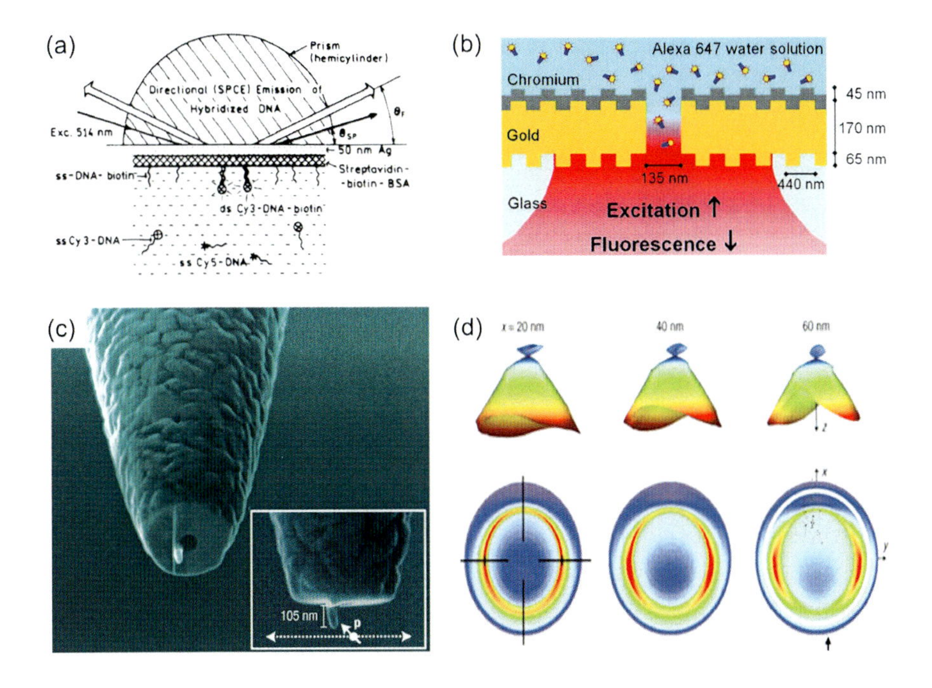

Figure 6.7: Emission directivity modified with metal nanostructures. (a) Directional fluorescence emission with a thin Ag film. (b) Numerical aperture surrounded with periodic corrugate. (c) Emission control with a metal nanorod. (d) Examples of emission pattern with respect to the distance between the emitter and antenna. Reprinted with permission. (a) Reference [18]. © 2003 American Chemical Society. (b) Reference [19]. © 2011 American Chemical Society. (c) and (d) Reference [20]. © 2008 Nature publishing group.

ily adjusted by controlling the distance between the emitter and antenna, as shown in Fig. 6.7(d) [20].

6.4 Fluorescence enhancement with dielectric nanostructures

The major drawback of metal nanostructures are the high losses inherent to the metallic materials at visible wavelengths that limits the quantum yield and the fluorescence enhancement. An alternative approach is to use nanostructures made of high refractive index dielectric materials such as silicon. These materials can also achieve optical confinement in their vicinities by utilizing high refractive index contrast compared to the surrounding medium. In addition, dielectric materials have much smaller losses in the visible range than metals. Eventually, they result in much higher quantum yields. Besides, most dielectric materials are compatible with the CMOS technology, hence reducing the fabrication cost. In this section, we will review the fluorescence enhancement techniques based on dielectric nanostructures.

6.4.1 *Photonic crystal microcavities*

A photonic crystal (PC) is a dielectric nanostructure whose refractive index changes periodically on the length scale comparable to the wavelength of light. Interference among multiple reflections traps light into a small volume and also prevents light propagation in a certain frequency range (stopband or photonic bandgap). PC cavities were first proposed by Purcell in 1946 to enhance the spontaneous emission of a dipole emitter. The ratio between the modified and free-space emission rates is known as the Purcell factor,

$$F_p = \frac{\Gamma_g}{\Gamma_0} = \frac{3Q\lambda^3}{4\pi^2 V_0}, \tag{6.3}$$

where λ is the wavelength of the transition, Q is the quality factor of the cavity, and V_0 is the cavity volume. A PC cavity could be designed so that its fundamental mode is resonant with the transition frequency of the dipole, and the dipole emission could be significantly enhanced as long as the emitter is located at the maximum field in the cavity and oriented with the polarization of the cavity mode. Optimized PC cavities can achieve a minimum volume of $0.1(\lambda_0/n)^3$, which is set by the diffraction limit of the periodic structure.

The early experiments demonstrated the feasibility of enhancing the transition in a single atom using PC cavities with very limited Purcell factor (around 1.02 in visible wavelengths), which subsequently was resolved using semiconductor structures such as quantum wells incorporated planar microcavities or pillars. Later, a better control of fluorescence emission was demonstrated [21, 22] with the well-known 'woodpile' PC structures and the 'inverse-opal' PC structures shown in Fig. 6.8(a) and (b), respectively. In recent years, researchers realized that the bandedge states become localized and lower in energy in the presence of localized defects or disorders, which can be viewed as a type of optical cavity with surrounding PC acting as a mirror in all directions. These defect cavities can be introduced into other PC structures such as woodpile and the emission can be coupled to the resonant mode of the cavities. The most common configuration consists of a high refractive index semiconductor membrane with an etched hole array. More details can be found in Ref. [8].

6.4.2 *Dielectric nanoantennas*

Dielectric nanoantennas are subwavelength nanostructures that have very similar configurations to metal nanoantennas (e.g., nanoparticle type) except for material composition [23]. Albella et al. proved that dielectric nanoantennas can provide comparable enhancements for radiative decay compared to metal counterparts. At the same time, they offer larger and more stable quantum yield due to much smaller optical losses [24]. In addition, dielectric nanoantennas exhibit both electric resonances

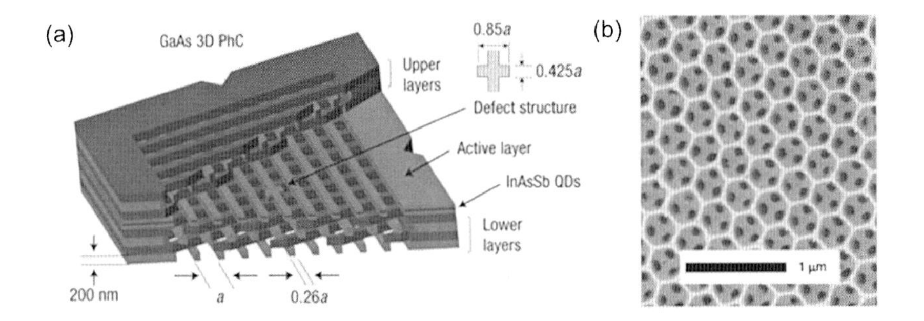

Figure 6.8: Fluorescence enhancement with photonic crystal microcavities. (a) Schematic of woodpile photonic crystal. (b) Three-dimensional inverse opal photonic crystal. Reprinted with permission. (a) Reference [21]. © 2008 Macmillan Publishers Limited. (b) Reference [22]. © 2004 Nature publishing group.

and magnetic ones, as shown in Fig. 6.9(a) [23], and this provides an additional degree of freedom in manipulating the spontaneous emission rates of fluorescent emitters. Electric and magnetic resonances observed for a single dielectric sphere [see Fig. 6.9(b)] can be coupled in a dimer configuration to provide a stronger field enhancements, as shown in Fig. 6.9(c). Despite these amazing features, the study on dielectric-enhanced fluorescence has just started, the actual performance is not yet clear, and the full potential has not yet been realized. We expect to see more dielectric nanostructures proposed in the near future.

6.4.3 Metal-dielectric hybrid structures

Besides pure metal and dielectric systems, fluorescence enhancement techniques have been developed using metal-dielectric hybrid structures. Aslan et al. reported the development of highly versatile fluorescent core-shell Ag@SiO_2 nanocomposites to incorporate any fluorophore to the outer silica shell to enhance its fluorescence signal up to 200 times, as shown in Fig. 6.10(a). When compared to a control sample of fluorescent nanoparticles, the core-shell architecture yielded up to 20-fold detectability [25]. Recently, Faggiani et al. investigated photonic devices with quantum emitters embedded in nanogaps for operation at visible and near-infrared frequencies, as shown in Fig. 6.10(b). The system can provide large spontaneous emission rate enhancements and good photon-radiation efficiencies, and quenching is thus effectively overcome [26]. The high decay rates found in planar nanogaps can be harnessed to realize tapered antennas offering strong decay rate enhancements ($\sim 10^2$ to 10^3) and large photon-radiation efficiencies. Livneh et al. introduced a new design for a hybrid metal-dielectric nanoantenna consisting of a metallic bullseye nanostructure and a dielectric waveguide layer, as shown in Fig. 6.10(c) [27]. Such configuration directs photon emission from colloidal QDs over a spectral range of $\sim$20 nm, compatible

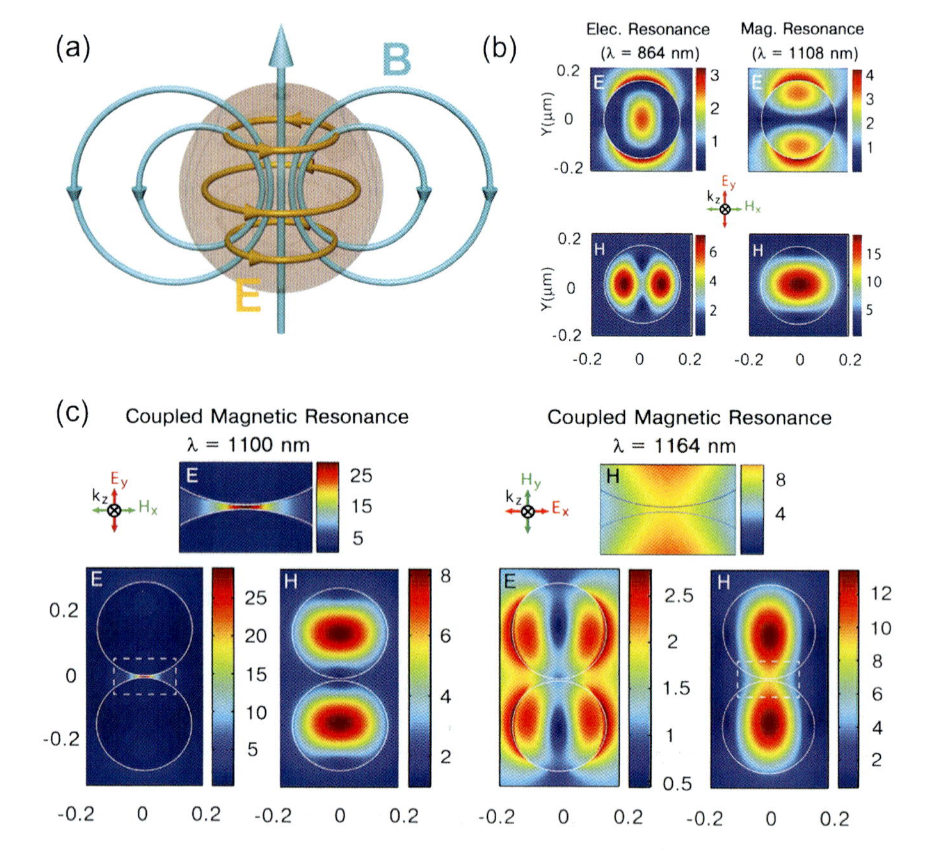

Figure 6.9: The principle of fluorescence enhancement with dielectric nanoantennas. (a) Schematic of magnetic resonance of a single dielectric particle. (b) Near-field enhancement of a Si particle at electric and magnetic resonance. (c) Near-field enhancement of a Si dimer at coupled electric and magnetic resonances. Reprinted with permission. (a) Reference [23]. © 2012 Nature Publishing group. (b) and (c) Reference [24]. © 2013 American Chemical Society.

with many modern QD spectral spreads at room temperature. The hybrid nanoantenna also presented a very small divergence angle of only 3.25 degrees along with broadband high directivity and high collection efficiency.

6.5 Summary

In this chapter, we discussed enhancement of fluorescence emission by using resonant nanostructures for detection and imaging of the molecular and cellular activities in biological tissues. We described the mechanism of fluorescence emission and showed that photonic and plasmonic nanostructures are able to enhance the flu-

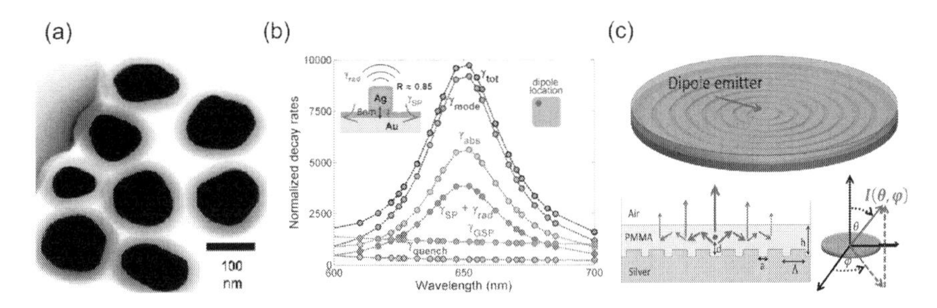

Figure 6.10: Fluorescence enhancement with metal-dielectric hybrid structures. (a) TEM images of Ag@SiO_2 core-shell particles. (b) Decay channels of nanocube antennas. (c) Schematic of the hybrid metal-dielectric nanoantenna. Reprinted with permission. (a) Reference [25]. © 2007 American Chemical Society. (b) Reference [26]. © 2015 American Chemical Society. (c) Reference [27]. © 2015 American Chemical Society.

orescence emission and manipulate its directivity due to their resonant nature. We briefly reviewed a few typical configurations (nanoparticles, thin films, periodic patterns, etc.) based on metals, high-index dielectric materials, and metal-dielectric hybrid structures used to enhance fluorescence emission up to several thousand times and/or constrain the directivity of the emission within a few degrees. Despite the remarkable progress, the full potential of resonant nanostructures in enhancement of fluorescence emission has not been realized yet. In the future, more effort will be focused on cost reduction, performance improvement (intensity, directivity, tunability), and an efficient way to bond the fluorophore specifically at the hot-spots of the nanostructures. In general, the research on fluorescence enhancement is still at the early stage of promises and opportunities.

Bibliography

[1] A. Wu and P. Senter, "Arming antibodies: prospects and challenges for immunconjugates," *Nat. Biotechnol.*, vol. 23, pp. 1137–1146, 2005.

[2] X. Gao, Y. Cui, R. Levenson, L. Chuang, and S. Nie, "In vivo cancer targeting and imaging with semiconductor quantum dots," *Nat. Biotechnol.*, vol. 22, pp. 969–976, 2004.

[3] K. Shah, A. Jacobs, X. Breakefield, and R. Weissleder, "Molecular imaging of gene therapy for cancer," *Gene Ther.*, vol. 11, pp. 1175–1187, 2004.

[4] P. Boer, J. Hoogenboom, and B. Giepmans, "Correlated light and electron microscopy: ultrastructure lights up!," *Nat. Methods*, vol. 12, pp. 503–513, 2015.

[5] W. Weber, J. Czernin, M. Phelps, and H. Herschman, "Technology insight: novel imaging of molecular target is an emerging area crucial to the development of targeted drug," *Nat. Clin. Pract. Oncol.*, vol. 5, pp. 44–54, 2008.

[6] X. Hao, C. Kuang, Z. Gu, Y. Wang, S. Li, Y. Ku, Y. Li, J. Ge, and X. Liu, "From microscopy to nanoscopy via visible light," *Light Sci. Appl.*, vol. 2, p. e108, 2013.

[7] U. Resch-Genger, M. Grabolle, S. Cavaliere-Jaricot, R. Nitschke, and T. Nann, "Quantum dots versus organic dyes as fluorescent labels," *Nat. Methods*, vol. 5, pp. 763–775, 2008.

[8] M. Pelton, "Modified spontaneous emission on nanophotonic structure," *Nat. Photon.*, vol. 9, pp. 427–435, 2015.

[9] P. Bharadwaj and L. Novotny, "Spectral dependence of single molecule fluorescence enhancement," *Opt. Express*, vol. 15, pp. 14266–14274, 2007.

[10] P. Anger, P. Bharadwaj, and L. Novotny, "Enhancement and quenching of single-molecule fluorescence," *Phys. Rev. Lett.*, vol. 96, p. 113002, 2006.

[11] M. Drobizhev, N. Makarov, S. Tillo, T. Hughes, and A. Rebane, "Two-photon absorption properties of fluorescent proteins," *Nat. Methods*, vol. 8, pp. 393–399, 2011.

[12] M. Drobizhev, N. Makarov, S. Tillo, T. Hughes, and A. Rebane, "Resonance enhancement of two-photon absorption in fluorescent proteins," *J. Phys. Chem. B*, vol. 111, pp. 14051–14054, 2007.

[13] D. Punj, M. Mivelle, S. Moparthi, T. Zanten, H. Rigneault, N. Hulst, M. F. Garcia-Parajo, and J. Wenger, "A plasmonic 'antenna-in-box' platform for enhanced single-molecule analysis at micromolar concentrations," *Nat. Nanotech.*, vol. 8, pp. 512–516, 2013.

[14] S. Huo, Q. Liu, S. Cao, W. Cai, L. Meng, K. Xie, Y. Zhai, C. Zong, Z. Yang, B. Ren, and Y. Li, "Surface plasmon-coupled directional enhanced Raman scattering by means of the reverse Kretschmann configuration," *J. Phys. Chem. Lett.*, vol. 6, pp. 2015–2019, 2015.

[15] P. Brooksby and W. Fawcett, "Determination of the electric field intensities in a mid-infrared spectroelectrochemical cell using attenuated total reflection spectroscopy with the Otto optical configuration," *Anal. Chem.*, vol. 73, pp. 1155–1160, 2001.

[16] Y. Wang, W. Knoll, and J. Dostalek, "Bacterial pathogen surface plasmon resonance biosensor advanced by long range surface plasmons and magnetic nanoparticle assays," *Anal. Chem.*, vol. 84, pp. 8345–8350, 2012.

[17] L. Zhou, F. Ding, H. Chen, W. Zhang, and S. Chou, "Enhancement of immunoassay's fluorescence and detection sensitivity using three-dimensional plasmonic nano-antenna array," *Anal. Chem.*, vol. 84, pp. 4489–4495, 2012.

[18] J. Malicka, I. Gryczynski, Z. Gryczynski, and J. Lakowicz, "DNA hybridization using surface plasmon-coupled emission," *Anal. Chem.*, vol. 75, pp. 6629–6633, 2003.

[19] H. Aouani, O. Mahboub, N. Bonod, E. Devaux, E. Popov, H. Rigneault, T. Ebbesen, and J. Wenger, "Bright unidirectional fluorescence emission of molecules in a nanoaperture with plasmonic corrugations," *Nano Lett.*, vol. 11,

pp. 637–644, 2011.

[20] T. Taminiau, F. Stefani, F. Segerink, and N. Hulst, "Optical antennas direct single-molecule emission," *Nat. Photonics*, vol. 2, pp. 234–237, 2008.

[21] K. Aoki, D. Guimard, M. Nishioka, M. Nomura, S. Iwamoto, and Y. Arakawa, "Coupling of quantum-dot light emission with a three-dimensional photonic-crystal nanocavity," *Nat. Photon.*, vol. 2, pp. 688–692, 2008.

[22] P. Lodahl, A. Driel, I. Nikolaev, A. Irman, K. Overgaag, D. Vanmaekelbergh, and W. Vos, "Controlling the dynamics of spontaneous emission from quantum dots by photonic crystals," *Nature*, vol. 430, pp. 654–657, 2004.

[23] A. Kuznetsov, A. E. Miroshnichenko, Y. H. Fu, J. Zhang, and B. Lukyanchuk, "Magnetic light," *Sci. Rep.*, vol. 2, p. 492, 2012.

[24] P. Albella, M. Poyli, M. Schmidt, S. Maier, F. Moreno, J. Saenz, and J. Aizpurua, "Low-Loss electric and magnetic field-enhanced spectroscopy with subwavelength silicon dimers," *J. Phys. Chem. C*, vol. 117, pp. 13573–13584, 2013.

[25] K. Aslan, M. Wu, J. Lakowicz, and C. Geddes, "Fluorescent core-shell Ag@SiO2 nanocomposites for metal-enhanced fluorescence and single nanoparticle sensing platforms," *J. Am. Chem. Soc.*, vol. 129, pp. 1524–1525, 2007.

[26] R. Faggiani, J. Yang, and P. Lalanne, "Quenching, plasmonic, and radiative decays in nanogap emitting devices," *ACS Photonics*, 2015.

[27] N. Livneh, M. Harats, S. Yochelis, Y. Paltiel, and R. Rapaport, "Efficient collection of light from colloidal quantum dots with a hybrid metal-dielectric nanoantenna," *ACS Photonics*, 2015.

Chapter 7

Chiral optics

Eng Huat Khoo
Institute of High Performance Computing, Singapore

Wee Kee Phua
Institute of High Performance Computing, Singapore

Yew Li Hor
Institute of High Performance Computing, Singapore

Yan Jun Liu
Institute of Materials Research and Engineering, Singapore

In this chapter, we will discuss the use of polaritonic resonances in chiral nanostructures for biosensing applications. Section 7.1 will describe how chiral optics emerged, including natural optical activity and the observed chiroptical effects. Section 7.2 will address the chiroptical effects in flat chiral nanostructures. Section 7.3 will discuss the application of chiral nanostructures in the realm of biosensing for detection of chiral biomolecules. Summary appears in Section 7.4.

7.1 Introduction and background

7.1.1 History of chiroptics

The physicochemical studies of proteins and nucleic acids began in 1815 and became a hot topic in the 1950s [1–3]. The discovery of two important components, α helix and β sheets [3] by L. Pauling and R. B. Corey led to insight into the first fundamental modeling of protein structures [4]. The α helix and β sheets are the secondary structures of proteins. The former is the helical conformation of a protein polymer while the latter is a laterally connected polymer. In 1951, F. Sanger, a biochemist, discovered the amino acid sequences [5]. Following this, J. Watson and F. H. C. Crick discovered that DNA has a double helix structure, winning applause from the chemistry community and a nobel prize. All this made it critical to demonstrate the existence of the α helix and β sheets in proteins.

The characterization of proteins was carried out using various physical methods such as intrinsic viscosity, osmotic pressure, sedimentation velocity, and equilibrium [6–8]. These techniques differ in terms of sensitivity, characterization time, and cost. However, optical methods are still preferred since they allow proteins and nucleic acids to be studied with higher sensitivity, speed, and at a much lower cost. The optical methods such as light scattering, infrared and ultraviolet spectroscopy, optical rotatory dispersion (ORD), and circular dichroism (CD) are popular for determining the structures and conformations of protein biopolymers [9, 10]. Among them, CD and ORD are more widely used due to their higher accuracy and sensitivity. CD and ORD also give information about the structural conformations of protein biopolymers up to the secondary level, which is crucial to determine the resultant handedness of the synthesized biopolymers.

The history of optical activity, also known as *chiroptics*, goes back more than 200 years ago, starting with the research of optical rotatory power. Pioneering research showed that the optical rotatory power consists of normal and anomalous types. The CD and anomalous types of ORD arise from the active absorbing band termed the *Cotton effect* [11] defined as the change in the CD or ORD in the absorption band of the material. For normal ORD, the rotation of light progressively increases with decreasing wavelength.

In 1848, Louis Pasteur found that the phenomenon of chiroptics resulted from molecular *dissymmetry* in compounds. It laid the foundation of *stereochemistry* [12]. Pasteur solved the puzzle of why some crystals of sodium ammonium tartrate compound were chiroptical, and some were not. He proposed that the molecular structures are three-dimensional and introduced the concept of molecular dissymmetry based on this postulation. In the past, scientists assumed that molecules are two-dimensional, and that their mirror images superimpose. Before Pasteur, no one saw that sodium ammonium tartrate was chiroptical at certain angles.

Nearly all optical activities focused on the ORD, where light rotates, as it travels through chiral molecules. The focus on the CD came after the discovery of a helical rotation in proteins. In 1957, Moffitt used modern spectropolarimeters to measure ORD and formulated the theory of ORD for a helix in proteins [13]. ORD has largely been replaced by CD in studies of biopolymer conformations mainly due to the ability to observe CD bands in the visible and ultraviolet spectra using advanced nano-optics methods.

7.1.2 Natural optical activity

Optical activity was first observed in the colors of sunlight that passed through the optical axis of a quartz crystal placed between crossed polarizers. Experiments carried by Biot in 1812 [14] showed that the colors were due to the two distinct effects: (i) optical rotation, i.e., the rotation of the polarization plane of a linearly polarized beam; and (ii) ORD, i.e., the unequal rotation of the polarization plane for light of different wavelengths. For ORD, the angle of rotation was inversely proportional to the square of the wavelength λ for a fixed path of the light through the quartz crystal.

Besides crystalline objects, optical activity was also found in organic liquids. The optical activity of fluids resides in the individual molecules and may be observed regardless of molecular orientation. The molten form of a crystalline solid does not show optical activity. It was realized that the source of natural optical activity in

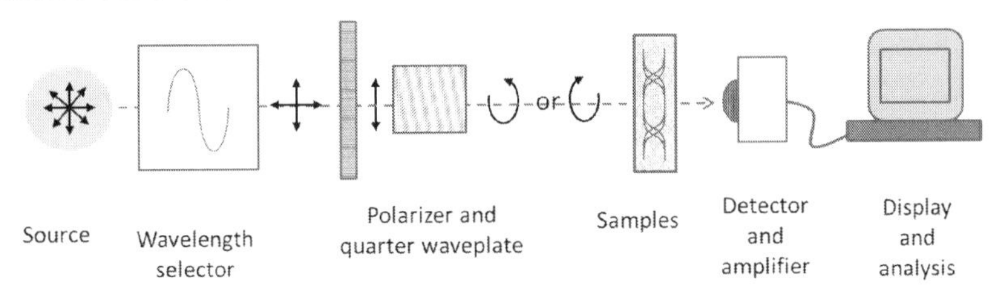

Figure 7.1: Conventional setup for CD measurement consisting of various optical components for generating circularly polarized light, detecting transmission signals, and analyzing absorption coefficients.

solids and liquids is due to a special class of molecular structures known as *chiral* structures that are not superimposable with their mirror images. They have two distinct forms that are the opposite absolute configurations featuring optical activities with equal magnitude but of opposite signs at a given wavelength.

Although a number of approaches are available for CD measurements, the setup in Fig. 7.1 can be used as a starting point. The fundamental CD measurement setup consists of a spectrometer, a light source, a polarizer, a quarter waveplate, and a conventional detector. Depending on the measurement requirements, each optical component can be modified. Early CD detection was based on the determination of the absorption difference in left (LCP) and right (RCP) circularly polarized light that defines a CD. This method was originally proposed in 1896 by the French physicist Aimé Cotton [11]. His setup consisted of a Nicol polarizing prism and a Fresnel rhomb quarter waveplate, and hence required frequent changing of achromatic Fresnel rhomb plates. Later, ellipticity measurement techniques were introduced [15]. They featured higher sensitivity and were in use until electronic modulated systems took over in the 1960s.

CD measurements place very strict constraints on the light source and other optical components to minimize the background absorption. The fact that CD measurement is inherently background-limited sets a minimum for the absorption difference to be detected. An alternative way is to use the transmission-based method demonstrated with a new setup configuration on a time resolved basis. However, the sensitivity of these measurements is still not sufficient to provide clear results for determining the handedness of the chiral molecules in the structures.

The main issues encountered in measuring CDs of molecular materials, nanomaterials, hybrid materials, bio-materials, and metamaterials are the low sensitivity and poor signal due to the short interaction time and complex interaction mechanisms. In order to measure chiral molecular materials with higher accuracy, the equipment must be scaled down to nanometer dimensions where unique physical effects arise from increased surface-to-volume ratios and pronounced resonances on localized and/or propagating polaritons.

7.1.3 *Chiroptical effects*

Measurements of optical activities in molecular materials, nanomaterials, and metamaterials are difficult due their small sizes, short interaction times, and poor signal-to-noise ratios. In addition, the chirality is a logical quantity, studied materials are either chiral or achiral. It makes the quantification of chirality very difficult. However, this problem can be overcome with improved interaction between light waves and chiral structures. A number of chiroptical effects can be measured, allowing the quantification of chirality and classification of chiral materials with different degrees of chirality.

Chiroptical effects occur when chiral molecules interact with electromagnetic waves with varying degrees of difference in refractive index and absorption coefficient. Two most dominant chiroptical effects are ORD, where the difference in refractive index for LCP (n_l) and RCP (n_r) light causes the rotation of the linearly polarized light,

$$\mathrm{ORD} = \frac{\pi}{\lambda}\left(n_l - n_r\right), \tag{7.1}$$

and CD, which arises from the absorption (or extinction) difference for LCP and RCP light,

$$\mathrm{CD} = A_l - A_r, \tag{7.2}$$

where A_l and A_r are the absorption coefficients of the material for LCP and RCP light, respectively.

In this chapter, we will focus on CD, as it can be obtained with a simple experimental setup. Typically, the measured CD signal is very weak (in the range of 10^{-7}) and poses a challenge for ultrasensitive detection. Several solutions exist to enhance the CD signal for sensing chiral molecules. Fluorescent or radioactive tags attached to biomolecules were proposed as a means to enhance the CD signal with polarized light. However, this method is not recommended due to photo-bleaching of fluorescent molecules and blocking of the active sites of target chiral molecules.

Another proposed solution is to use nanoparticles exhibiting strong resonances on localized absorption polaritons. These polaritons feature high fields confined inside the nanoparticles and can provide strong absorption (see Chapter 3). If the nanoparticles are arranged helically, they can generate resonant *superchiral near fields*, prolonging and enhancing chiral interaction between molecules and light waves. Nanoparticles can also be assembled into a giant spiral filament via a molecular chiral scaffold. These hybrid chiral materials can be tailored to be as small as clusters or as large as filaments such as DNA and protein. The nanoparticle trimer, which is a composition of nanoparticles arranged asymmetrically along the planar axis, is also shown to exhibit localized optical chirality and generate superchiral near fields.

While superchiral fields can be generated by nanoparticles with unique arrangements, the generation of superchiral fields in flat chiral nanostructures has attracted much attention in the development of biosensors for detection of chiral macromolecules and their hierarchical structures. Flat chiral nanostructures possess larger surface area than nanoparticles, producing continuous superchiral fields of wider coverage. As a result, the localized chiral fields can distinctly interact with the accumulated molecules on the nanostructure surface. These accumulated molecules are analogous to a thin layer of dielectric material; they change the environment on the nanostructure surface and produce a shift in the CD spectrum, signaling detection of chiral molecules.

Flat gammadion is a common design for the generation of superchiral fields because it has a large dissymmetry factor. Other flat chiral nanostructures include the circular disk array and G-shaped structures. In addition, there are also three-dimensional spiral structures and multi-layer flat achiral nanostructures that mimic the three-dimensional properties of optical chirality.

7.2 Flat chiral nanostructures

Flat chiral nanostructures are the most studied systems that provide a high dissymmetry factor together with wide coverage critical for sensing applications. Superchirality in such structures has been demonstrated in many designs of single- and multi-layer systems made of metallic and dielectric materials. They are all based on the same principles but rely on different polariton resonances. For example, single-layer systems exhibit strong resonances on horizontally coupled polaritons while multi-layer systems can utilize resonances on the vertically coupled eigenmodes. In addition, metallic and dielectric chiral systems react to LCP and RCP differently: metallic nanostructures create superchiral fields mainly through electric resonances while dielectric systems can use both electric and magnetic resonances. In this section, we will consider peculiarities of single- and multi-layer flat chiral systems with focus on metal structures.

7.2.1 Single-layer chiral systems

To overview typical features of single-layer chiral systems, we consider the most studied gammadion-shaped structures. They have attracted much attention in the scientific community due to their obvious handedness. Generally, gammadion-shaped systems provide high dissymmetry that can sufficiently improve CD measurements of chiral molecules and filaments. We can further improve their dissymmetry factor by simply tapering gammadion's arms. The idea of arm tapering arises from References [17, 18] showing that the overall dissymmetry factor of nanostructures was modified to get higher electric fields for nano-focusing. With such tapering, gammadion design provides dissymmetry sufficient to detect small chiral molecules and their associated chiral filaments.

The schematic layouts of left-handed tapered gammadion arrays[1] are shown in Fig. 7.2(a). The ends of the gammadion arms can be tapered via symmetric and asymmetric configurations. In symmetric tapering, the width is narrowed symmetrically along the arm axis to produce an enhanced localized field, as shown in Fig. 7.2(b).

[1] We focus on the left-handed gammadion to demonstrate the effectiveness of different tapering designs. The right-handed structure yields similar results.

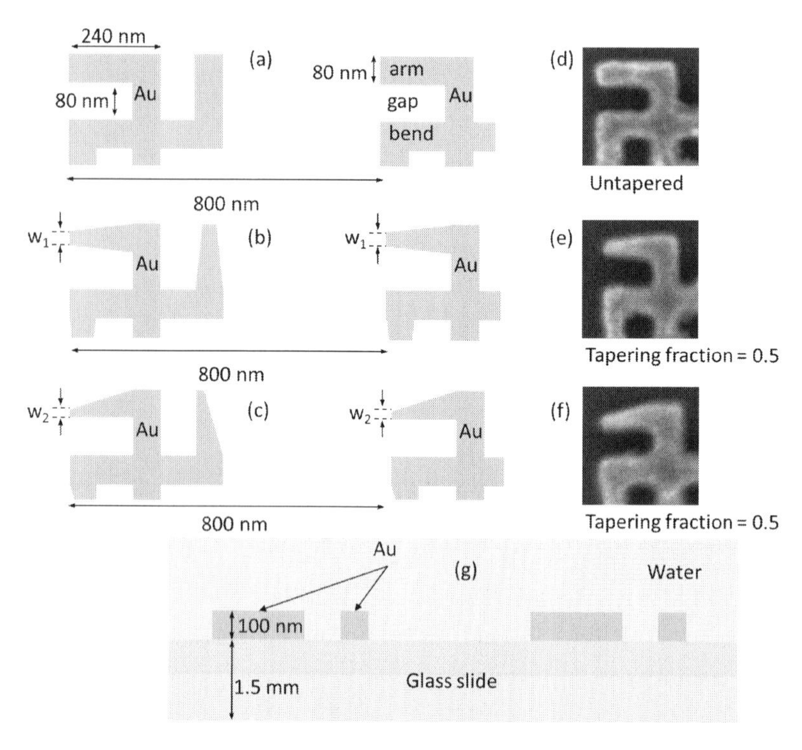

Figure 7.2: Schematic layout for designs of two-dimensional left-handed gold gammadion arrays. (a) Untapered, (b) symmetrically tapered, and (c) asymmetrically tapered gammadion designs. Symmetric tapering ensures the symmetry along the horizontal axis, while asymmetric tapering performs tapering only on one side, keeping the gap between the arm and bend constant. (d)through(f) Scanning electron microscopy (SEM) images of gammadion arms for different designs. The gammadions are fabricated using electron beam lithography and photo-resin lift-off, (d) untapered gammadion, (e) symmetrically tapered gammadion, (f) asymmetrically tapered gammadion. (g) Cross-sectional layout of the gammadion design. The gammadion is fabricated on the glass slides with a thickness of 1.5 mm. Reprinted with permission from Ref. [16].

Asymmetric tapering means that only one side of the gammadion arm is tapered and the distance between the arm and bend, known as the arm-bend gap, remains constant as shown in Fig. 7.2(c). Tapering is characterized with the tapering fraction (TF) defined as the ratio of the width w_1 (or w_2) to w_0 where w_0 is the width of the untapered arm as shown in Figs. 7.2(b) and (c). Figures 7.2(d) through (f) show the fabricated gammadion structures for a TF of 0.5. Figure 7.2(g) demonstrates the cross-sectional layout when the sample is immersed in pure water for calibration purposes; the biomolecules and filaments are dissolved in water for testing.

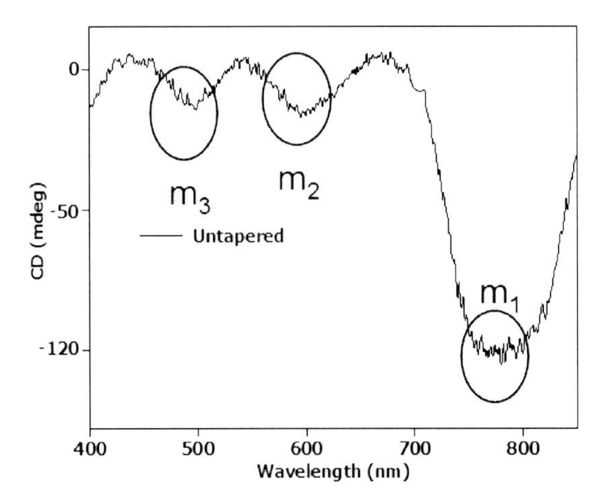

Figure 7.3: Experimental extinction difference (ED) spectrum for an untapered gammadion. Reprinted with permission from Ref. [16].

CD analysis

To find the CD of a gammadion structure, the extinction spectra were obtained with a dark-field microscope within the wavelength range of 400 to 850 nm. Consisting of both absorption and scattering, extinction has line shapes similar to absorption, thereby allowing the extinction difference (ED),

$$\mathrm{ED} = \mathrm{Ext}_{\mathrm{LCP}} - \mathrm{Ext}_{\mathrm{RCP}}, \tag{7.3}$$

to be used as a substitute for CD. The ED spectrum for the untapered gammadion is shown in Fig. 7.3 for the case of plain water. We can see a number of distinct dips. They are inherent to any single-layer chiral system. Typically they consist of chiral resonances (i) on the Bloch polaritons brought by the array periodicity (marked in Fig. 7.3 as m_1) and (ii) on the localized eigenmodes of the unit cell (depicted in Fig. 7.3 as m_2 and m_3).

Figure 7.4 demonstrates the ED spectra for symmetrically tapered gammadion at different TFs. It reveals that ED at the the m_1 resonance is weaker for the symmetrically tapered gammadion than for the untapered one. This implies that the respective CD is weaker too. Figure 7.4 also shows that the dip m_3 redshifts, while the dip m_2 becomes weaker and disappears as TF decreases. Finally, at the TF of 0.1, only the resonances m_1 and m_3 can be distinguished.

From Fig. 7.4(c), the enhanced localized fields at the lower part of the arm and its opposing bend become weaker and less distributed for the TF of 0.1. These localized fields are also known as the gap fields. Tapering of the gammadion arms increases the gap distance between the arm and bend and makes the arm narrower. Coupling of

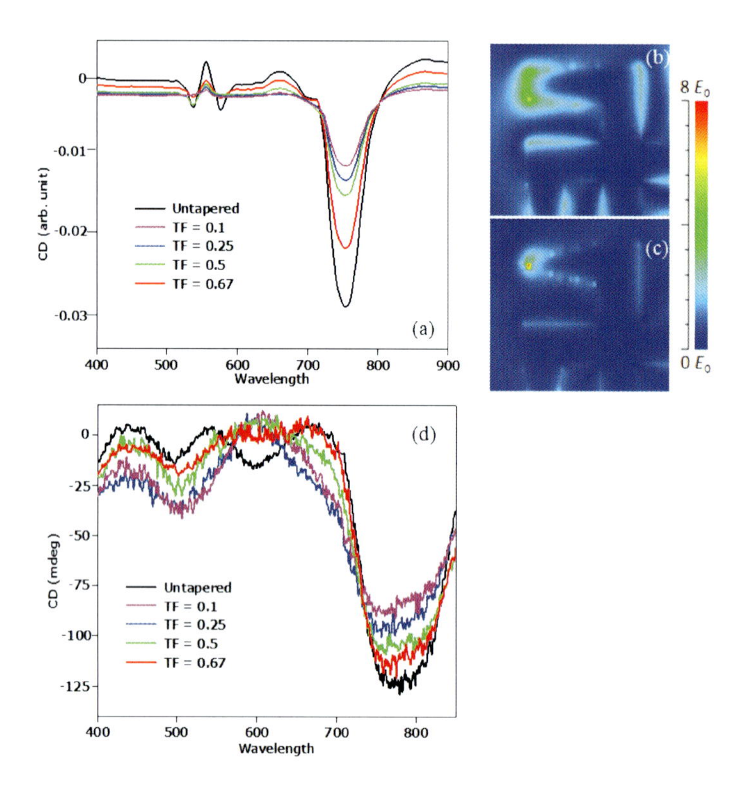

Figure 7.4: (a) Simulated ED for symmetrically tapered gammadions. (b) and (c) Simulated field distribution at the resonance m_1 for the gammadion arms with TFs of (b) 0.67 and (c) 0.1. (d) Experimental results of the ED spectrum for the symmetrically tapered gammadion nanostructures. Reprinted with permission from Ref. [16].

localized fields between the arm and bend decreases for larger gap distance, resulting in weaker and less spatially distributed fields for smaller TF at the resonance m_1. The resonances m_2 and m_3 are due to the gammadion eigenmode hybridization [19] inside the gap between the arm and bend. With larger gap distance, coupling between individual eigenmodes weakens and the resonance m_3 redshifts. The resonance m_2 becomes weaker and disappears at TF 0.1.

Figure 7.5(a) shows ED spectra for the asymmetrically tapered designs at different TFs. For the dip m_1, ED is larger for the asymmetrically tapered gammadion at TFs of 0.67 and 0.5 compared to the untapered one. This difference becomes apparent from Fig. 7.5(b) showing the local field distribution that spreads over a larger part of the asymmetrically tapered gammadion. In addition, the gap field does not weaken, as the gap distance remains unchanged. In contrast to Fig. 7.4(d), the ED spectra in Fig. 7.5(d) show stronger EDs at TFs of 0.67 and 0.5 for all three resonances.

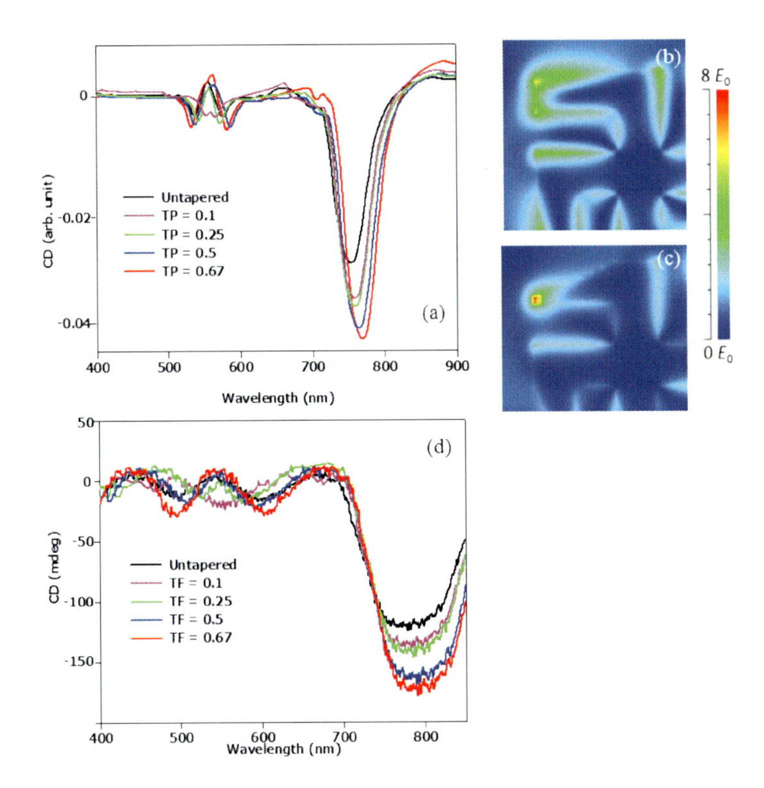

Figure 7.5: (a) Simulated ED asymmetrically tapered gammadions. (b) and (c) Simulated field distribution at the resonance m_1 for the gammadion arms with TFs of (b) 0.67 and (c) 0.1. (d) Experimental results of the ED spectrum for the asymmetrically tapered gammadion nanostructures. Reprinted with permission from Ref. [16].

Enhanced-field distribution analysis

In order to correlate TF and enhenced-field spread, the surface-enhanced field ratio (SEFR) is introduced,

$$\text{SEFR} = \frac{P(z = 50, E_z \geq 5E_0)}{P_{\text{total}}(z = 50)}, \tag{7.4}$$

where the surface perimeter P represents the length along the gammadion edge at $z = 50$ nm above the substrate with enhancement five times more than the incident field. P_{total} represents the total surface perimeter of gammadion respectively at $z = 50$ nm above the substrate. The parameter z is the vertical position of the gammadion. A larger SEFR indicates stronger field absorption and resonance, which in turn indicates a greater circularly polarized light absorption difference or ED [20].

Figures 7.6(a) and (b) show the SEFRs at resonances m_1 and m_3 for the symmetri-

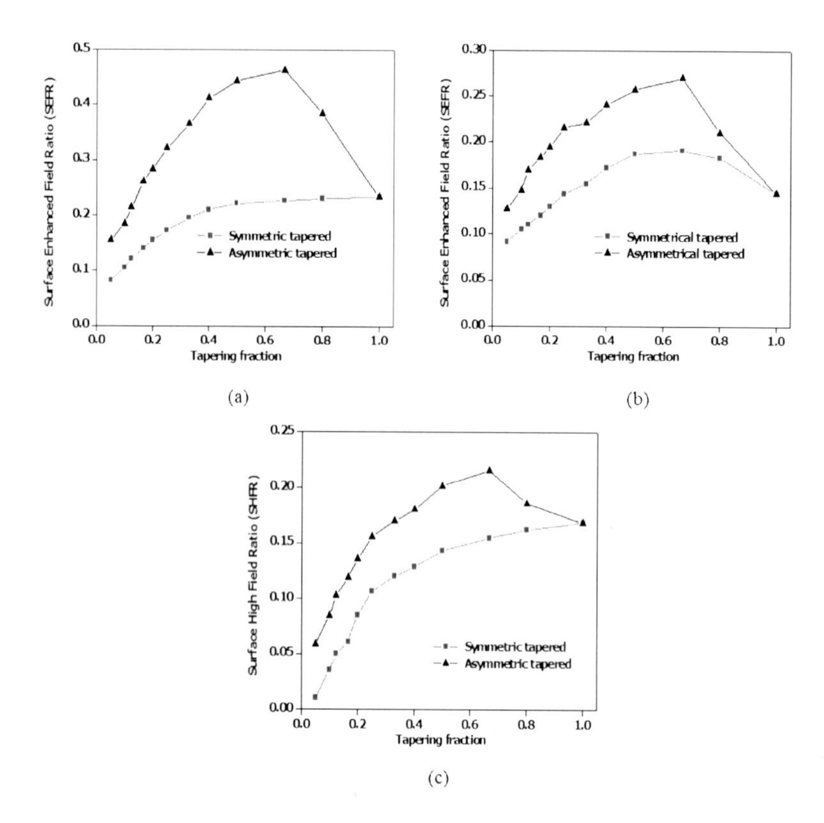

Figure 7.6: SEFR results for different gammadions for (a) resonance m_1, and (b) resonance m_3. (c) Surface high-field ratio in the gap. Reprinted with permission from Ref. [16].

cally and asymmetrically tapered gammadions at different TFs, respectively. For the symmetrically tapered gammadions, SEFR decreases after tapering is introduced due to the reduction of the high-field region at lower TF. At TFs lower than 0.6, the fields at the top and bottom of the gammadion arm merge together. This results in weaker gap fields and smaller field spread, indicating weaker absorption. For resonance m_3, the ED increases slightly before decreasing.

In the case of the asymmetrically tapered gammadion, SEFR increases as the TF decreases from 1 to 0.5 for the resonance mode m_1. It is caused by an increase of the amplitude and spread of the gap fields between the arm and bend. At TFs lower than 0.5, the SEFR drops significantly, as the reduction of the enhanced-field region at the tapered arm is larger than the field at the arm-bend gap, as shown in Fig. 7.5(c).

The results in Fig. 7.6(b) demonstrate the comparison of SEFR for the resonant mode m_3 at various TFs, indicating a similar trend. Figure 7.6(c) shows the comparison of SEFR at the gap for both the symmetrically and asymmetrically tapered

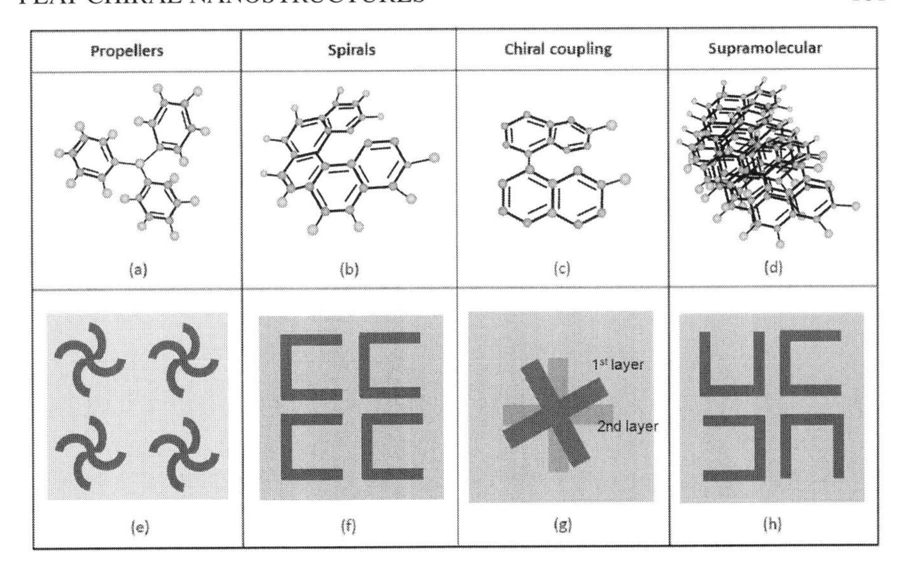

Figure 7.7: Naturally occurring chiral molecules and their corresponding man-made structures.

gammadions. For the asymmetrically tapered gammadion, the field at the arm-bend gap increases, as the field distribution and amplitude increase at the lower part of the arm due to the tapering [seen in Fig. 7.5(b), but not in Fig. 7.5(c)]. Hence, the arm tapering to a smaller width is a disadvantage even though the field density per unit area is increased at the tapered end.

7.2.2 Multi-layer chiral systems

The nanoengineering strategies developed for chiral structures can be related to the concepts originated in chiral molecules, as shown in Fig. 7.7 [21]. Many of the concepts associated with chirality initially developed for molecules are also applicable to nanostructures. Figures 7.7(a) through (d) show the common structures of the chiral molecules that exist in nature while (e) through (h) are the corresponding nanostructures that resemble them. For example, Figs. 7.7(a) and (b) are helical chiral molecules in the shape of a propeller and a spiral. Correspondingly, the propeller- and spiral-shaped nanostructures shown in Figs. 7.7(e) and (f) exhibit similar chirality [22, 23].

Chirality can also be achieved by coupling two or more achiral molecules as illustrated in Fig. 7.7(c). Similarly the coupling mechanism occurs in the superposition of achiral crosses [25] as shown in Fig. 7.7(g). Another well-known chiral structure is the supramolecule represented in Fig. 7.7(d). In this case, the building blocks alone are chiral and chirality is achieved through coupling of the building blocks. Conveying this concept onto a nanostructure, the pattern in Fig. 7.7(h) shows four two-dimensional spirals arranged at different angles. This arrangement gives rise to

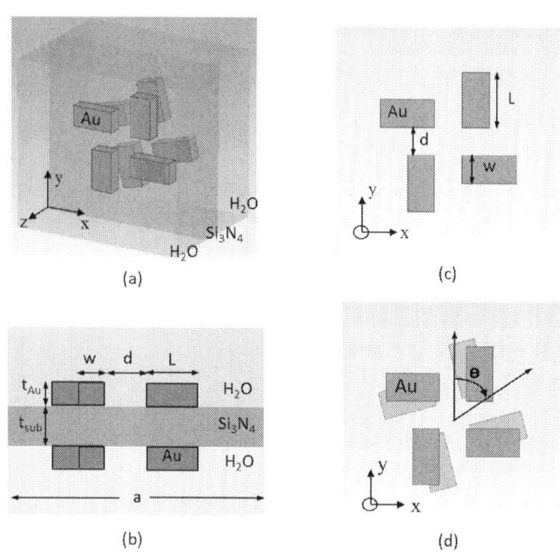

Figure 7.8: Schematic layout of the double-layer nanostrip chiral structure. (a) Unit cell in three-dimensions with a pitch of 250 nm. (b) Side view of the structure. (c) Top view of the single layer nanostructure layout. (d) Top view of the double layer with the second layer rotated by θ degrees with respect to the first layer. The thickness of the gold (t_{Au}) layer is 20 nm and for Si_3N_4 (t_{sub}) is 30 nm. Each gold layer consists of four nanostrips in a windmill arrangement, with the length of nanostrip, $L = 60$ nm, separation of nanostrip, $d = 30$ nm, and width of nanostrip, $w = 30$ nm. Reprinted with permission from Ref. [24].

strong nonlinear CD attributed to chiral coupling of the nanostructures [23]. Translating this concept to supramolecular detection, a fourfold symmetric structure with double layer coupling is a good candidate. The double layer allows coupling and manipulation of the chiral effects when one of the layers is rotated with respect to the planar axis of the other layer.

The chiral sub-wavelength nanostructure that will be discussed in this section consists of two layers of gold strip arrays with a sandwiched layer of thin silicon nitrate [26, 27]. Figure 7.8 shows the unit cell of this array with four sub-wavelength strips at each layer in a windmill orientation and stacked on top of each other [24]. The windmill orientation forms a fourfold rotational symmetry that describes the helical chiral phenomenon, while the double layers of the structure represent the coupling effect between the layers. In this structure, light normally incident onto the nanostructure experiences strong resonances on the localized absorption polaritons.[2] At the resonant wavelengths, strong absorption improves the CD measurement.

[2]The discussed metal strip arrays exhibit strong electric resonances on surface absorption polaritons. A similar response can be observed if we make the strips of any high-index dielectric material such as silicon. In this case, incident light can experience both electric and magnetic resonances on the volume absorption polaritons.

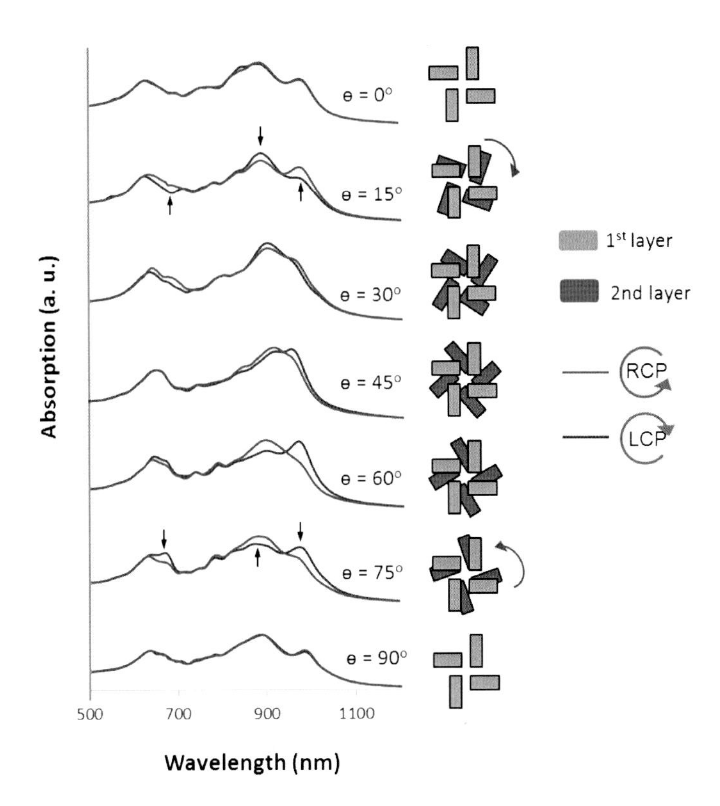

Figure 7.9: (Left) spectral absorption of the bilayer nanostrip chiral structure with the second layer rotated by θ. (Right) nanostructure with the second layer rotated by θ. Black arrows indicate a switch of the absorbance peak in RCP and LCP for $\theta = 15$ degrees and 75 degrees. Reprinted with permission from Ref. [24].

CD analysis

In this design, the chiral response strictly depends on the rotational angle θ of two arrays. Figure 7.9 shows the spectral absorbance of LCP and RCP light for different values of θ. Reversal of the absorbance peak for the RCP and LCP light is observed when θ changes from 0 degrees to 90 degrees, with 45 degrees being the turning point. For example, the arrows at Fig. 7.9 indicate the reversal of the absorbance peak for $\theta = 15$ degrees and $\theta = 75$ degrees that occurs at the particular wavelength, e.g., at 980 nm, when the RCP absorbance is dominant for $\theta = 15$ degrees, while the LCP absorbance dominates for $\theta = 75$ degrees. Apparently, the reversal is due to the change in handedness of the nanostructure. From $\theta = 0$ degrees to 30 degrees the nanostructure is left-handed, and from $\theta = 45$ degrees to 75 degrees the nanostructure is right-handed.

The sign of CD signal depends on the extent by which LCP light is absorbed

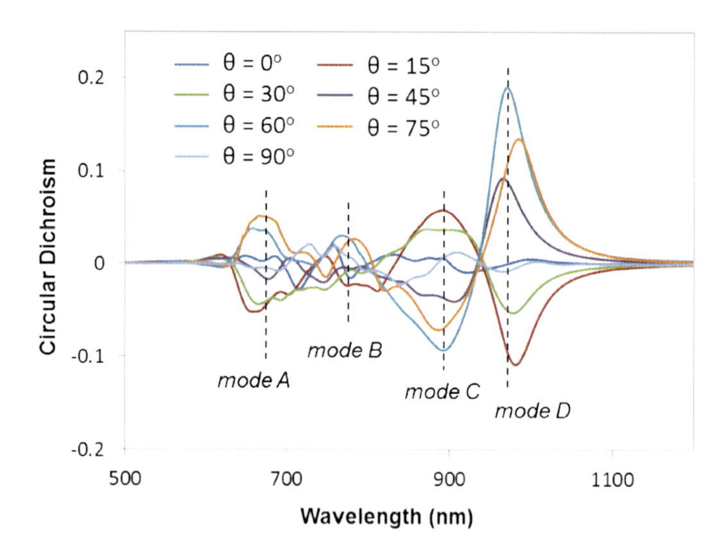

Figure 7.10: CD spectrum of the fourfold symmetry nanostrips. Reprinted with permission from Ref. [24].

more or less than RCP light. CD is positive when the LCP absorbance is larger than RCP, and vice versa. Figure 7.10 shows the profiles when CD switches between positive and negative as a function of wavelength. The CD profile at each angle θ may exhibit both positive and negative peaks and dips in different wavelength regions. As observed, the CD peaks or dips fall around 630 nm, 690 nm, 900 nm, and 980 nm. These CD features are denoted as the resonant modes A, B, C, and D. The modes A and D exhibit negative CD for θ = 15 degrees and 30 degrees, and positive CD for $\theta = 60$ degrees, 75 degrees, while in $\theta = 0$ degrees, 45 degrees, 90 degrees, the sign of the CD is undefined. The CD signature for mode C is opposite the modes A and D. The modes C and D exhibit opposite chiral effects, and a bisignate CD appearance when the spectra consist of peaks next to dips (or vice versa).

The CD sign flipped indicates a switching behavior due to a change in handedness of the nanostructure. Figure 7.11 shows the switching effects for the three modes within rotation angle range from 0 degrees to 180 degrees. It reveals that the mode D provides the most significant switching effect among the four modes.

Resonant mode analysis

The modes A through D can be categorized as longitudinal and transverse modes, depending on the dipole moment orientation. In the longitudinal mode, the dipole moments are oriented along the strips, while the transverse modes have dipole moments oriented perpendicularly to the strip. Figure 7.12 shows the LCP and RCP electric field distribution for the first and second layers of the designed structures for modes A, B, C, and D at the transmittance profile. The nanostructure responses

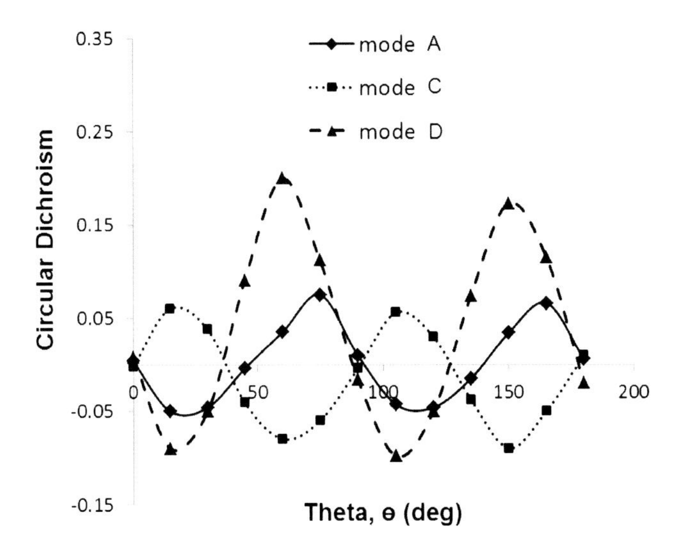

Figure 7.11: The switching effect of CD for modes A, C, and D, when the second layer rotates from 0 degrees to 180 degrees with respect to the first layer. The mode B is not included because it is of less significance. Reprinted with permission from Ref. [24].

to LCP and RCP light are identical except for the switch of the positive and negative behaviors in the effective region. Surface charge distributions are illustrated in Fig. 7.12 for every mode. The field distribution plots clearly show that the modes A and B are transverse, while C and D are longitudinal.

For the resonant mode A, the surface charges are distributed along the transverse direction and accumulated along the edges. Therefore, we designated them as the transverse edge modes. Similarly for the mode B, the charges are also transversely separated and accumulated at the corners, so we designate them as transverse corner modes [28]. The resonant modes C and D, are longitudinal with different magnitudes of the x and y components of the RCP and LCP [19]. The mode C has pure x or y components of the LCP and RCP, while the mode D has hybridized x and y components. Therefore, for the mode C, we observe the field distribution only in the two nanostrips, while for the mode D, the field distribution is approximately identical in all four nanostrips.

The role of the second layer is to impart handedness to the bilayer system. For example, in the mode D, the electric field distribution of the second layer with RCP light is stronger than the electric field distribution with LCP light. It indicates that the right-handedness of the nanostructure dominates.

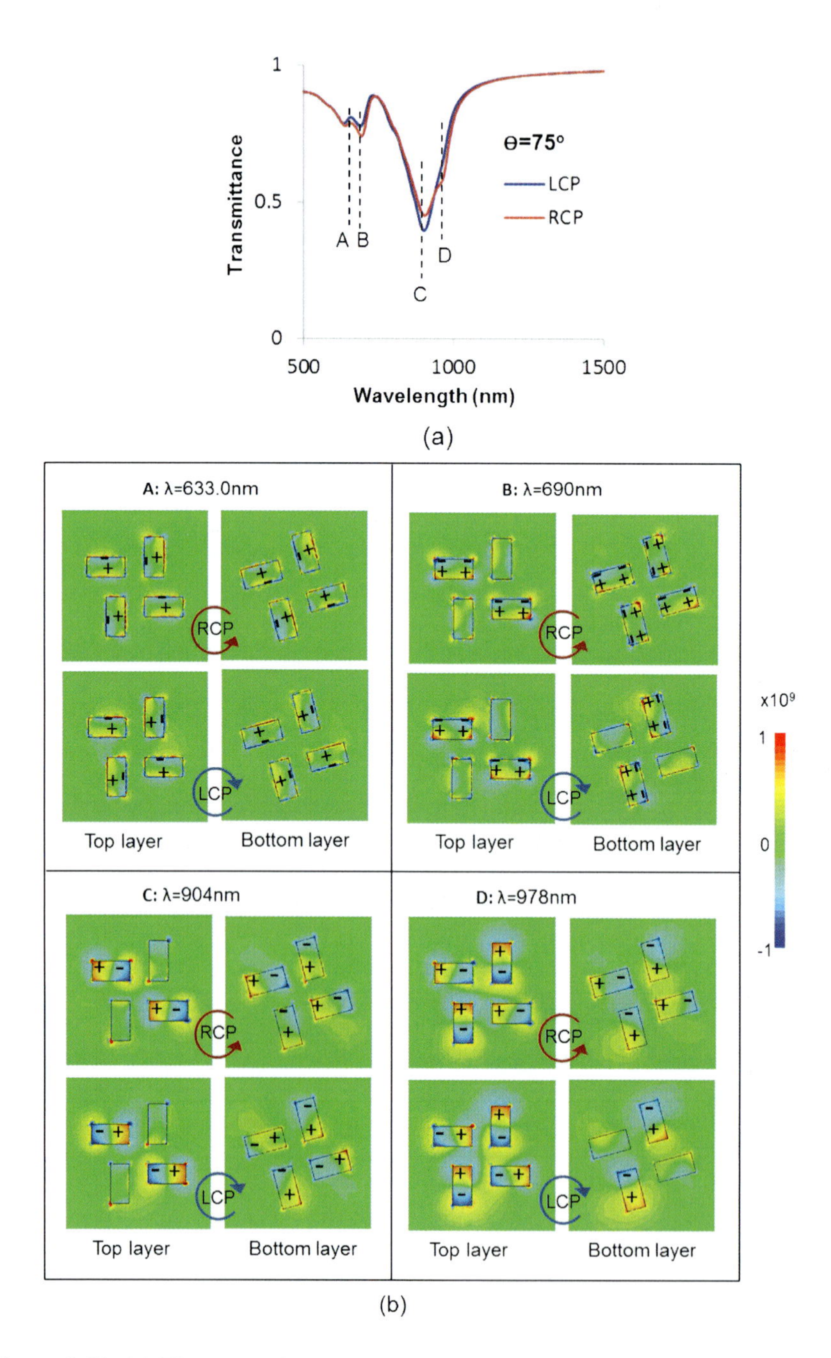

Figure 7.12: (a) The transmittance profile indicating the modes A, B, C, and D; (b) Electric field response of the individual layers under RCP and LCP light illumination at resonance peaks indicated in (a). Reprinted with permission from Ref. [24].

7.3 Biosensing with flat chiral systems

Actin [29–32] is a multi-functional protein found in eukaryotic cells. It is an important protein that takes part in many important biological processes, such as muscle contraction, cell division, and cytokinesis [33]. Actin exists as a free monomer known as G-actin (globular), shown in Fig. 7.13(a). G-actin has an α-helix fold, which is asymmetric. In the presence of ATP, G-actin can polymerize with other actin monomers to form an actin filament known as F-actin, shown in Fig. 7.13(a). F-actin is a unique right-handed dissymmetric single-stranded helix filament.

To demonstrate detection of actin proteins with chiral systems, we will use the gammadion-shaped nanostructures considered in Section 7.2.1 with both left- and right-handedness. The detection is based on the wavelength shift produced in the ED spectrum and mainly attributed to the adsorption of molecules in the high-field region [17, 34–36]. The adsorbed G- and F-actins contain tryptophan, whose carboxyl group allows binding of the actin molecules to the surface of the gammadion nanostructures and formation of a single flat chiral layer [34, 37]. G- and F-actins adopt geometries with a well-defined orientation axis with respect to the surface of the gammadion and are randomly oriented in the plane parallel to the gammadion surface.

7.3.1 Sensing of G-actin

G-actin samples were prepared and stirred to ensure that they would not polymerize to F-actin. Figure 7.13(b) shows the ED spectra for the symmetrically and asymmetrically tapered gammadion nanostructures immersed in a solution of G-actin molecules. From there, we can observe a larger wavelength shift for the asymmetrically tapered gammadion, indicating great sensitivity to the G-actin monomers. For the symmetrically tapered gammadion, the average wavelength shift was very small and almost similar to the untapered gammadion in water. Figures 7.13(c) and (d) show a comparison of the average wavelength shifts for different TFs. From there, we see that the gammadion resonance mode m_3 shows larger wavelength shifts than the mode m_1.

On the molecular level, G-actin has absorption energies closer to the resonance mode m_3 [30]. Although the mode m_1 has larger SEFR, its ED wavelength does not match the energy level of G-actin. Eventually it causes a larger wavelength shift for mode m_3 compared to m_1. Note that the largest wavelength shift occurs at TFs of 0.67 and 0.5. As observed in Figs. 7.6(a) and (b), lower TF results in smaller SEFR and hence smaller wavelength shift. The results plotted in Figs. 7.13(c) and (d) together with the SEFR results in Figs. 7.6(a) and (b) show a correlation between SEFR and the sensitivity of chiral biomolecules detection.

7.3.2 Sensing of F-actin

To demonstrate the effect of F-actin, we added KCl to a solution of G-actin to polymerize it to F-actin. To prevent F-actin from dissociating back into G-actin, we stirred the solution for 2 to 3 minutes to ensure homogeneity in the solution. Figures 7.14(a) and (b) show the average wavelength shift produced by polymerizing G-

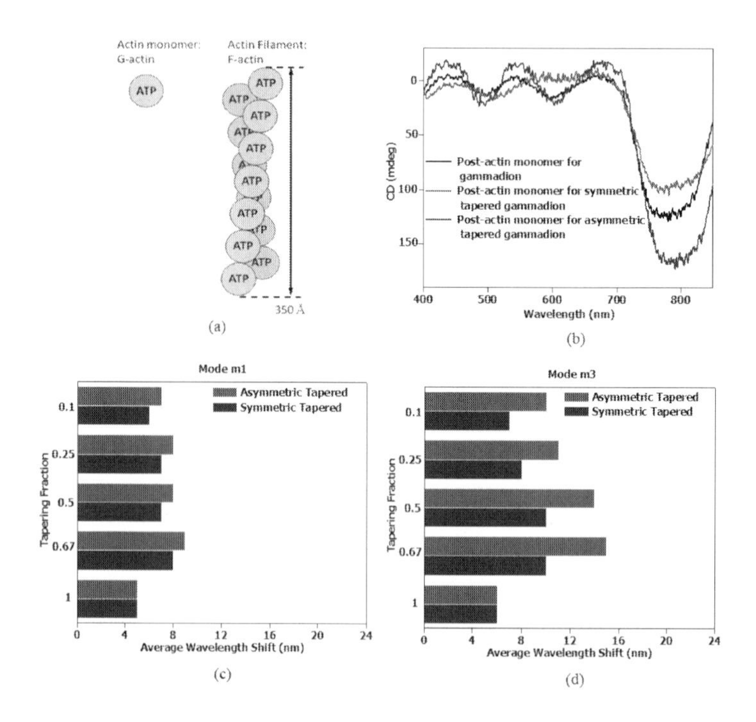

Figure 7.13: (a) Schematic representation of G-actin monomer and F-actin filament. The polymerization of G-actin monomers forms an F-actin, which is a double helix right-handed filament. (b) ED spectra of the left-handed gammadion designs with TF of 0.67 after submersion in a solution of G-actin. (c) Average wavelength shift for the gammadion resonance mode m_1. (d) Average wavelength shift for the gammadion resonance mode m_3. Reprinted with permission from Ref. [16].

actin to F-actin. Unlike Fig. 7.14(c), we observed a larger average wavelength shift in Fig. 7.14(a) for the resonant mode m_1. This is due to the larger size of F-actin, which has a length of 700 nm (20× pitch of 35 nm).

To further demonstrate the effects of F-actin on mode m_1, another sample with similar gammadion size and height and a period of 650 nm was fabricated. Obviously, its wavelength shift should be different from the sample with the period of 800 nm. Figures 7.14(c) and (d) support this hypothesis and show the wavelength shift produced by the symmetrically and asymmetrically tapered gammadion structures for modes m_1 and m_3 at TF 0.67. The wavelength shift is reduced by approximately 12% for the TF of 0.67 and mode m_1. However, for mode m_3, the change in period does not affect the wavelength shift much.

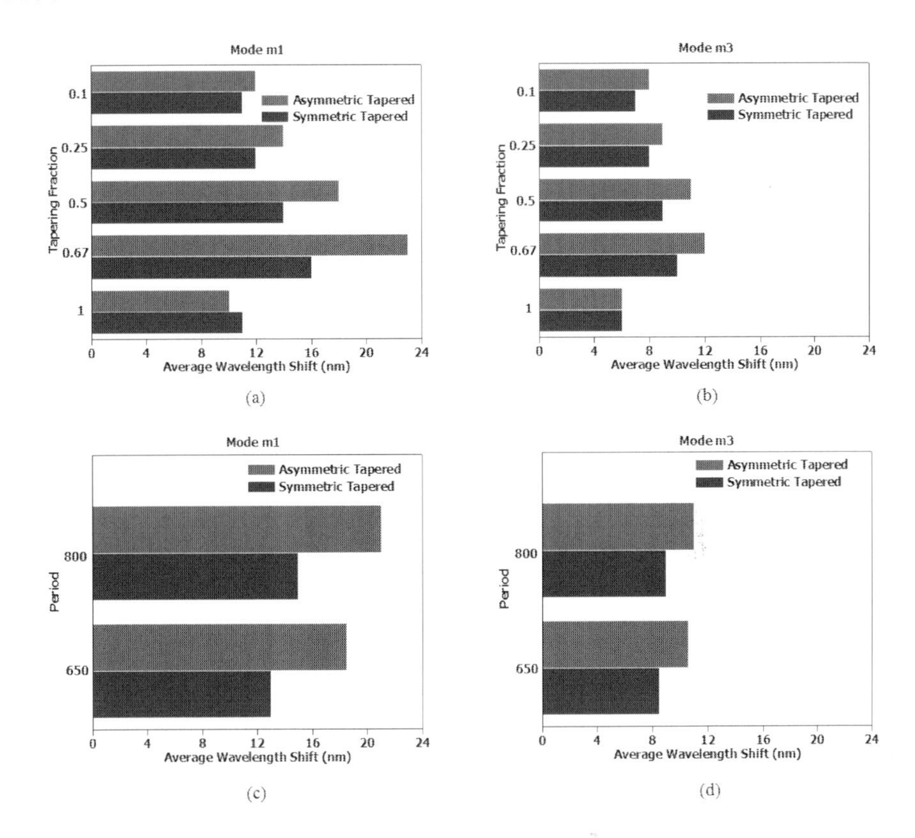

Figure 7.14: Wavelength shift of the symmetrically and asymmetrically tapered gammadion for the resonant modes (a) m_1 and (b) m_2 using F-actin filaments. (c) and (d) Response of F-actin on the different gammadion structures with different periods. Reprinted with permission from Ref. [16].

7.3.3 Effects of superchiral fields

The above results highlight the importance of local field enhancement and its spatial distribution. These findings were applied to chiral molecule solutions. However, a theoretical explanation of the sensitivity improved by using the concept of optical chirality will further demonstrate the importance of superchiral field interaction with chiral molecules.

Following Ref. [38], the optical chirality can be expressed as

$$\xi = \frac{\varepsilon_0}{2}\mathbf{E}\cdot\nabla\times\mathbf{E} + \frac{1}{2\mu_0}\mathbf{B}\cdot\nabla\times\mathbf{B}, \tag{7.5}$$

where ε_0 and μ_0 are the electric and magnetic constants and $\mathbf{E}$ and $\mathbf{B}$ are the time-varying electric field and magnetic induction. Eq. (7.5) describes the degree of chiral dissymmetry, which is proportional to ξ. This implies that chiral dissymmetry increases the rate of interaction of chiral molecules with enhanced fields.

SEFR given by Eq. (7.2) is directly proportional to the rate of interaction with chiral molecules. Thus, optical chirality ξ supports the previously obtained results, where the enhanced fields from the tapered gammadion increase the interaction with chiral molecules and enhance the observed wavelength shift.

For incident circularly polarized light, the optical chirality ξ can be simplified [39] and written as

$$\xi = \pm \frac{2U_e \omega}{c}, \tag{7.6}$$

where c and ω represent the speed of light and angular frequency; the $\pm$ sign indicates LCP and RCP lights. With the time-average density U_e of the electric field energy (for details, see Section 1.2.1),

$$U_e = \frac{\varepsilon_0}{4}|E|^2, \tag{7.7}$$

we can evaluate the optical chirality of the near fields generated by the asymmetrically tapered gammadion nanostructures. Figures 7.15(a) through (c) show the local optical chirality for asymmetrically tapered gammadion at TFs of 1, 0.67, and 0.1 for the resonant mode m_1. It suggests that the optical chirality for the TF of 0.67 is the strongest. Its local optical chirality spreads over a larger area compared to the untapered and asymmetrically tapered gammadions at the TF of 0.1. From Fig. 7.15(c), we also observe that the optical chirality of the asymmetrically tapered gammadion is continuous compared to the untapered design.

As we mentioned above, optical chirality ξ correlates with SEFR. When the SEFR is large, the localized field amplitude is very strong and the distribution is large. This results in a large optical chirality, as ξ is directly proportional to the modulus of the electric field. A large ξ also indicates higher field rotation amplitude, as described in Eq. (7.6). The higher field rotation amplitude results in stronger chiral field generation, creating a larger wavelength shift and improving the sensitivity of molecular detection.

The optical chirality ξ affects the rate of interaction with chiral molecules according to the gammadion TF. As observed from Figs. 7.13 and 7.14, the rate of interaction improves for TFs ranging between 0.5 and 0.75. This range of TF produces large ξ to excite many chiral molecules, resulting in a larger change of the refractive index locally around the nanostructures. It gives a larger wavelength shift and improved sensitivity.

Note the ω dependence of the optical chirality further increases the rate of interaction with chiral molecules for incident light of higher frequency. As a result, we can observe that the wavelength shift for mode m_3 is larger because it occurred at shorter wavelength and higher frequency for G-actin molecules, as shown in Fig. 7.14. However, for larger F-actin molecules, a larger wavelength shift is seen for mode m_1.

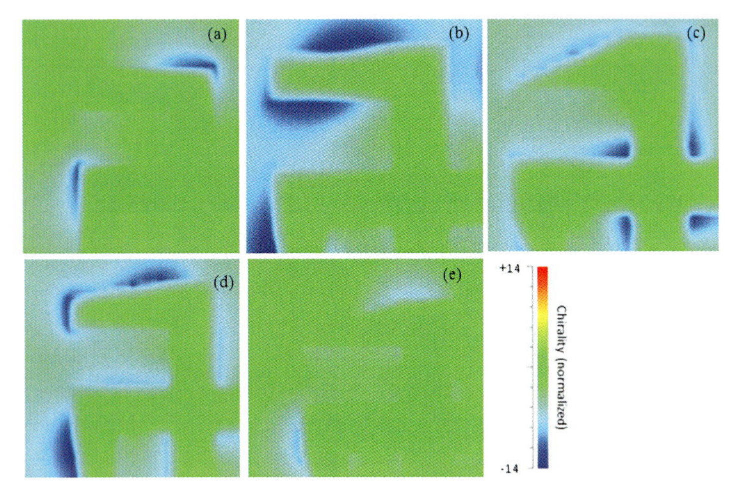

Figure 7.15: Optical chirality plot of the left-handed gammadion structures incident with left circularly polarized light for different designs of gammadion arms. (a) Untapered gammadion. (b) Asymmetrically tapered gammadion with TF of 0.67, mode m_1. (c) Asymmetrically tapered gammadion with TF of 0.1, mode m_1. (d) Asymmetrically tapered gammadion with TF of 0.67, mode m_3. (e) Asymmetrically tapered gammadion with TF of 0.1, mode m_3. Reprinted with permission from Ref. [16].

7.4 Summary

In conclusion, chiral optics emerged as a subfield of wave optics due to the finding of natural dissymmetry in molecular compounds and observation of their optical activity. To detect the chiroptical effects from optical activity of small molecules, we need to amplify those effects. This amplification can be provided by chiral nanostructures, which exhibit polaritonic resonances with superchiral fields that can substantially improve the detection of chiral biomolecules.

Bibliography

[1] L. Pauling and R. B. Corey, "A proposed structure for the nucleic acids," *Proc. Natl, Acad. Sci.*, vol. 39, pp. 84–93, 1953.

[2] L. Pasteur, "Researches on the molecular dissymmetry of natural organic product," *Am. J. Pharm.*, vol. 34, pp. 1–16, 1860.

[3] J. D. Waston and F. H. C. Crick, "Molecular structure of nucleic acids," *Nature*, vol. 171, pp. 737–738, 1953.

[4] L. Pauling, R. B. Corey, and H. R. Branson, "The structure of protein, two hydrogen bonded helical conformation of the polypeptide chain," *Proc. Natl, Acad. Sci.*, vol. 37, pp. 205–211, 1951.

[5] F. Sanger and H. Tuppy, "The amino acid sequence in the phenylalanine chain of insulin. the investigation of peptides from enzyme hydrolysates," *Biochem.*

J., vol. 49, pp. 481–490, 1951.

[6] G. Weber, "Polarization of the fluorescence of macromolecules," *Biochem. J.*, vol. 51, pp. 155–167, 1952.

[7] E. Gavrilesco, E. Barbu, and M. Macheboeuf, "Gelification of proteins, changes with ph in the viscosity of solutions of serum albumin too low in protein concentration to form gels," *Bull. Soc. Chim. Biol.*, vol. 32, pp. 924–933, 1950.

[8] D. D. Fitts and J. G. Kirkwood, "The rotatory power of helical molecules," *Proc. Natl, Acad. Sci.*, vol. 42, pp. 33–36, 1956.

[9] G. Holzwarth, "Circular dichroism measurements to 185 mm in a commercial recording spectrophotometer," *Rev. Sci. Instrum.*, vol. 36, pp. 59–63, 1965.

[10] L. Velluz, M. Legrand, and M. Grosjean, *Optical Circular Dichroism. Principles, Measurements and Applications*. Academic Press, New York, 1965.

[11] Y. J. Cassim and J. T. Yang, "A computerized calibration of circular dichrometer," *Biochemistry*, vol. 8, pp. 1947–1951, 1969.

[12] J. March, *Advanced Organic Chemistry: Reactions, Mechanisms, and Structure*. Wiley, New York, 3rd ed., 1985.

[13] W. Moffitt, D. D. Fitts, and J. G. Kirkwood, "Critique of the theory of optical activity of helical polymers," *Proc. Natl, Acad. Sci.*, vol. 43, pp. 723–730, 1957.

[14] J. S. Balcerski and E. S. Pysh, "Optical rotatory dispersion and vacuum ultraviolet circular dichroism of polysaccharide," *J. Am. Chem. Soc.*, vol. 97, pp. 6274–6275, 1975.

[15] J. T. Yang and T. Samejima, "Optical rotatory dispersion and circular dichroism of nucleic acids," *Prog. Nucleic Acid Res. Mol. Biol.*, vol. 9, pp. 223–300, 1969.

[16] E. H. Khoo, E. S. P. Leong, S. J. Wu, W. K. Phua, Y. L. Hor, and Y. J. Liu, "Effects of asymmetric nanostructures on the extinction difference properties of actin biomolecules and filaments," *Sci. Rep.*, vol. 6, 2016.

[17] E. Hendry, T. Carpy, J. Johnston, M. Popland, R. V. Milhaylovskiy, A. J. Lapthorn, S. M. Kelly, L. D. Barron, N. Gadegaard, and M. Kadodwala, "Ultrasensitive detection and characterisation of biomolecules using superchiral fields," *Nat. Nanotechnology*, vol. 5, p. 783, 2010.

[18] A. Papakostas, A. Potts, D. M. Bagnall, S. L. Prosvirnin, H. J. Coles, and N. I. Zheludev, "Optical manifestations of planar chirality," *Phys. Rev. Lett.*, vol. 90, no. 10, p. 107404, 2003.

[19] W. K. Phua, Y. L. Hor, E. S. P. Leong, Y. J. Liu, and E. H. Khoo, "Study of circular dichroism modes through decomposition of planar nanostructures," *Plasmonics*, vol. 11, no. 2, pp. 449–457, 2016.

[20] Z. Fan and A. O. Govorov, "Plasmonic circular dichroism of chiral metal nanoparticle assemblies," *Nano. Lett.*, vol. 10, p. 2580, 2010.

[21] V. K. Valev, J. J. Baumberg, C. Sibilia, and T. Verbiest, "Chirality and chiroptical effects in plasmonic nanostructures: fundamentals, recent progress, and outlook," *Adv. Mat*, vol. 25, no. 18, pp. 2517–2534, 2013.

[22] V. K. Valev, B. D. Clercq, X. Zheng, D. Denkova, E. J. Osley, S. Vandendriessche, A. V. Silhanek, V. Volskiy, P. A. Warburton, G. A. E. Vandenbosch, and M. Ameloot, "The role of chiral local field enhancements below the resolution limit of second harmonic generation microscopy," *Opt. Express*, vol. 20, pp. 256–264, 2012.

[23] V. K. Valev, N. Smisdom, A. V. Silhanek, B. D. Clercq, W. Gillijns, M. Ameloot, V. V. Moshchalkov, and T. Verbiest, "Plasmonic ratchet wheels: switching circular dichroism by arranging chiral nanostructures," *Nano. Lett.*, vol. 9, no. 11, pp. 3945–3948, 2009.

[24] Y. L. Hor, W. K. Phua, and E. H. Khoo, "Chirality switching via rotation of bilayer fourfold meta-structure," *Plasmonics*, pp. 1–5, 2016.

[25] M. Decker, "Strong optical activity from twisted-cross photonic metamaterials," *Opt. Lett.*, vol. 34, no. 11, p. 2501, 2009.

[26] J. Olofsson, T. M. Grehk, T. Berlind, C. Persson, S. Jacobson, and H. Engqvist, "Evaluation of silicon nitride as a wear resistant and resorbable alternative for total hip joint replacement," *Biomatter*, vol. 2, no. 2, pp. 94–102, 2012.

[27] M. Mazzocchi and A. Bellosi, "On the possibility of silicon nitride as a ceramic for structural orthopaedic implants. part i: Processing, microstructure, mechanical properties, cytotoxicity," *J. Mater. Sci. Med*, vol. 19, no. 8, pp. 2881–2887, 2008.

[28] Y. Huang, L. Wu, X. Chen, P. Bai, and D. H. Kim, "Synthesis of anisotropic concave gold nanocuboids with distinctive plasmonic properties," *Chem. Mater.*, vol. 25, no. 12, pp. 2470–2475, 2013.

[29] Y. Kimori, E. Katayama, N. Morone, and T. Kodama, "Fractal dimension analysis and mathematical morphology of structural changes in actin filaments imaged by electron microscopy," *J. Struct. Biol.,*, vol. 176, p. 1, 2011.

[30] K. C.Holmes, D.Popp, W.Gebhard, and W.Kabsch, "Atomic model of actin filament," *Nature*, vol. 347, p. 44, 1990.

[31] K. Xu, G. Zhong, and X. Zhuang, "Actin, spectrin, and associate proteins form a periodic cytosleletal structure in axons," *Science*, vol. 339, p. 452, 2013.

[32] T. Yanagida, M. Taniguchi, and F. Oosawa, "Conformational changes of f-actin in the thin filaments of muscle induced in vivo and in vitro by calcium ions," *J. Mol. Biology*, vol. 90, p. 509, 1974.

[33] G. J. Doherty and H. T. McMahon, "Mediation, modulation and consequences of membrane-cytoskeleton interactions," *Annu. Rev. Biophys.*, vol. 37, p. 65, 2008.

[34] W. P. Hall, J. N. Anker, Y. Lin, J. Modica, M. Mrksich, and R. P. V. Duyne, "A calcium-modulated plasmonic switch," *J. Am. Chem. Soc.*, vol. 130, p. 5836, 2008.

[35] S. Link and M. A. El-Sayed, "Spectral properties and relaxation dynamics of surface plasmon electronic oscillations in gold and silver nanodots and nanorods," *J. Phys. Chem. B*, vol. 103, p. 8410, 1999.

[36] C. G. dos Remedios and D. D. Thomas, *Molecular Interactions of Actin: Actin Structure and Actin-Binding Proteins*. Springer Heidelberg, 2012.

[37] X. Zhao, R. Zhao, and W. S. Yang, "Self-assembly of l-tryptophan on the cu(001) surface," *Langmuir*, vol. 18, no. 2, p. 433, 2002.

[38] D. Lipkin, "Existence of a new conservation law in electromagnetic theory," *J. Math. Phys.*, vol. 5, p. 696, 1964.

[39] Y. Tang and A. E. Cohen, "Optical chirality and its interaction with matter," *Phys. Rev. Lett.*, vol. 14, p. 163901, 2010.

Chapter 8

Localized polariton-based sensors

Ping Bai
Institute of High Performance Computing, Singapore

Xiaodong Zhou
Institute of Materials Research and Engineering, Singapore

Ten It Wong
Institute of Materials Research and Engineering, Singapore

Lin Wu
Institute of High Performance Computing, Singapore

Song Sun
Institute of High Performance Computing, Singapore

In this chapter, we consider the main principles of design and operation of localized polariton-based sensors. The operation principles will be reviewed in Section 8.1, while the design of two configurations based on metal nanoparticles and patterned films will be discussed in Section 8.2. The effects of light illumination and sensor materials will be addressed in Sections 8.3 and 8.4, respectively. Nanochip fabrication and characterization aspects will be covered in Section 8.5. Point-of-care sensing systems will be considered in Section 8.6. The chapter will conclude with summary presented in Section 8.7.

8.1 Operation principles

Polaritonic sensors are able to monitor binding events on a sensor surface in real time, providing a distinct advantage over endpoint detection methods. They are typically based on the spectral shift of polariton resonances or their intensity change induced by the molecule binding. The traditional propagating polariton-based sensors [1] are generally bulky, while the localized polariton-based sensors offer similar functionalities at smaller dimensions. In particular, sensors based on localized surface polariton resonance (LSPR) utilize the sub-wavelength interaction between light and nanostructures, largely relaxing the resonance conditions. For instance, strong LSPR may be realized with a nanohole array patterned on a metal film with ordinary white light illumination on under the normal conditions. LSPR promises to develop a low-cost point-of-care sensing system. Another advantage of the LSPR sensor is the short penetration depth of evanescent fields, making the sensor less sensitive to bulk dielectric environment changes induced by temperature fluctuations in the solvent far from the sensor surface.

LSPR of metal nanoparticles exhibits fields that are highly localized at the nanoparticle surface and decay rapidly away from it into the dielectric background. Coupling of light with the localized surface absorption polaritons (see Chapter 3) results in strong scattering and absorption by nanoparticles at LSPR. Field enhancement and localization are the two important characteristics of LSPR, which define the key sensing parameters such as sensitivity and spatial resolution.

To quantify the performance of a sensor, the most common parameter is sensitivity, which indicates the sensor signal change responding to the measured quantity change. In LSPR-based sensing, spectral peak wavelength, intensity, and phase have been explored as the sensing signals, among which the peak wavelength is the most frequently used. Typically, the LSPR peak wavelength depends on the surrounding dielectric environment; to quantify it, the *refractive index sensitivity* S per refractive index unit (nm/RIU) is introduced,

$$S = \frac{\Delta\lambda}{\Delta n}, \tag{8.1}$$

where $\Delta\lambda$ is the wavelength shift of the LSPR peak, and Δn is the change of refractive index in the surrounding environment. As the precision that can be achieved with respect to changes in the refractive index depends on the resonance linewidth as well, the use of sensitivity alone does not provide an adequate performance quantification. To overcome this issue and take the resolution into account, a figure of merit (FOM) obtained by dividing the sensitivity by the resonance linewidth is commonly used [2],

$$\mathrm{FOM} = \frac{S}{\mathrm{FWHM}}, \tag{8.2}$$

where FWHM is the full width at half-maximum of the corresponding LSPR peak.

Besides the peak wavelength, field intensity is often used to evaluate the performance of the LSPR sensors. For instance, strong field enhancements also accompany polariton resonances used for performing surface-enhanced spectroscopies or surface-polariton field-enhanced fluorescence spectroscopy (SPFS).

8.2 Sensing structures

LSPR can be efficiently realized with sub-wavelength metallic structures such as nanoparticles and patterned films. Metallic nanoparticles and periodic nanostructures have commonly been employed to LSPR sensors. In this section, we will focus on these two configurations.

8.2.1 Nanoparticles

The sensitivity of nanoparticle-based LSPR sensors to local changes in refractive index is highly dependent on resonance characteristics such as spectral linewidth,

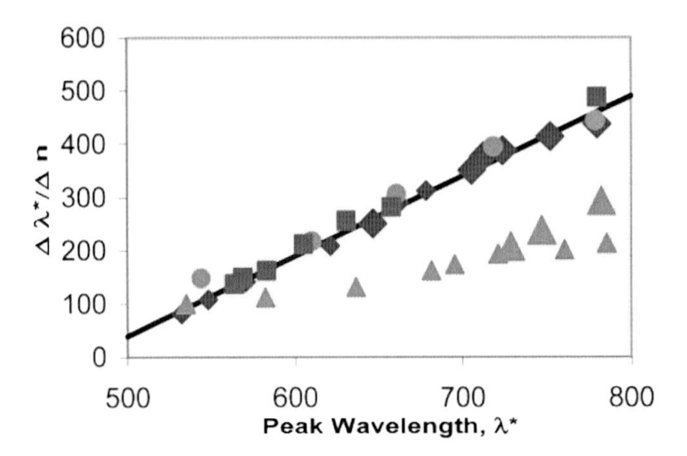

Figure 8.1: Theoretical predictions of refractive index sensitivity for gold nanocylinders (circles), nanodisks (squares), hollow nanoshells (diamonds, larger diamonds showing larger hollow nanoshells), and solid nanoshells (triangles, larger triangles for larger particles). Reprinted with permission from Ref. [4]. © 2005 American Chemical Society.

extinction intensity, electromagnetic-field strength, and decay length [3]. All these characteristics are determined by the material composition, size, and shape of the nanoparticle and also the substrate to support the nanoparticle and the surrounding environment. To optimize the sensitivity of nanoparticle sensors, we need to understand how these characteristics affect LSPR properties.

First, let us look at the nanoparticle size effect. According to the Mie theory, the scattering cross section of a sub-wavelength spherical particle of radius R is proportional to R^6, while the absorption is proportional to R^3 [see Chapter 5, Eq. (5.6)]. For smaller nanoparticles, LSPR extinction is dominated by absorption. As the particle size increases, scattering takes over. For gold nanospheres, this transition occurs around a diameter of 80 nm. More importantly, the surface-polariton resonant wavelength is dependent on the particle size [see Chapter 5, Fig. 5.4]. It redshifts with the nanosphere diameter. The LSPR wavelengths of gold nanospheres can be changed over 60 nm by varying particle diameter from 10 to 100 nm. Generally, the longer the resonant wavelength is, the higher the refractive index sensitivity of the nanoparticle can be achieved, following the linear dependence shown in Fig. 8.1 [4].

The aspect ratio of elongated nanoparticles such as nanorods also affects sensitivity. Increasing nanoparticle aspect ratio has been shown to redshift the LSPR wavelength and increase the electromagnetic field decay length. Figure 8.2 shows that the resonant wavelength of gold nanorods redshifts from 550 to 800 nm, when the aspect ratios change from 1.35 to 4.42. It also shows that the larger the aspect ratio, the better the sensitivity. For instance, the sensitivity of a gold nanorod with the radius of 10 nm increases from 157 to 497 nm/RIU when the aspect ratio increases from 1.0 (sphere) to 3.4 (rod) [5]. For the nanorods of the same aspect ratio

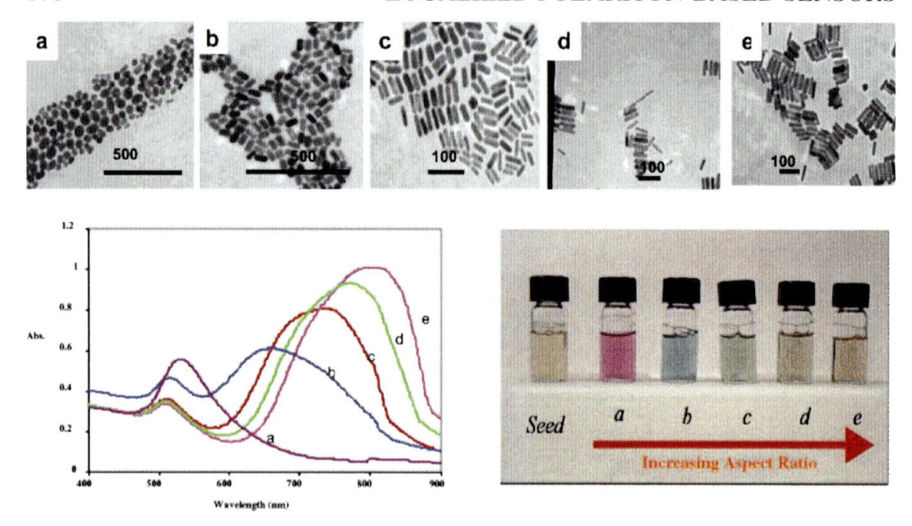

Figure 8.2: Transmission electron micrographs (top), optical spectra (bottom left), and photographs of aqueous solutions of gold nanorods of various aspect ratios (bottom right). Seed sample: aspect ratio 1; samples (a), (b), (c), (d), and (e) have the aspect ratios of 1.35, 1.95, 3.06, 3.50, and 4.42, respectively. Reprinted with permission from Ref. [6]. © 2005 American Chemical Society.

and different sizes, the larger nanorods possess higher refractive index sensitivity. It is caused by the redshift of resonant wavelength when the aspect ratio or radius of the nanorod increases, leading to a sensitivity increase.

The nanoparticle shape plays a large role in determining sensitivity. For instance, silver nanoparticles with similar volumes, but different shapes, such as spheres, triangles, and cubes, have distinctive scattering spectra, as shown in Fig. 8.3 [7]. The silver nanotriangles demonstrated higher sensitivity (350 nm/RIU) than the spheres (160 nm/RIU) [7]. In addition, nanoparticles with sharp tips such as nanotriangles and bipyramids exhibit especially high refractive index sensitivities. One effect of the sharp tips is a redshift of LSPR, improving the refractive index sensitivity. Moreover, a sharp tip causes high field enhancements, offering an additional advantage for molecular detection at the microscopic level.

Another key parameter is the nanoparticle material; it has a strong effect on the LSPR wavelength. Silver and gold are the most commonly used nanoparticle materials that feature LSPR within the visible range. LSPR is possible in many other metals, alloys, and semiconductors, exhibiting dielectric permittivity with a sufficiently negative real part and small imaginary part.

8.2.2 *Periodic nanostructures*

Sensitivity of nanoparticle sensors is dominated by spectral position of LSPR and dielectric property. A similar claim is valid for the case of periodic nanostructure-based

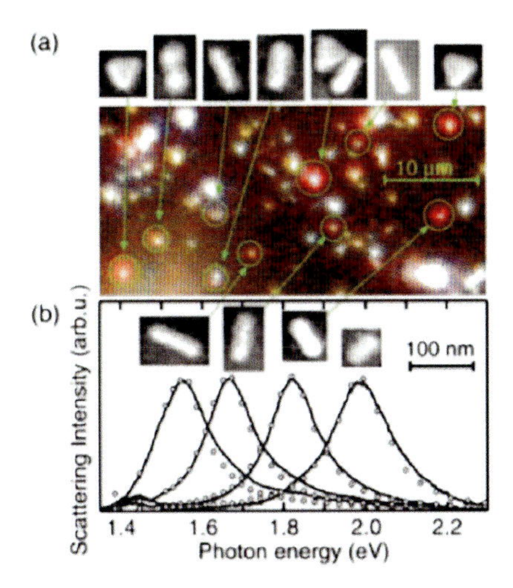

Figure 8.3: (a) Dark-field microscopy image and corresponding scanning electron microscope (SEM) images and (b) light scattering spectra of Au nanocrystals of different shapes. Reprinted with permission from Ref. [7]. © 2003 American Institute of Physics.

sensors. By exploiting the geometry of periodic nanostructures, surface-polariton extinction spectra with intense, narrow resonances and high refractive index sensitivities can be designed. Among numerous nanostructures, nanoslit and nanohole arrays have been the most intensively investigated for LSPR sensing.

Nanoslit array

To implement nanoslit array-based sensors, a metal film can be deposited and patterned on a glass substrate with the analytes dissolved in water (e.g., buffer, human urine or serum) and bound onto the array surface. To excite the surface polariton resonance, the nanoslit array should be illuminated from either the analyte or substrate side. By measuring the transmitted or scattered light, the density of the analyte bound to the surface can be detected.

Here we consider a simple gold nanoslit array fabricated on a glass substrate. Figure 8.4(a) is a cross-sectional view of one unit cell of the nanoslit array in water medium. The left panel displays the case of no analyte bound to the nanoslit array, while the right panel depicts analyte presence. Figure 8.4(b) shows the calculated reflectance in the nanoslit structures. It demonstrates three polaritonic resonances labeled α, β, and γ. After 50 nm of PLK1 mRNA molecules (refractive index: 1.343) are bound to the metallic structures, all three resonances shift by different values, among which the α resonance shifts the most.

To understand why the α dip has the largest shift, we should determine the nature of all resonances. To do so, we calculate the electric field distribution at each resonance. Figure 8.5 shows the field distributions at the α, β, and γ resonance modes

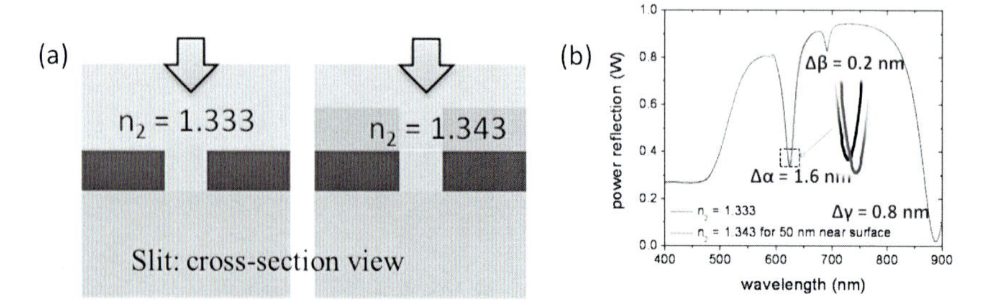

Figure 8.4: (a) Cross-sectional view of a unit cell of a nanoslit array. (b) Simulated reflection spectrum for the nanoslit array with and without the analytes bound, where the pitch, width, and thickness of the slit array are 400, 250, and 100 nm, respectively. The sensitivity is calculated for the bulk refractive index of 1.333 (water) and the 50 nm analyte of refractive index of 1.343. Reprinted with permission from Ref. [8]. © 2013 IEEE.

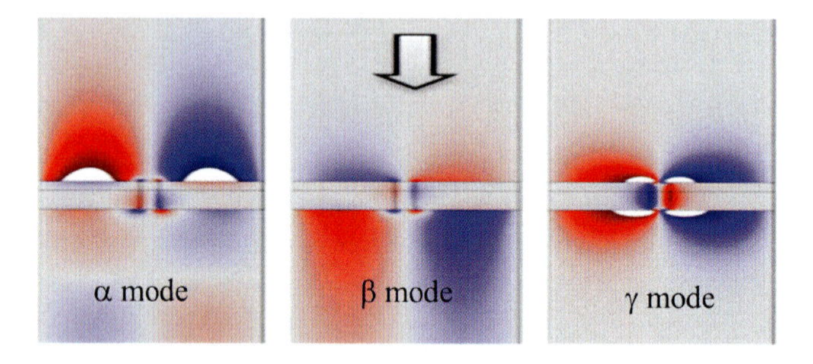

Figure 8.5: Simulated electric-field distributions for the α, β, and γ resonances across the slit structure on a glass substrate. Light is incident from top of the nanoslit array. Reprinted with permission from Ref. [8]. © 2013 IEEE.

across the slit structure. From these distributions, we discern that the α resonance is the surface mode with fields excited at the water-metal surface; the β resonance is the surface mode with fields excited at the substrate-metal interface; and the γ resonance is the edge mode with fields excited at the substrate-metal interface close to the slit edges. When the analytes are bound to the surface of the nanoslit array, the refractive index change occurs at the water-metal interface, where the fields of α resonance originate. The α resonance is the most sensitive to the refractive index change at the water side, because its electric fields have the largest overlap with the analyte bound area. Conversely, the fields of β resonance are concentrated at the metal-glass interface having the least overlap with the bound analytes; thus, its resonant wavelength shifts the least.

To further study the effects of nanoslit array parameters on sensitivity, we consider the pitch, slit width, and thickness of the metal film. The effect of pitch change is the most prominent, as presented in Fig. 8.6. All three polariton resonances red-

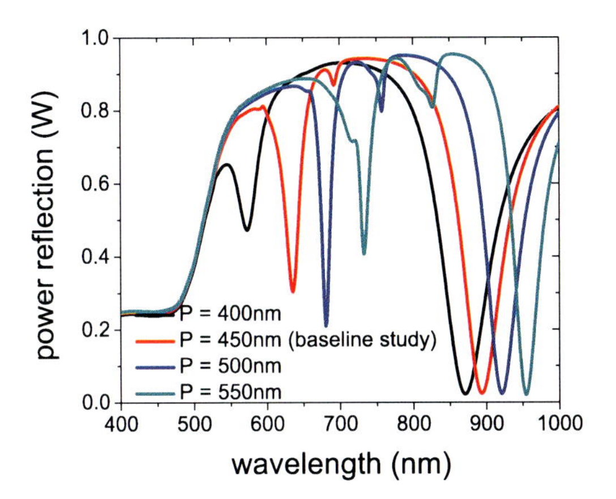

Figure 8.6: Pitch effects on the polariton resonances in the nanoslit arrays where the slit width is 37.5 nm and the metal film thickness is 150 nm. Reprinted with permission from Ref. [8]. © 2013 IEEE.

shift when the pitch increases. The maximized sensitivity is achieved for the nanoslit array of 450 nm pitch, 20 nm width, and 150 nm thickness.

Nanohole array

Compared to nanoslits, nanohole arrays show lower wavelength shift sensitivity caused by the oscillations of polarization charges restricted in the two planar directions for a nanohole array. This restriction is preferable for surface plasmon-polariton field-enhanced fluorescence spectroscopy (SPFS) in which strong fields are used to enhance the excitation rates of fluorescent molecules [9]. The LSPR wavelength should match the excitation wavelength of the fluorescent dyes. For a nanohole array, its resonant wavelength may be tuned by changing the pitch, diameter, and depth (or the thickness of the metal film) of the hole. The resonant wavelength and magnitude of the LSPR in a nanohole array depend on the size, shape, composition, and local dielectric environment. For a two-dimensional square array of nanoholes with a pitch p, the LSPR wavelength λ_p at normal incidence can be expressed by [10]

$$\lambda_m = \frac{P}{\sqrt{i^2 + j^2}} \sqrt{\frac{\varepsilon_m' \varepsilon_d}{\varepsilon_m' + \varepsilon_d}}, \tag{8.3}$$

where ε_d is the dielectric constant of the medium, and ε_m is that of the metal. The integers i and j indicate diffraction orders. Following this formula, the pitch strongly affects the LSPR wavelength.

To demonstrate the pitch effect, we consider a nanohole array patterned in a silver film suspended in air, where the two metal-air interfaces are identical [11]. Figures 8.7(a) and (b) show the simulated transmission and absorption spectra for nanohole

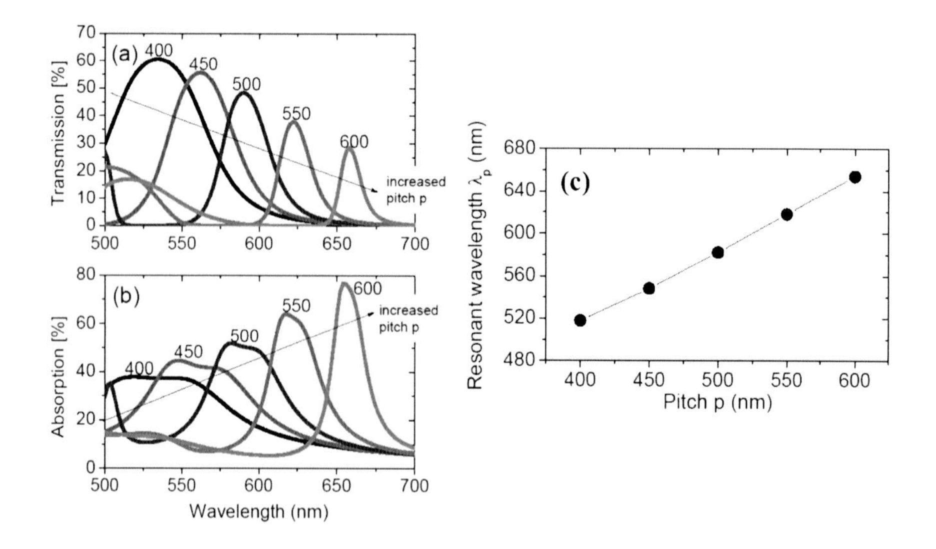

Figure 8.7: Calculated (a) transmission, (b) absorption spectra, and (c) resonant wavelength as functions of the array pitch for normally incident light. The nanoholes have a diameter of 250 nm and depth of 400 nm. Reprinted with permission from Ref. [11]. © 2012 The Optical Society of America.

array with pitches between 400 nm and 600 nm, where the holes have a diameter of 250 nm and a depth of 400 nm. As the pitch increases, the transmission is reduced, whereas the absorption is increased and the peak redshifts. The resonant wavelength linearly increases with the pitch, as shown in Fig. 8.7(c). The results agree well with the prediction of the above formula. An increment of 100 nm in the pitch leads to an increment of 60 to 70 nm in the resonant wavelength.

Compared with the pitch effect, the diameter and depth of the holes have lesser influence on the LSPR wavelength. An increment of 100 nm in diameter leads to an increase of the LSPR wavelength up to 30 to 40 nm; a similar increment in the hole depth causes a resonance shift up to 10 to 20 nm [11].

Among all geometrical parameters, the film thickness has the weakest effect on the resonant wavelength. However, it has significant influence on field enhancement. Figure 8.8 shows the electric field enhancement at the edge of the nanohole along the z direction (the hole axis direction) for three thicknesses t: 100, 400, and 600 nm at the resonant wavelengths of 534, 582, and 592 nm, respectively. For a thin film (e.g., t of 400 nm), the field enhancements at the two surfaces are comparable. As the silver film gets thicker, the field enhancement increases at the front surface and decreases at the back one. The incident optical power primarily excites fields at the front surface, with some power losses for reflection. Less optical power could pass through the metal film and excite strong fields at the back surface. For a thick film

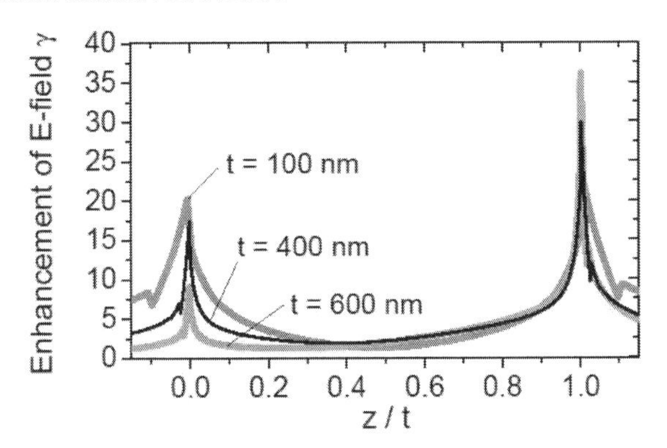

Figure 8.8: Electric-field enhancement at the edge of the hole along the z direction (the hole axis direction) for three thicknesses: t = 100 nm (λ_p = 534 nm), 400 nm (λ_p = 582 nm), and 600 nm (λ_p = 592 nm), where the z coordinate is normalized with the film thickness t. Reprinted with permission from Ref. [11]. © 2012 The Optical Society of America.

structure (e.g., t of 600 nm), there is little power to excite fields at the back surface; therefore the electric field enhancement there is very small.

By increasing the pitch and decreasing the hole diameter, we can increase the absorption power inside the nanohole array and hence improve the electric field enhancement as well. This can be understood by using a physical picture. Let us assume that we have plenty of holes of the same dimensions and the electric field enhancement is given by how we arrange the holes. Larger pitch means that we can put fewer holes and hence have more silver in the array. As a result, more incident optical power can be absorbed inside the array. Similarly, smaller holes also allow more incident optical power to penetrate a film. Consequently, more optical power may be absorbed by silver and the fields are enhanced more at the LSPR resonance. However, compared with film thickness variation, the pitch and hole diameter have fewer effects on the field enhancement.

8.3 Light illumination effects

In the previous section, we studied a symmetric nanohole silver film suspended in air with two identical metal-air interfaces. In reality, the nanohole metal film is normally fabricated on a substrate and analytes are bound on the top surface of the metal film. In this case, the two surfaces of the metal film contact different materials: the substrate and analytes. As a result, light incident from the analytes (front) or the substrate (rear) sides may produce different effects. In addition, the light could illuminate the array under oblique incidence, which also affects the sensitivity. We now consider these aspects and study how they influence the analyte detection.

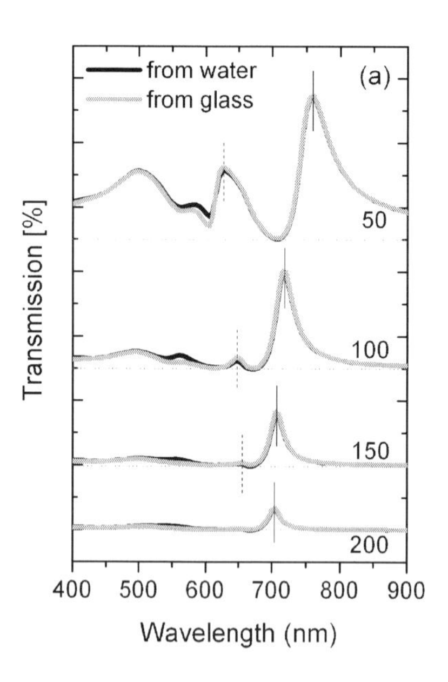

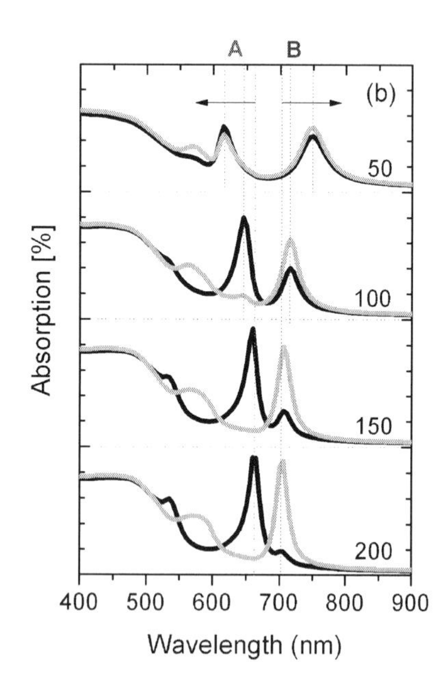

Figure 8.9: (a) Transmission and (b) absorption spectra of nanohole arrays (pitch: 400 nm, diameter: 150 nm), milled in gold films with thicknesses of 50, 100, 150, and 200 nm, where the light illuminates the arrays from the water (black) or glass (grey) sides. Reprinted with permission from Ref. [12]. © 2012 IEEE Photonics Society.

8.3.1 Front and rear illumination

Consider a water-gold-glass structure, in which a nanohole array is patterned in a gold film deposited on top of a glass substrate, with aqueous analytes applied to the top surface of the array. A square lattice nanohole arrangement can be fully characterized by array pitch, hole diameter, and gold film thickness. Figure 8.9 shows the calculated transmission and absorption spectra for nanohole arrays with different film thicknesses when light illuminates from the analyte (front) or glass (rear) sides [12]. Two major absorption peaks are observed: peak A at about 610 to 660 nm and peak B at 700 to 750 nm. For a thicker film of 200 nm, peak A is dominant if light illuminates from the front side, whereas peak B dominates for the rear side illumination. However, for a smaller film thickness such as 50 nm, both peaks are present regardless of the illumination scheme.

The field distributions plotted in Fig. 8.10 demonstrate that the fields at resonance A are generally stronger for illumination from the water side. It confirms that peak A is an LSPR excited primarily at the water-gold interface. Similarly, peak B is an LSPR excited primarily at the glass-gold interface due to the extraordinary optical transmission (EOT).

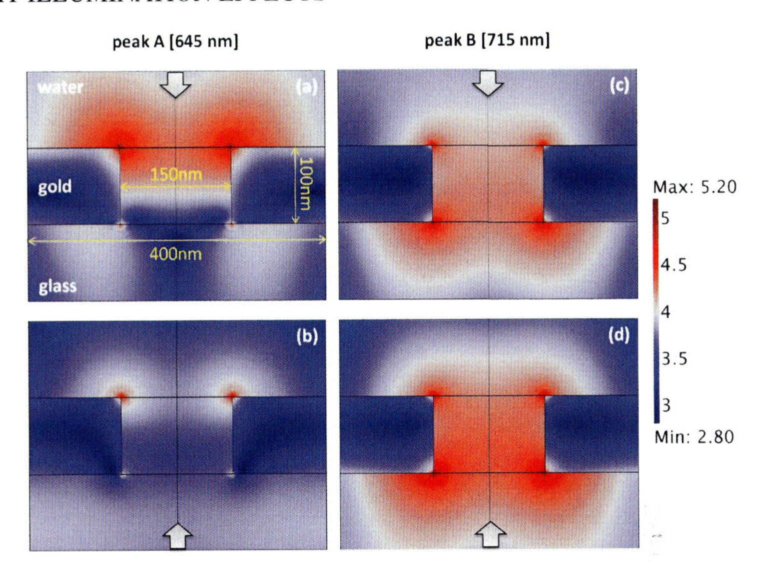

Figure 8.10: Field distributions across a nanohole array for two cases where light illuminates from the water [(a) and (c)] or glass [(b) and (d)] side at two peak wavelengths: peak A [(a) and (b)] or B [(c) and (d)]. The nanohole array under analysis has pitch of 400 nm, diameter of 150 nm, and hole depth of 100 nm. Reprinted with permission from Ref. [12]. © 2012 IEEE Photonics Society.

As peak A is excited at the water-gold interface where the analytes are located, the field enhancement for peak A is more important for detection of the analytes. Figure 8.11 shows the field enhancement of peak A at the rims of holes at the water-gold interface with light illuminating from either the water or glass side. When light illuminates from the water side, the field enhancement increases rapidly with the film thickness. It saturates around a value of 32 with a variation of 3.5%, when a critical film thickness of 150 nm is reached. This can be understood by analyzing the optical power flow, as illustrated in the inset of Fig. 8.11(a). When the metal film is thinner, more optical power will tunnel through the nanohole array and less power will be available to excite the LSPR at the water-gold surface. Consequently, a relatively thick film (of 150 nm) is required to reduce the tunneling power and retain enough optical power to excite peak A.

On the other hand, the field enhancement at peak A increases with film thickness until $t < 60$ nm, when light illuminates from the glass side, as shown in Fig. 8.11(b). This can be easily understood. The increased field enhancement is due to more optical power tunneling through the nanohole array. Beyond 60 nm, when film thickness is away from the skin depth of gold in the visible spectrum (around 30 nm), field enhancement starts to drop. We find that a thinner film allows more optical power to reach the gold surface layer near the water. However, most of the reached optical power turns into loss instead of excitation of the LSPR, as the surface-layer power absorption difference between the Au nanohole array and the similar Au film is re-

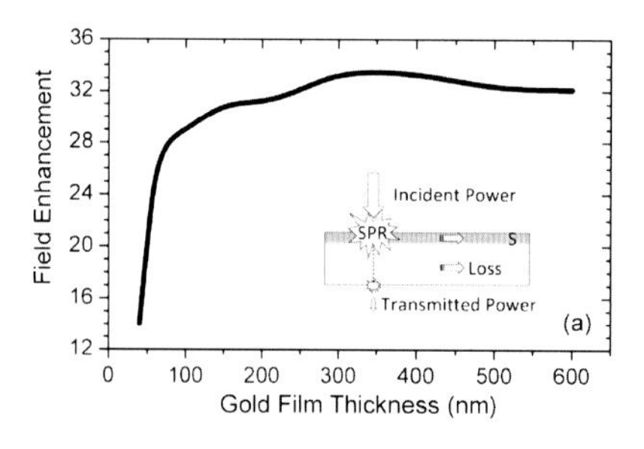

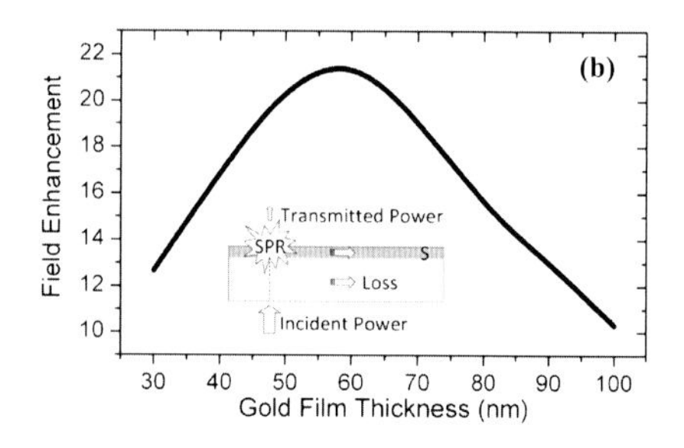

Figure 8.11: Field enhancement in resonance A at the rims of holes on the water-gold interface as a function of gold film thickness with light illumination from (a) the water side and (b) the glass side. The insets show the power flow. Reprinted with permission from Reference [12]. © 2012 IEEE Photonics Society.

duced when the film thickness decreases from 60 nm to 30 nm [12]. In other words, for the glass side illumination, the gold film should be thin enough, but larger than the skin depth (60 nm in the considered case).

8.3.2 Oblique illumination

In some cases, such as in dark-field microscopy, the sample is peripherally illuminated at large oblique angles. Such tilted illumination may influence the LSPR wavelength and change the field enhancement. Here we explore the possible effects of oblique incidence in two configurations: (i) a gold nanohole array on a glass substrate (Au-hole array) and (ii) a gold-coated photoresist nanohole array (Au/PR-hole array) [13].

Figure 8.12(a) shows the Au-hole array (pitch: 400 nm, diameter: 40 nm, and thickness: 50 nm) to be investigated for fluorescence excitation of Alexa 647. At normal incidence, the simulated absorption spectra of the gold nanohole array in water medium exhibit peaks around 626 nm and 750 nm marked in Fig. 8.12(b) as α and β. The α resonance is excited on the top rims (water side) of the gold nanoholes, while the β peak is excited at the bottom rims (glass side) of the gold nanoholes.

The simulated spectra in Fig. 8.12(b) indicate that the variations of the incident angle from 0° to 80° cause redshift of the α resonance by around 5 nm, whereas no shift is observed for the β mode. The field enhancements for the wavelengths of 590, 600, 610 and 620 nm, as shown in Fig. 8.12(c) also indicate small variations upon changing the incident angles. For example, a field enhancement of 37 is obtained at the wavelength of 610 nm with a variation within 10% for all incident angles.

Figure 8.13(a) shows the Au/PR-hole array that can be fabricated by imprinting a nickel mold (with nanopillars) on a 220 nm thick photoresist layer, and evaporating 50 nm of gold on it. Unlike the Au-hole array, the α and β resonant modes of the Au/PR-hole array blueshift to 608 nm and 726 nm at normal incidence because of the lower refractive index of the 220 nm thick photoresist as compared to glass, as shown in Fig. 8.13(b). The absorption peak of the α mode is more sensitive to changes in

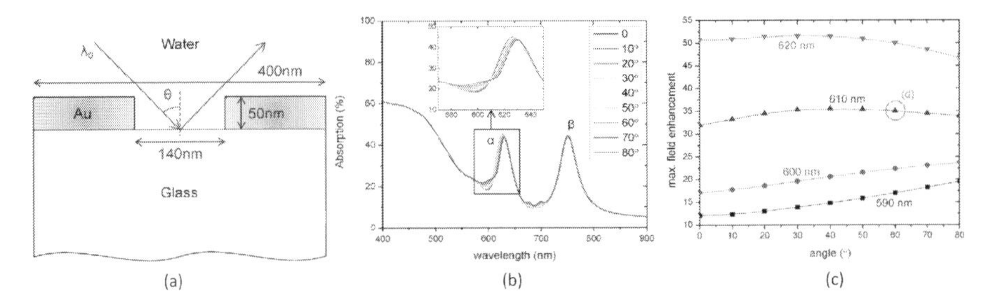

Figure 8.12: (a) Simulation model of the Au-hole array ($P = 400$ nm, $D = 140$ nm, $T = 50$ nm) on glass at an incident angle of θ. (b) Absorption spectra simulated for the Au-hole array in water at various incident angles. (c) Simulated maximum field enhancement versus the incident angle at wavelengths of 590, 600, 610, and 620 nm. Reprinted with permission from Ref. [13]. © 2013 Elsevier B.V.

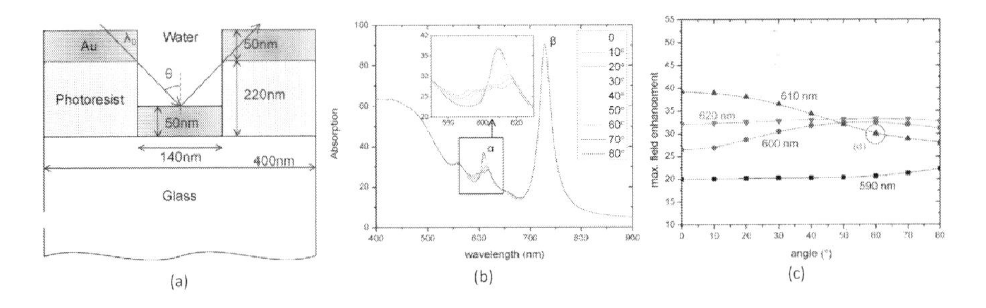

Figure 8.13: (a) Cross-sectional view of a unit cell of the Au/PR-hole array with 50 nm of gold coated on a 220 nm thick photoresist nanohole array. (b) Simulated absorption spectra of the Au/PR-hole array in water at various incident angles. (c) Maximum field enhancement versus incident angle at wavelengths of 590, 600, 610, and 620 nm. Reprinted with permission from Ref. [13]. © 2013 Elsevier B.V.

the incident angle on the Au/PR-hole array compared with the Au-hole array. As the incident angle increases, the peak associated with the α mode is broadened, slightly redshifted, appears with reduced intensity, and further splits into three separate features. It is likely due to the presence of two more interfaces (glass-photoresist and photoresist-Au), which also contribute to the appearance of an additional peak around 575 nm. The angle-dependent field distribution plotted in Fig. 8.13(c) indicates that the field enhancements at the wavelengths of 600 and 610 nm are more dependent on the light incidence angle than those at 590 and 620 nm, with a variation up to 33% over the entire angular span.

8.4 Effects of nanostructure materials

Gold is so far the most commonly used material in nanostructure-based LSPR biosensors due to its excellent polaritonic properties in the longer wavelength visible light range (600 to 800 nm) and chemical inertness in solutions. However, in some scenarios, fluorescent dyes with excitation wavelength below 600 nm may be preferred, where gold is not a good candidate for LSPR. Here we explore the possibilities of using various metals such as Au, Ag, Cu, and Al or their composites Ag-Au, Cu-Au, and Al-Au in the nanohole array-based SPFS biosensors, aiming to excite different fluorescent dyes (emitting within different wavelength ranges) with satisfactory performance [14].

First, let us consider how much power from incident light may be used to excite LSPR. The dimension of the nanohole array will be fixed with the pitch of 400 nm, hole diameter of 150 nm, and metal thickness of 100 nm. White light with incident optical power of 1 nW is assumed to be normally incident from the glass substrate side. Among the power absorbed within the nanohole array structure, some part is used to excite LSPR either at the glass-metal or water-metal interface, while the rest is absorbed in the metal as bulk losses. We will discuss bare metal films as the references for absorbed powers attributing to bulk losses.

For example, the Au nanohole array supports three surface polariton resonances with wavelengths of 540, 640 and 715 nm, respectively. A similar Au film showed no resonance and feature absorption monotonically decreasing with wavelength within the range of 400 to 600 nm. It was caused by a large intrinsic absorption of Au in the mentioned range with a characteristic electron inter-band transition around 500 nm. The total optical power used to excite LSPR can be estimated by determining the area difference of the absorbed powers between the Au film and the Au nanohole array in the absorption spectra. We found that of the 1 nW incident powers, about 0.104 nW was used to excite the three LSPRs of the nanohole array, which is around 10.4% of the power efficiency [14].

Cu possesses optical data very similar to that of Au. The Cu nanohole array has surface polariton resonances mainly in the wavelength range of 600 to 900 nm, and it naturally absorbs light in the spectrum of 400 to 600 nm. By analyzing the Cu film and Cu nanohole array power absorption, we estimate that 11.5% of the incident power is actually used for LSPR excitation.

For Ag and Al films, much lower bulk loss was observed over the whole visible spectrum. Therefore, we perceived both Ag and Al films approached white color in contrast to Au and Cu films of yellow and orange color. The Ag nanohole array supports a few resonances at 460, 525, 610, and 695 nm, respectively. The Al nanohole array exhibited LSPRs in a wide wavelength range from 405, 455, 560, to 630 nm. To excite these resonances, Ag (or Al) consumes about 10.8% (or 7.83%) of the incident power. Compared with Au and Cu, Ag and Al provide better tunability of the resonant wavelength across the whole visible spectrum, especially in the range of 400 to 600 nm.

In SPFS, the field enhancement near the excitation wavelength of the fluorescent dyes plays the most important role. We will analyze below the field enhancement for all possible modes. The resonant wavelengths and corresponding field enhancements are summarized in Table 8.1 for different metals. The italic values represent the primary water side resonance and the rest indicate the water side LSPRs caused by the primary glass side resonances. The results suggest that Au is preferred when the excitation wavelength of the fluorescent dyes is larger than 600 nm; Ag performs better when the excitation wavelength is below 600 nm; Cu is a cheaper alternative to Au with slightly degraded field enhancement; Al provides the smallest field enhancements, but it brings the resonant wavelength down to 405 nm.

	400 to 500 nm	500 to 600 nm	600 to 700 nm	700 to 800 nm
Au	-	540 nm (13.68)	*640 nm (25.28)*	715 nm (18.12)
Ag	*460 nm (54.27)*	525 nm (15.66)	*610 nm (19.55)*	695 nm (18.30)
Cu	-	-	*625 nm (17.29)*	710 nm (15.71)
Al	*405 nm (6.79)*	*560 nm (6.88)*	630 nm (4.06)	-

Table 8.1: Surface polariton resonances (field enhancements at water interface) for different metals [14]. Italic values are caused by resonances from the water-metal interface and the rest from the glass-metal interface.

Combining different metals could achieve better performance, especially in the range of 500 to 600 nm. Three examples are explored by coating a layer of 10 nm thick Au on the top and sidewall of 90 nm thick Ag core, Cu core, or Al core. The results are summarized in Table 8.2. The combination of 90 nm Ag and 10 nm Au would achieve a higher field enhancement for the 535 nm (22.86) and 700 nm (20.35) resonances than using pure Ag (15.66, 18.30) or pure Au (13.68, 18.12). The combination of Al and Au surprisingly increased the field enhancement of the LSPR at 645 nm to 28.10.

	400 to 500 nm	500 to 600 nm	600 to 700 nm	700 to 800 nm
Ag	455 nm (2.11)	535 nm (22.86)	*620 nm (24.42)*	700 nm (20.35)
Cu	-	535 nm (12.05)	*630 nm (22.01)*	710 nm (17.14)
Al	*420 nm (2.62)*	525 nm (10.63)	645 nm (28.10)	-

Table 8.2: Surface polariton resonances (field enhancements at water interface) for different metal composites (metal (90nm)+ Au (10 nm)) [14]. Italic values are caused by resonances from the water-metal interface and the rest from the glass-metal interface.

Thus, the Ag nanohole array is suitable for excitation of 400 to 500 nm fluorescent dyes. The Au to coated Al nanohole array supports very good resonances for fluorescence excitation at 600 to 700 nm, whose field enhancement even exceeds that of pure Au. The combination of Au and Ag is superior over pure Au or Ag for 500 to 600 nm or 700 to 800 nm fluorescent dyes.

8.5 Nanochip fabrication and characterization

As the typical LSPR sensors feature sub-wavelength nanostructures, electron-beam (e-beam) lithography is commonly used to fabricate nanochips. However, it is not

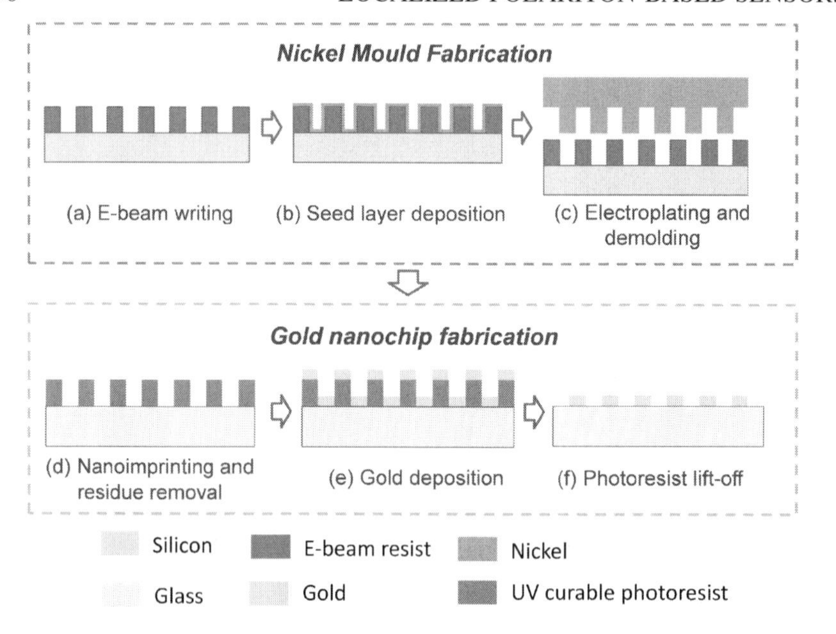

Figure 8.14: Fabrication processes of nickel mold [(a) through (c)] and gold nanohole array on a 4″ glass wafer [(d) through (f)]. Adapted with permission from Ref. [16]. © 2013 The Royal Society of Chemistry.

suitable for mass fabrication. Therefore, nanoimprinting is adopted [15]. Nanoimprinting of gold nanoholes or nanopillars array fabrication is presented in Fig. 8.14 [16]. Two major steps involved in nanostructure fabrication are nickel mold fabrication and nanoimprinting of the mold. This process can be used for mass production of gold nanostructures by arranging tens of nanochip designs on one wafer.

The fabricated gold nanochips were surface modified for biosensing. Gold nanostructures were used to enhance the fluorescent labels in a sandwich bioassay in the form of capturing antibody (cAb) biomarker detection antibody (dAb). The fluorescent labels were established on the surface of the gold nanoholes and nanopillars arrays with the steps illustrated in Fig. 8.15(a) [15].

In Fig. 8.15(a), the PSA antibody was cleft into a half (noted as cleft cAb) and immobilized onto a clean gold nanostructure surface [step (i)] via the gold-SH link commonly used in bioassays. Bovine serum albumin (BSA) was added to passivate the other binding sites to reduce the non-specific binding [step (ii)]. Then PSA in buffer was dripped on the gold nanostructures through which the PSA was captured by cleft cAb. Later the biotinylated PSA detection antibody (noted as biotinlyated dAb) was added followed by applying of the streptavidin conjugated quantum dot (QD) solution [step (iii)] so that the cleft cAb-PSA-biotinylated dAb-QD with streptavidin was established on the gold nanostructures. The AFM images in Figs. 8.15(b) and (c) show the nanopillar and nanohole array surfaces after each immobilization step. After the cleft cAb immobilization, the gold nanopillars were slightly enlarged,

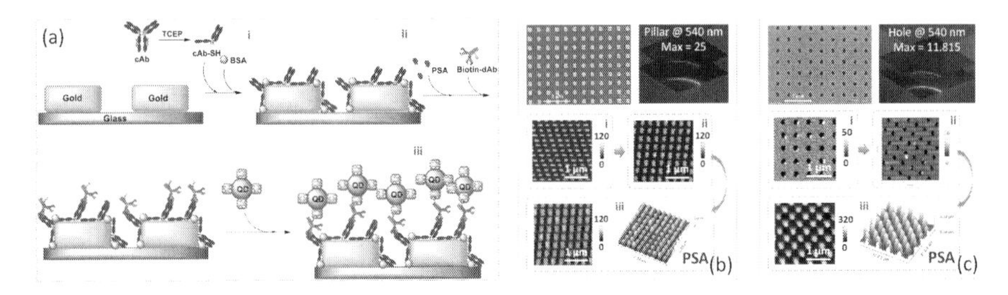

Figure 8.15: (a) The sandwich assay of cleft cAb-PSA-biotinylated dAb-QD with streptavidin established on the gold surface of a nanochip. (i), (ii), and (iii) are the steps showing the pristine gold nanochip surface, cleft cAb assembled on the gold surface, and the final sandwich assay, respectively. (b) and (c) SEM images of the 140 nm × 140 nm gold nanopillar (pitch is 320 nm) and nanohole (pitch is 400 nm) arrays, their near field simulations and AFM images of the gold surface after the immobilization steps (i), (ii), and (iii). AFM images for step (iii) taken with PSA concentration of 100 ng/ml. The metal layer is 5 nm of chromium and 50 nm of gold. Adapted with permission from Ref. [15]. © 2015 The Royal Society of Chemistry.

while the gold nanoholes slightly shrunk, indicating that a thin cleft cAb layer was immobilized on the surface. The AFM images show that for the gold nanopillar and nanohole arrays, the QD aggregation heights are around 40 and 240 nm, which are reasonable as the QDs are not at the heights of close-pack status yet.

The surface-modified nanochip was investigated under a Nikon Ti Eclipse fluorescent microscope illustrated in Fig. 8.16 [15]. For the gold nanopillar array, the bright- and dark-field images of the nanochips with sandwich assay established at various PSA concentrations are presented in Fig. 8.17(a). The bright-field images are similar for different PSA concentrations. However, in dark-field images with 100 ms integration, QD emission is still observed clearly for 5 ng/ml PSA, and the reference with 0 ng/ml PSA [blank in Fig. 8.17(a)] shows no specific binding-caused QD emission. The QD emission spectra in Fig. 8.17(b) present the same results as the images in the spectral form.

The QD emission spectra at shorter integration times of 20 and 50 ms were also taken for comparisons. Based on the peak count at the wavelength of 655 nm for QD-655 emission, the characterization curves for the PSA detection at different optical integration durations are plotted in Fig. 8.17(e). The Y-axis data were obtained by normalizing the counting of the QD peak emission for different PSA concentrations to 100 ng/ml. The characterization curves are nonlinear, due to the aggregation of the QDs in the bioassay and the polariton tweezers effect [17] more prominent for low PSA concentrations and long integration times. Based on the three-time noise level, the limits of detection (LOD) for the gold nanopillar array at 20 and 100 ms of integration are 100 pg/ml and 10 pg/ml, respectively. Longer integration time always increases the sensitivity for low fluorescent signal detection [15].

As the sandwich assay in the case of nanohole array is established on both the

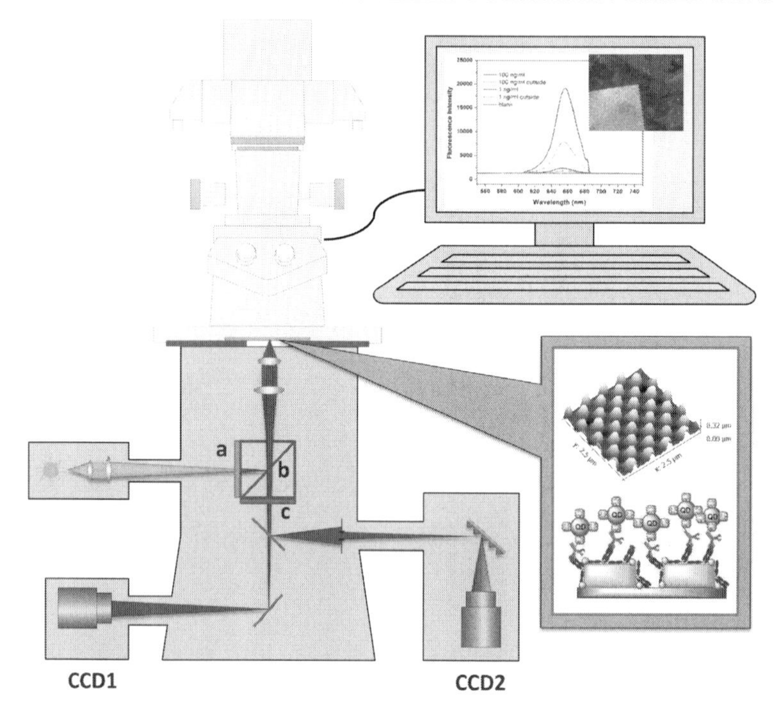

Figure 8.16: Microscope and spectrograph for nanochip detection. The inset square represents a beam splitter filter cube (a is a band-pass filter, b is a dichroic beam splitter, and c is a band-pass filter). CCD1 records the images, and CCD2 detects the spectrum of the light passing through an incoming slit 500 μm wide. Adapted with permission from Ref. [15]. © 2015 The Royal Society of Chemistry.

array and gold film surfaces, these two surfaces can be compared. The gold film surface provides strong fields to enhance the QD emission. However, as shown in Fig. 8.17(c), the areas with gold nanoholes are obviously much brighter than the gold film surfaces. The QD emission spectra in Fig. 8.17(d) for the gold film (denoted 'out') and the nanohole array (denoted 'in') indicate the same results as Fig. 8.17(c) in the spectral form. According to the three-time noise level of the characterization curves in Fig. 8.17(f) for the PSA detection with gold nanohole array, the LODs are 10 ng/ml and 1 ng/ml at 20 ms and 100 ms of optical integration, respectively.

The gold nanopillar array is found to be more sensitive for PSA detection, because its near field at the wavelength of 540 nm is two times higher than that of the nanohole array. Meanwhile, the nanopillar array has higher surface and volume coverage of QDs within the high-field area. Although the atomic force microscopy (AFM) images in Figs. 8.15(b) and (c) reveal that many more QDs are piled on the gold surface (they are piled to 40 nm and 240 nm in height for the gold nanopillar and nanohole arrays, respectively), most of them are far from the near field and not enhanced with polaritonic resonance. This indicates that polaritonic enhancement

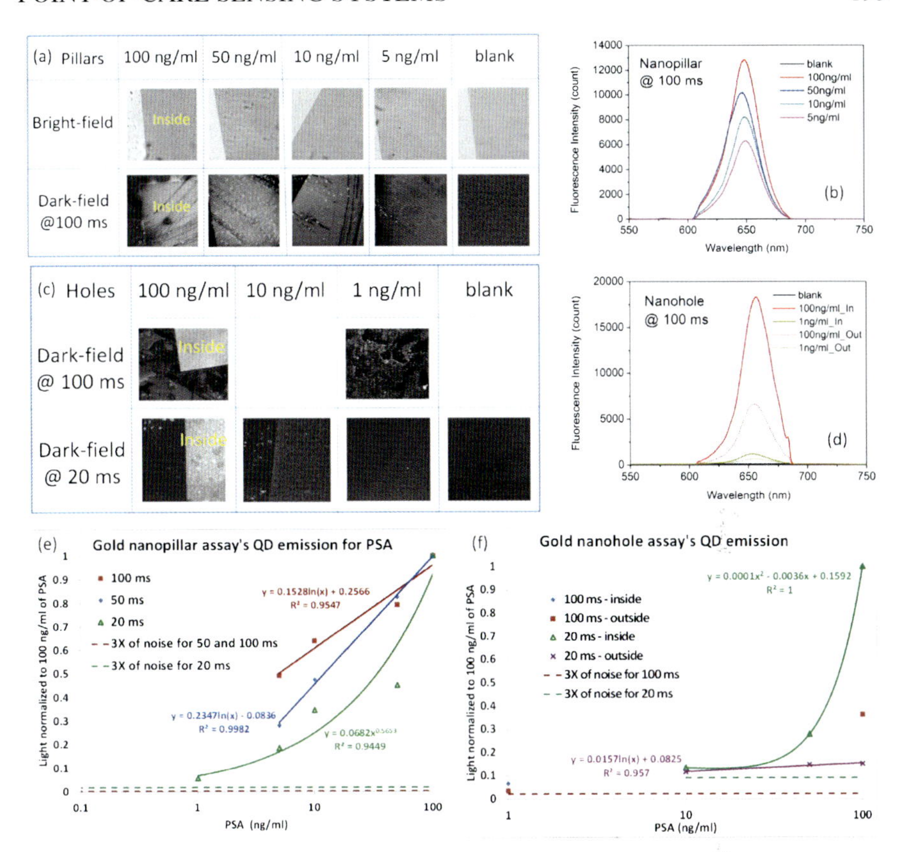

Figure 8.17: Bright- and dark-field images (right is the gold nanostructure area) of the QD assay for (a) gold nanopillar (the left in the images is glass) and (c) gold nanohole (the left in the images is gold film) arrays. Fluorescent spectra of QDs for (b) gold nanopillar and (d) nanohole arrays. Characterization curves for PSA detection with (e) gold nanopillar and (f) nanohole arrays. Optical integration time is shown in the figures. Blank is without PSA (i.e., 0 pg/ml of PSA). Reprinted with permission from Ref. [15]. © 2015 The Royal Society of Chemistry.

is the key factor to increase the sensitivity of the sandwich bioassay for biomarker detection.

8.6 Point-of-care sensing systems

Point-of-care (POC) sensing systems have attracted much attention due to their promise for medical detection. They also could greatly reduce the diagnostic delays due to the heavy detection loads in diagnostic laboratories. POC systems are supposed to achieve low cost and fast detection, and may be operated by untrained

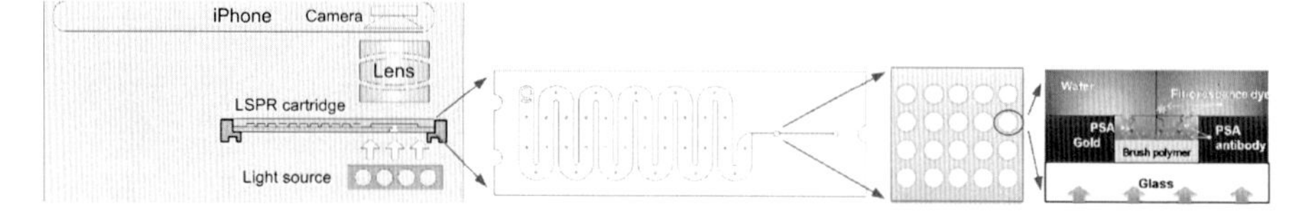

Figure 8.18: A scheme of a simple LSPR POC system including the microfluidic device, nanochip, and bioassay. Adapted with permission from Ref. [18]. © 2014 Springer.

users. An LSPR nanochip is a good candidate for achieving a POC system featuring high sensitivity, compact size, and low cost.

8.6.1 System configuration

An LSPR POC system can be configured with the components shown in Fig. 8.18. Light illuminates the LSPR cartridge where the microfluidic chip with the integrated nanochip is placed, and a camera detects the polariton-enhanced fluorescent signal from the bioassay for the quantitative detection of the analyte. Overall, the POC system is a cost-effective and small footprint substitution for a microscope. In this POC system, the major processes are the actuation of microfluidics, optical detection, and software control.

The microfluidic device in POC system is usually disposable to eliminate the cross-contamination of measurements for different patients and thus should be inexpensive. The microfluidic device is mostly fabricated from a polymer. It is desirable to have multiple microfluidic channels in the LSPR cartridge in which some channels could be used for reference and characterization to increase the accuracy of the detection; some could be used to test several biomarkers simultaneously for one specimen.

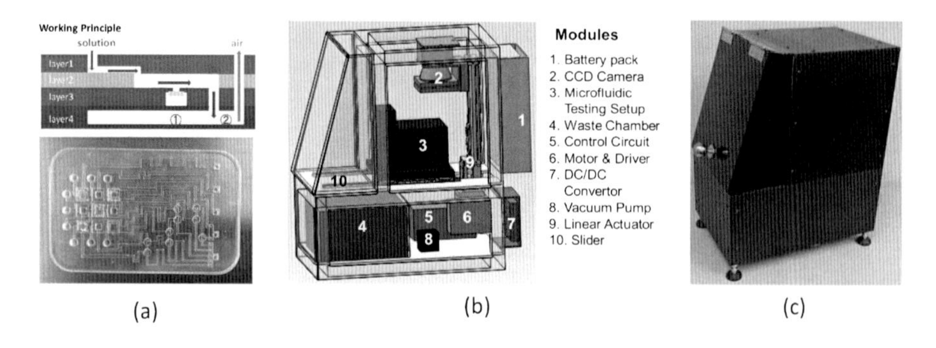

Figure 8.19: (a) The working principle (top) and fabricated nine-chambers microfluidic device (bottom). (b) The principle and (c) the fabricated POC based on a microscope camera for LSPR nanochip detection. Reprinted with permission from Ref. [18]. © 2014 Springer.

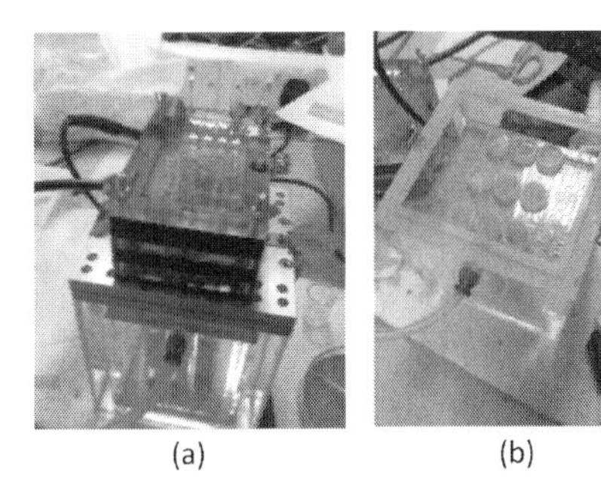

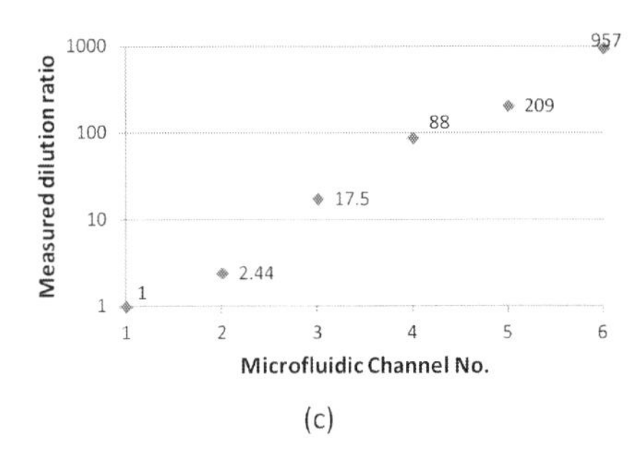

Figure 8.20: (a) Dilution and flow-rate control jig in the POC. (b) Diluted flow coming from a nine-chamber microfluidic device collected by nine test tubes (for test purposes only). (c) Dilution ratios of the six diluted channels. Reprinted with permission from Ref. [18]. © 2014 Springer.

An LSPR cartridge with nine-chambers (each chamber has a nanochip) microfluidic device was designed and fabricated, as shown in Fig. 8.19(a). The LSPR cartridge is of credit card size and has four layers of polymethyl methacrylate (PMMA) [18].

Figures 8.19(b) and (c) show a designed and fabricated POC system of size 200 mm (long) $\times$ 300 mm (width) $\times$ 350 mm (height) [18]. LEDs provide the necessary broadband light source for the fluorescent label excitation, while an optical filter is attached to select the excitation wavelength. A dedicated microscope camera with control software is integrated into the system for fluorescent gain control. The system can handle dilution of multiple channels. A large chamber collects the waste solution for long-time use without need for cleaning. A larger air pump is integrated to quickly draw and control the microfluidics during the sample dilution and bioassay establishment.

The POC system relies on the control software, adjustable mechanical system, and air pump to control analyte injection, dilution, bioassay establishment, calibration, and reading of the fluorescent counts. The system features a control panel in a laptop computer where results are processed and presented on the screen.

8.6.2 *System characterization*

The dilution and flow rate control jigs were tested for dilution effect in the POC, because the nine-chambers microfluidic device is meant to handle the specimen dilution. To conduct this calibration, the liquid coming from the channels of the nine-chambers microfluidic device [mounted on the dilution jig, as shown in Fig. 8.20(a)] is collected by nine plastic test tubes in the waste chamber of the POC [see Fig. 8.20(b)]. The amount of the liquid from each channel is measured to find the dilution rate in each channel; an exponential function is assumed for satisfactory sample dilution.

In the test, dilution of six channels is targeted. In real applications, the six diluted channels can either be diluted specimens (for the three leftover channels, two are

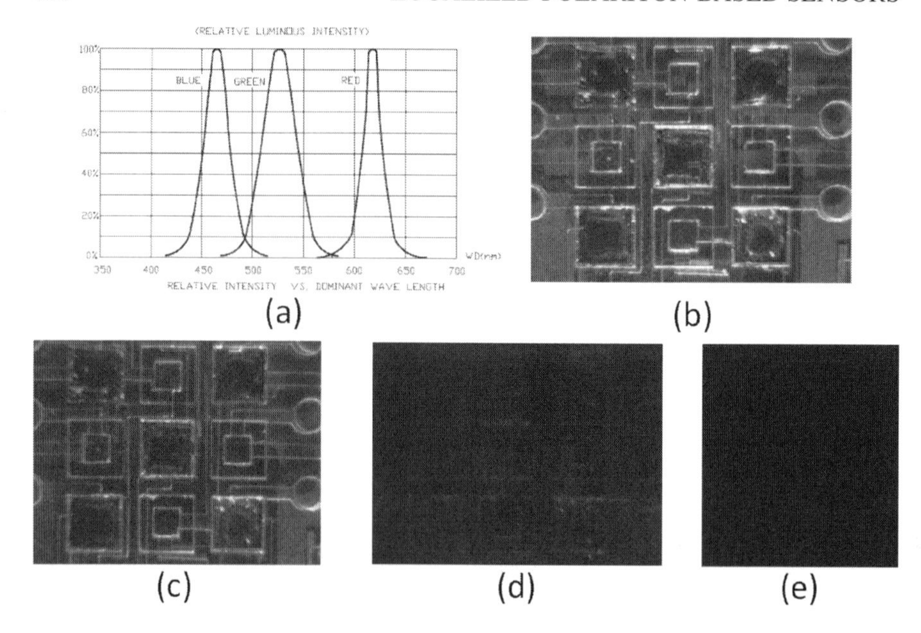

Figure 8.21: (a) The spectra of red, green, and blue light from three LEDs incorporated in the POC. (b) through (e) Images of the microfluidic device viewed from the top. (b) through (d) Images with the illuminating LED white light passed through (b) no filter, (c) 725 long pass filter, (d) 590 nm long pass filter. (e) Fluorescent image with 5.5 seconds signal integration. Reprinted with permission from Ref. [18]. © 2014 Springer.

used for reference and one is for background calibration) or be filled with diluted reference samples for calibration (for the three leftover channels, two are used for analyte and one is for background calibration). In Fig. 8.20(c), the dilution ratios for the six channels indicate that the dilution of the microfluidic device in the control jig has only small deviations from the designed exponential response.

The POC system can take images in the transmission mode (light illuminates from the bottom of the LSPR nanochip) or in the reflection mode (light sheds from the top of the nanochip). The images are analyzed to quantify the light intensity for deriving the biomarker concentrations.

For fluorescence dye excitation, the light source in the POC system is a combination of three LEDs of different wavelengths: red (619 to 624 nm), green (520 to 540 nm), and blue (460 to 480 nm), as presented in Fig. 8.21(a). Filters are integrated in the fluorescent excitation and emission system; the optical subsystem is collimated.

Figures 8.21(b) through (e) microscopic images without and with filters, as well as dark-field noise when the light sheds from the top of the chip. Figure 8.21(e) shows the signal noise saturation if the images are integrated at 5.5 seconds. If the exposure time is 0.3 second, the averaged signal can effectively reduce the noise.

8.7 Summary

This chapter introduced the design and operation principles for LSPR-based sensors. We explored two platforms for LSPR sensing elements based on nanoparticles and patterned films. Also, we investigated the effects of geometrical and material parameters on detection in different illumination schemes. Finally, we demonstrated a POC system enabling fluorescence enhancement for accurate sensing. This kind of highly sensitive LSPR biosensor shows great promise for replacing the bulky and expensive immunosystems in medical diagnostic laboratories and clinics. With the further development, LSPR biosensors may find ample applications in medical, chemical, agricultural, and environmental detections.

Bibliography

[1] J. Homola, S. S. Yee, and G. Gauglitz, "Surface plasmon resonance sensors: review," *Sensors and Actuators B: Chemical*, vol. 54, no. 12, pp. 3 – 15, 1999.

[2] L. J. Sherry, S.-H. Chang, G. C. Schatz, R. P. V. Duyne, B. J. Wiley, and Y. Xia, "Localized surface plasmon resonance spectroscopy of single silver nanocubes," *Nano Letters*, vol. 5, no. 10, pp. 2034–2038, 2005.

[3] J. N. Anker, W. P. Hall, O. Lyandres, N. C. Shah, J. Zhao, and R. P. Van Duyne, "Biosensing with plasmonic nanosensors," *Nature Materials*, vol. 7, no. 6, pp. 442–453, 2008.

[4] M. M. Miller and A. A. Lazarides, "Sensitivity of metal nanoparticle surface plasmon resonance to the dielectric environment," *The Journal of Physical Chemistry B*, vol. 109, no. 46, pp. 21556–21565, 2005.

[5] K.-S. Lee and M. A. El-Sayed, "Gold and silver nanoparticles in sensing and imaging: Sensitivity of plasmon response to size, shape, and metal composition," *The Journal of Physical Chemistry B*, vol. 110, no. 39, pp. 19220–19225, 2006.

[6] C. J. Murphy, T. K. Sau, A. M. Gole, C. J. Orendorff, J. Gao, L. Gou, S. E. Hunyadi, and T. Li, "Anisotropic metal nanoparticles: Synthesis, assembly, and optical applications," *The Journal of Physical Chemistry B*, vol. 109, no. 29, pp. 13857–13870, 2005.

[7] H. Kuwata, H. Tamaru, K. Esumi, and K. Miyano, "Resonant light scattering from metal nanoparticles: Practical analysis beyond rayleigh approximation," *Applied Physics Letters*, vol. 83, no. 22, pp. 4625–4627, 2003.

[8] P. Bai, L. Wu, and E. P. Li, "Patterning metallic films to enhance plasmonic modes for mrna detection," in *Nanoelectronics Conference (INEC), 2013 IEEE 5th International*, pp. 43–45, Jan 2013.

[9] S. Sun, L. Wu, P. Bai, and C. E. Png, "Fluorescence enhancement in visible light: dielectric or noble metal?," *Physical Chemistry Chemical Physics*, vol. 18, pp. 19324–19335, 2016.

[10] H. F. Ghaemi, T. Thio, D. E. Grupp, T. W. Ebbesen, and H. J. Lezec, "Surface

plasmons enhance optical transmission through subwavelength holes," *Physical Reviews B*, vol. 58, pp. 6779–6782, Sep 1998.

[11] L. Wu, P. Bai, and E. P. Li, "Designing surface plasmon resonance of subwavelength hole arrays by studying absorption," *Journal of the Optical Society of America B*, vol. 29, pp. 521–528, Apr 2012.

[12] L. Wu, P. Bai, X. Zhou, and E. P. Li, "Reflection and transmission modes in nanohole-array-based plasmonic sensors," *IEEE Photonics Journal*, vol. 4, pp. 26–33, Feb 2012.

[13] Y. Wang, L. Wu, X. Zhou, T. I. Wong, J. Zhang, P. Bai, E. P. Li, and B. Liedberg, "Incident-angle dependence of fluorescence enhancement and biomarker immunoassay on gold nanohole array," *Sensors and Actuators B: Chemical*, vol. 186, pp. 205 – 211, 2013.

[14] L. Wu, X. Zhou, and P. Bai, "Plasmonic metals for nanohole-array surface plasmon field-enhanced fluorescence spectroscopy biosensing," *Plasmonics*, vol. 9, no. 4, pp. 825–833, 2014.

[15] H. Y. Song, T. I. Wong, A. Sadovoy, L. Wu, P. Bai, J. Deng, S. Guo, Y. Wang, W. Knoll, and X. Zhou, "Imprinted gold 2d nanoarray for highly sensitive and convenient psa detection via plasmon excited quantum dots," *Lab Chip*, vol. 15, pp. 253–263, 2015.

[16] T. I. Wong, S. Han, L. Wu, Y. Wang, J. Deng, C. Y. L. Tan, P. Bai, Y. C. Loke, X. D. Yang, M. S. Tse, S. H. Ng, and X. Zhou, "High throughput and high yield nanofabrication of precisely designed gold nanohole arrays for fluorescence enhanced detection of biomarkers," *Lab Chip*, vol. 13, pp. 2405–2413, 2013.

[17] M. L. Juan, M. Righini, and R. Quidant, "Plasmon nano-optical tweezers," *Nature Photon*, vol. 5, pp. 349–356, Jun 2011.

[18] X. Zhou, T. I. Wong, H. Y. Song, L. Wu, Y. Wang, P. Bai, D.-H. Kim, S. H. Ng, M. S. Tse, and W. Knoll, "Development of localized surface plasmon resonance-based point-of-care system," *Plasmonics*, vol. 9, no. 4, pp. 835–844, 2014.

Chapter 9

Metasurfaces for flat optics

Zhengtong Liu
Institute of High Performance Computing, Singapore

In this chapter, we will discuss the use of resonant nanoparticles for creation of metasurfaces enabling efficient control of light at nanoscale. Section 9.1 will give a brief introduction to the field. Then, Section 9.2 will overview different types of metasurfaces, based on the type of used building blocks (meta-atoms). The chapter will conclude with summary in Section 9.3.

9.1 Introduction and background

9.1.1 History of metasurface development

The term *metasurface* refers to textured surfaces with sub-wavelength structures. The study of metasurfaces is a natural extension of research on metamaterials. The idea of metamaterials was first proposed by Pendry in 2000 [1]. In his seminal work, Pendry discussed a group of imaginary materials with both negative permittivity and permeability. He predicted that the refractive index of such materials would be negative, and a thin slab of such materials would act as a perfect lens that could focus both propagating and evanescent waves to create a perfect image of a source. But natural materials with both negative permittivity and permeability do not exist – in the optical ranges the magnetic responses of natural materials are practically nonexistent. To realize negative refractive index, metamaterials made of periodic artificial atoms (*meta-atoms*) – was proposed. Meta-atoms are sub-wavelength structures that have magnetic or electric (or both) resonances. Due to their sub-wavelength nature, meta-atoms' electromagnetic responses can be homogenized and the metamaterials can be treated as homogeneous entities instead of complex structures. The key features of meta-atoms are sub-wavelength and resonance.

The scientific community was excited by the new and exotic properties of metamaterials and the novel applications they promised. In the following years, research on metamaterials surged. Many designs of metamaterials were proposed, among which the split-ring resonator and fishnet structure are probably the most famous [2, 3]. The properties and applications of metamaterials have been extensively studied [4]. However, after about a decade of research researchers started to realize some intrinsic limitations of current metamaterial designs. Most metamaterials were made of metallic meta-atoms using silver (Ag), gold (Au), and aluminum (Al). These metals have high ohmic losses at optical frequencies, which are usually undesirable, and the devices using those designs had low efficiencies. Fabrication is another limiting factor. The optical wavelengths are sub-micron, so the size of an meta-atom can be no more than 200 to 300 nm, and the smallest feature is usually less than 50 nm. To fabricate such small structures, researchers had to resort to electron-beam lithography (EBL) or focused ion beam (FIB) lithography. Both EBL and FIB are expensive and time-consuming processes, and even with them it is not easy to produce consistent and good-quality samples. A necessary condition of a true metamaterial is that the design has to be three-dimensional, meaning that it has to have multiple layers of meta-atoms so the homogenization can be valid. A single layer of meta-atoms is already lossy and difficult to fabricate, and a multi-layer metamaterial can only make it worse, so very few examples have been demonstrated[1] [5].

To overcome those limitations, researchers explored various possible solutions. For example, Xiao and Shalaev tried to put optical gain material inside metamaterials to reduce the loss [6]. Boltsevar and Shalaev have explored alternatives to Ag and Au; they showed that TiN, AlZnO, and some other transition metal nitrides can be used to replace lossy Ag and Au [7, 8]. These materials can produce the necessary resonances while keeping the loss low. Although substantial progress has been made on those frontiers, three-dimensional (3D) metamaterials remain challenging, and optical metamaterials are still far from realizing their full potential.

Some researchers took a different approach – instead of trying to remove the difficulties in metamaterials, they tried to bypass them. Since multi-layer structures are difficult, they decided to use only a single layer of metamaterials, later called metasurface. However a metasurface is not just a single layer metamaterial. While each unit cell of a metasurface is still sub-wavelength, large-scale spatial variations have been introduced to realize rich functionalities, so metasurface cannot be treated as a homogeneous layer like metamaterials, and usually cannot be described by effective permittivity and permeability. This is obvious in the famous work by Yu and Capasso [9], who created a metasurface to change the refraction angle and proposed the generallized Snell's law. Although earlier works on metasurface existed [10–14], this work generated huge interests among researchers and the field of metasurface boomed.

9.1.2 Generalized Snell's law

One interesting function of metasurface is to control phase distribution of a light beam that may change the conventional reflection and refraction at interfaces. Snell's law relates the reflected and refracted light to the incident one, as shown in Fig. 9.1 [15]. At the interface between two media with refractive indices n_i and n_t, the law

[1] Note that the discussion here applies only to optical metamaterials. At microwave and radio frequencies the situation is different.

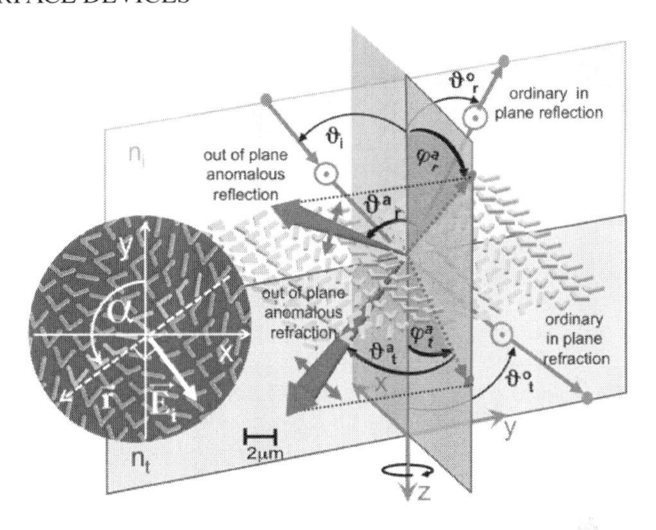

Figure 9.1: Generalized Snell's law at an interface with abrupt phase changes. Reprinted with permission from Ref. [15]. © 2012 American Chemical Society.

states that

$$\theta_r = \theta_i, \tag{9.1}$$

$$n_i \sin(\theta_i) = n_t \sin(\theta_t), \tag{9.2}$$

where θ_i is the incident angle, θ_r is the reflection angle, and θ_t is the refraction angle.

In the presence of a metasurface at the interface, an abrupt phase change is introduced and, if the phase change is space-variant, the conventional Snell's law does not hold. If the phase change has a constant gradient along a particular direction, Snell's law can be modified as:

$$n_t \sin(\theta_t) - n_i \sin(\theta_i) = \frac{\lambda_0}{2\pi} \frac{\mathrm{d}\Phi}{\mathrm{d}x}, \tag{9.3}$$

where λ_0 is the vacuum wavelength and $\mathrm{d}\Phi/\mathrm{d}x$ is the gradient of phase change. Similarly the refraction law is modified as:

$$\sin(\theta_r) - \sin(\theta_i) = \frac{\lambda_0}{2\pi n_i} \frac{\mathrm{d}\Phi}{\mathrm{d}x}. \tag{9.4}$$

Thus, through metasurface-induced phase change, we can control refraction and reflection, which are the two fundamental processes of light-matter interaction.

9.2 Metasurface devices

A large variety of metasurface have been proposed for different applications working at different frequencies. Here we focus on optical metasurface devices, sometimes called "flat optics" due to their planar geometries. In the following discussion, we classify metasurface devices according to the meta-atoms used. Once meta-atoms are chosen, many optical devices with different functionalities can be built.

9.2.1 Metasurfaces using rods as meta-atoms

One of the simplest meta-atoms is a dipole antenna, which in practice is usually realized as a rectangular rod. If the antenna is optically small ($l/\lambda \ll 1$, where l is the length of the antenna, and λ is the wavelength), its charge distribution instantaneously follows the incident field. Therefore, the incident and scattered fields are π out of phase. At antenna resonance ($l/\lambda \approx 1/2$), the incident field is in phase with the current at the center of the antenna, and the phase difference is $\pi/2$. For an antenna with the length equal to the wavelength ($l/\lambda = 1$), the antenna impedance is primarily inductive, so the scattered and incident fields are almost in phase. As the length of the antenna increases, the impedance changes from capacitive to resistive and then to inductive, and the phase shift changes from 0 to π [16]. In other words, a single rod provides only $[0-\pi]$ phase change for a linearly polarized beam, which does not meet the phase change requirement of $[0-2\pi]$ for general metasurfaces. To bypass this limitation, several designs have been proposed.

One way to realize $[0-2\pi]$ phase change with rods is to use circularly polarized incident waves [17–19]. Huang et al. showed that when a beam of circularly polarized light is incident onto a dipole antenna, the scattered wave is partially converted into the opposite handedness of circularly polarized light with a phase change determined solely by the orientation of the dipole [18]. Within a certain range of incident angle around the surface normal, a circularly polarized beam is primarily scattered into waves of the same polarization as that of the incident beam without phase change, and waves of the opposite circular polarization with a phase change $\Phi = 2\phi$ double the angle formed between the dipole and the x-axis [see Fig. 9.2(a)]. Since the ϕ may vary from 0 to π, Φ covers the whole $[0-2\pi]$ range. Using this concept, Zhang's group at the University of Birmingham realized optical vortex beams and a dual-polarity metalens for circularly polarized light [17, 18]. The dual-polarity metalens consists of gold nanorods with varying orientations along one horizontal direction, as shown in Fig. 9.2(a). Depending on the handedness of the incident wave (left-handed or right-handed circular polarization), the metalens can function as a convex lens or a concave lens (simulated light intensity shown in the figure). The optical vortex beams are realized with the structure shown in Fig. 9.2(b), where the measured optical intensity patterns at different wavelengths are demonstrated as well. The orientation of the rods and thus the abrupt phase change vary with the polar angle, creating an optical vortex. Using the same meta-atoms with coded phases, Zhang's group also demonstrated three-dimensional holograms [19].

Another way to realize $[0-2\pi]$ phase change is to use metallic rods with a reflector (usually a metallic ground plate) [20–23]. The metasurface and the ground plate form a Fabry-Perot cavity, and the incident waves undergo multiple reflections inside the cavity, so the accumulated phase change covers the whole $[0-2\pi]$ range. Such a device works in the reflection mode. This approach has been implemented

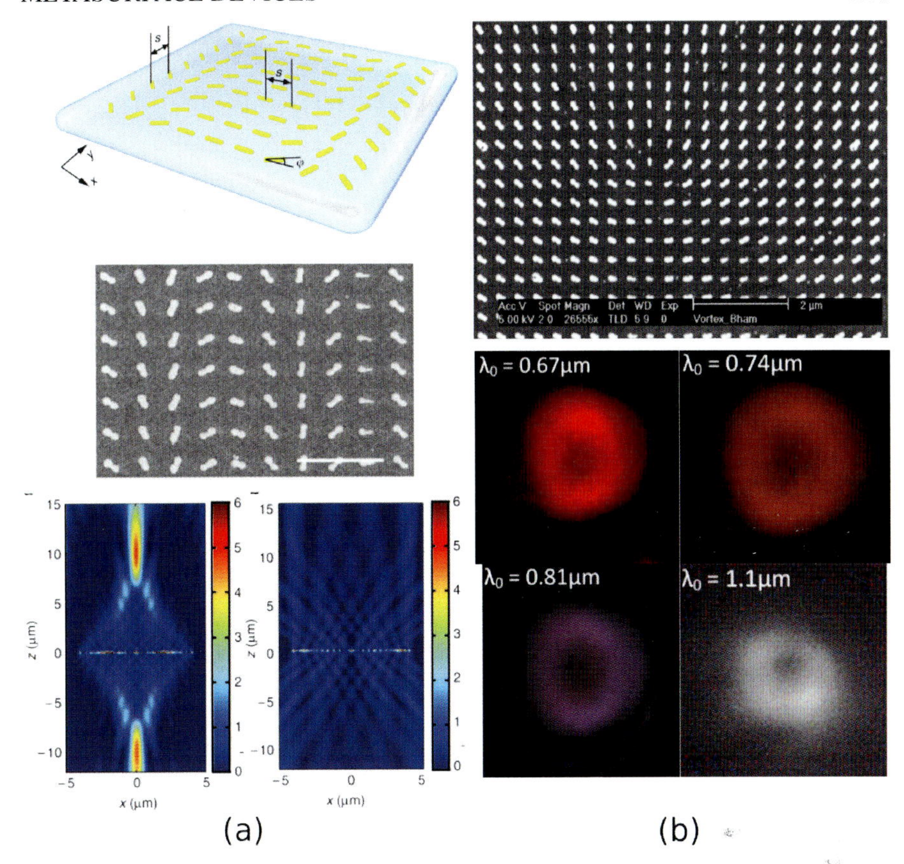

Figure 9.2: Metasurface using rods with circularly polarized light. (a) Metalens for circularly polarized light. The metasurface is a convex lens for right-handed circularly polarized light and a concave lens for left-handed circularly polarized light. Reprinted with permission from Ref. [17]. © 2012 Nature Publishing Group. (b) Metasurfaces made of nanorods generate optical vortex. Reprinted with permission from Ref. [18]. © 2012 American Chemical Society.

by several groups. Sun et al. used the design shown in Fig. 9.3(a) to create a surface phase-change gradient to realize anomalous reflection [20]. The phase gradient was created by changing the lengths of the rods. Because there are no transmitted waves, the efficiency of their device can reach 80%. Pors et al. further showed that if both the length and the width of the rods are adjusted, the reflections for two polarizations (TE and TM) can be tailored independently [see Fig. 9.3(b)] [21]. Jiang et al. discovered that if rods are placed close to one another, the electromagnetic near-field coupling between them can also be used; subsequently, half-wave plate and quarter-wave plate were created using the design shown in Fig. 9.3(c) [22]. To address the loss issue caused by localized polariton resonances in metallic structures, Yang et al. tried dielectric rods as building blocks [23]. They used silicon rods to make polariza-

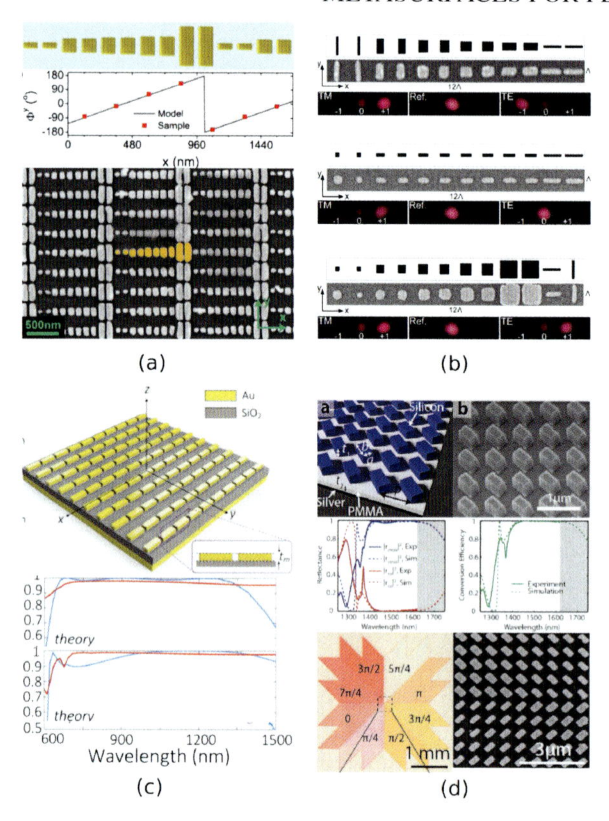

Figure 9.3: Metasurfaces using rods and reflectors. (a) Nanorod metasurface for anomalous refraction. Reprinted with permission from Ref. [20]. © 2012 American Chemical Society. (b) Both the length and width of the rods are tailored, so the metasurface works differently for the TE and TM polarizations. Reprinted with permission from Ref. [21]. © 2013 Nature Publishing Group. (c) Broadband half-wave plate and quarter-wave plate that make use of inter-rod coupling. Reprinted with permission from Ref. [22]. © 2014 Nature Publishing Group. (d) Metasurface for optical vortex using dielectric rods. Reprinted with permission from Ref. [23]. © 2014 American Chemical Society.

tion converters and an optical vortex and demonstrated that the efficiencies of those devices can approach 100% [see Fig. 9.3(d)].

9.2.2 *Metasurfaces using V-shaped meta-atoms*

Although metasurfaces using rods are simple to analyze and implement, their intrinsic limitation of $[0-\pi]$ phase change requires extra design considerations. If a rod can provide $[0-\pi]$ phase change, two rods together may give the full range of $[0-2\pi]$ phase change. Yu et al. have shown that if two rods are joined to form a V-shape, $[0-2\pi]$ phase change can be achieved [9]. The physics of the V-shape meta-atoms have been discussed in detail in Refs. [16, 24]. The V-shape meta-atoms consist of two arms of equal length l that support symmetric and anti-symmetric eigenmodes. The resonance on the symmetric mode is excited with electric fields parallel to the antenna symmetry axis, and the current distribution in each arm is same, similar to a single-rod antenna with first-order resonance occurring at $l \approx \lambda/2$. The resonance on anti-symmetric mode is excited by electric fields perpendicular to the antenna symmetry axis. In this case, the two arms act as one, and the equivalent length is $2l$, so the condition for the first-order resonance is $l \approx \lambda/4$. When the excitation electric field is parallel or perpendicular to the antenna symmetry axis, the scattered fields largely preserve the polarization of the excitation field. If the excitation electric field is not strictly parallel or perpendicular to the antenna symmetry axis, then both the symmetric and anti-symmetric resonances are excited, and the scattered fields represent a combination of the two states with tunable polarization. Usually the incident radiation is chosen to be polarized at 45° to the anteanna symmetry axis, and the output has polarization perpendicular to the incident wave. The remaining direct radiation after the structure can be filtered out by a crossed polarizer.

Using the V-shaped meta-atoms as building blocks, Yu et al. demonstrated anomalous refraction predicted by the generalized Snell's law using the design shown in Fig. 9.1 [15]. The experimentally measured refraction angles matched the theoretical prediction perfectly. Genevet et al. also created a metasurface using the V-shape meta-atoms to generate an optical vortex; the device and measured optical intensity are shown in Fig. 9.4(a) [25]. Using similar elements but different arrangements, they created a quarter-wave plate operating in a broad wavelength range (roughly from 6 to 10 μm), as shown in Fig. 9.4(b) [26]. Lens is one of the most common optical elements, and Aieta el al. used V-shaped meta-atoms to create flat convex lenses and axicons shown in Fig. 9.4(c) [27]. Convex lenses and axicons have similar requirements for phase change distribution, although the exact distribution profiles are different. By arranging the V-shaped meta-atoms according to the lens profiles, the predicted performances were achieved.

Ni et al. went one step further and encoded a hologram using V-shaped meta-atoms. To increase the signal-to-noise ratio (SNR), they used V-shaped grooves instead of V-shaped rods. According to Babinet's principle, the diffraction pattern from an opaque body is identical to its complementary aperture, except for the overall intensity, therefore the V-shaped grooves produce the same diffraction pattern (the hologram) as the V-shaped rods, but of different intensity. The advantage of the Babinet-inverted structure is that it effectively blocks the co-polarized light component from being transmitted. Ni et al. designed the hologram to show the word 'PURDUE' and fabricated the mask with V-shaped grooves in a 30 nm thick gold film by focused ion beam. At 10 μm above the metasurface, they successfully observed the image under a microscope, as shown in Fig. 9.4(d) [28].

Although metasurfaces using V-shaped meta-atoms have been successfully implemented to realize many optical elements, they suffer from a major drawback: limited efficiency. All the functionalities of those metasurfaces are realized by the gen-

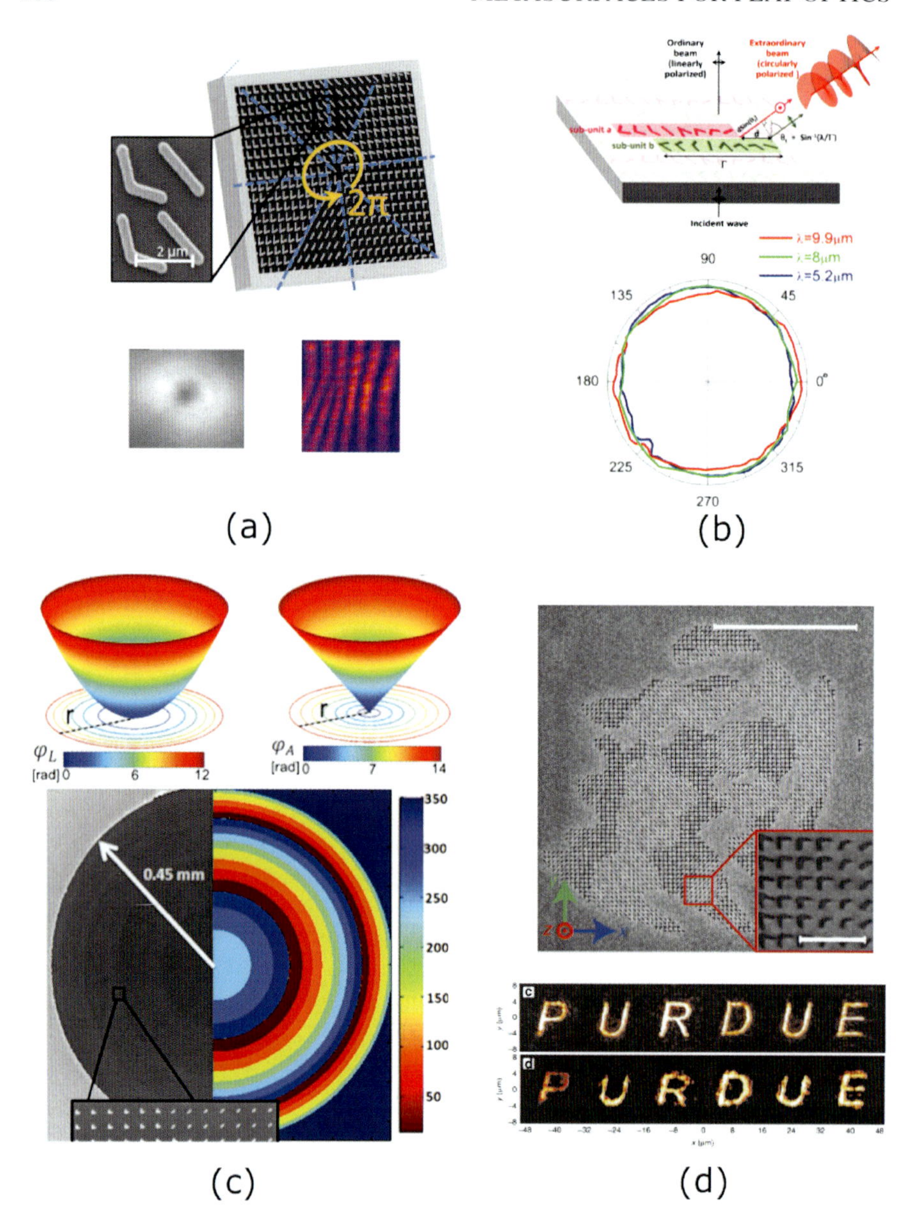

Figure 9.4: Metasurface using V-shaped meta-atoms. (a) Metasurface for anomalous refraction and optical vortex. Reprinted with permission from Ref. [25]. © 2012 AIP Publishing LLC. (b) Quarter-wave plate. Reprinted with permission from Ref. [26]. © 2012 American Chemical Society. (c) Lens and axicon. Reprinted with permission from Ref. [27]. © 2012 American Chemical Society. (d) Hologram (with V-shaped grooves). Reprinted with permission from Ref. [28]. © 2013 Nature Publishing Group.

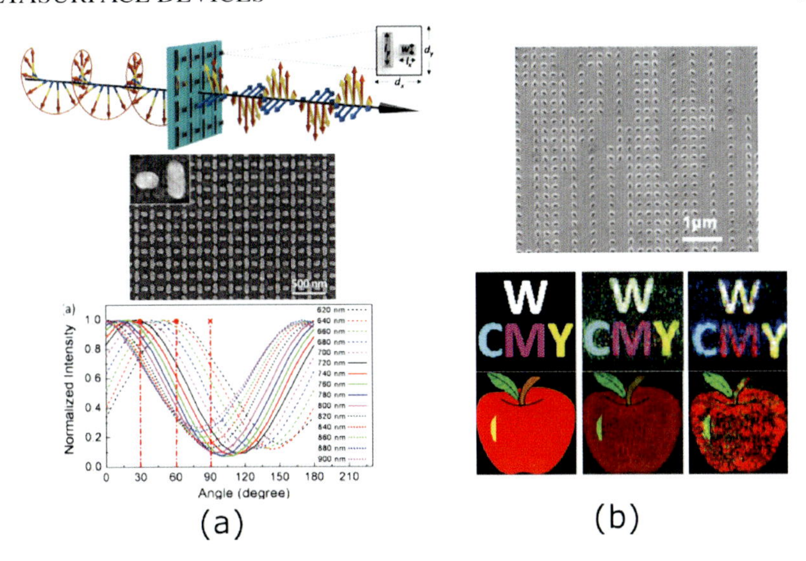

Figure 9.5: (a) A quarter-wave plate with two orthogonal rod arrays. Reprinted with permission from Ref. [31]. © 2013 American Chemical Society. (b) High resolution full-color holograms using nanoslits. Reprinted with permission from Ref. [33]. © 2016 American Chemical Society.

erated cross-polarized light. Monticone et al. demonstrated that for the device based on cross-polarization scheme, the theoretical limit of efficiency is only 25% [29].

9.2.3 Metasurfaces with other meta-atom shapes

Along with the rods and V-shaped meta-atoms, many other shapes and structures have been explored and proposed. Some use rods with different configurations. Zhang et al. used orthogonal arrays of rods (and their Babinet complementaries) to make quarter-wave plates [30, 31], as shown in Fig. 9.5(a). The lengths of the two orthogonal rods were chosen so that the scattered fields of the two rods have a phase difference of $\pi/2$. Therefore, when a circularly polarized light passes through the metasurface, the two orthogonal linear components experience a phase shift difference of $\pi/2$ and the output light will be linearly polarized. By tuning the dimensions of the rods, a quarter-wave plate worked in a broad wavelength range (600 – 900 nm), although at different wavelengths, the orientations of the output linearly polarized light differed due to the transmittance dispersion. They fabricated the sample using silver nanorods and demonstrated that the device worked and was broadband. A similar device was proposed by Khanikaev et al. [32].

Wan et al. used nanoslits (the Babinet complementary of nanorods) for full-color holograms [33]. By tuning the encoded phase shifts, the incident angles of light, and the NA of objective lens, full-color holographic images of high quality were demonstrated [see Fig. 9.5(b)].

Some researchers used more complicated shapes and configurations for metasurfaces. Memarzadeh et al. used loop antennae as meta-atoms to build a light concentrator shown in Fig. 9.6(a) [34]. Chen et al. used nanorods, nanocross, and nanodiscks to design a hologram mask (although the underlying principle is still based on nanorods), as illustrated in Fig. 9.6(b) [35]. Pfeiffer et al. proposed that multiple layers of metasurface can be cascaded, as shown in Fig. 9.6(c), to improve performance [36]. Apart from the electric resonances generated by the meta-atoms, the inter-layer coupling also supports magnetic resonances, providing additional tunability. Because the dielectric spacers between layers are sub-wavelength, the whole structure can still be considered a thin metasurface. Following this idea, a three-layer polarization converter was designed and fabricated, as shown in Fig. 9.6(d) [37]. Three layers of metasurface were fabricated with 28 nm thick gold and 200 nm thick SU-8 dielectric layers were deposited between the gold layers. This three-layer metasurface converted left-handed circularly polarized light into right-handed circularly polarized light with 50% efficiency.

9.2.4 *Material selections for metasurfaces*

The aforementioned metasurface devices were designed based on excitation of polaritonic resonances in metal structures. At optical frequencies, they are accompanied by strong ohmic losses that reduce the overall efficiency and impair device performance. Therefore, there is a strong motivation to eliminate metal parts and use resonant dielectric structures with low optical losses.

Recent studies showed that dielectric nanoparticles support both electric and magnetic resonances, and under certain conditions they can overlap [38, 39]. Decker et al. studied silicon nanodisks and found out that they support electric dipole resonance (on quasi-TM polaritons) and magnetic dipole resonance (on quasi-TE polaritons) at the same time [38]. When both resonances are present, the nanodisks can induce phase change covering the whole $[0-2\pi]$ range, while keeping the transmittance almost constant at 100%, which is a very desirable feature for metasurface design. Yu et al. have done similar work and reached the same conclusion [39]. They also designed and fabricated a metasurface composed of silicon nanodisks of various sizes. This metasurface demonstrated anomalous diffraction predicted by the generalized Snell's law.

While size change alone produces phase shift of $[0-2\pi]$, shapes and orientations of nanodisks may also be adjusted to provide additional tunability. Arbabi et al. used elliptical silicon nanodisks as meta-atoms and demonstrated that tuning of the size, shape, and orientation of the nanodisks gives complete control over the scattered light [40], as shown in Fig. 9.7. Using this platform, polarization beam splitters, polarization-dependent holograms, and optical vortices were demonstrated.

Nanodisks are not the only choices for designing metasurface. Lin et al. used silicon nanobeams as metasurface building blocks [41]. Like metallic nanorods, the nanobeams provide only $[0-\pi]$ phase change, so Lin et al. had to use circularly polarized light as input. With the silicon nanobeams, they demonstrated planar convex lenses and axicons.

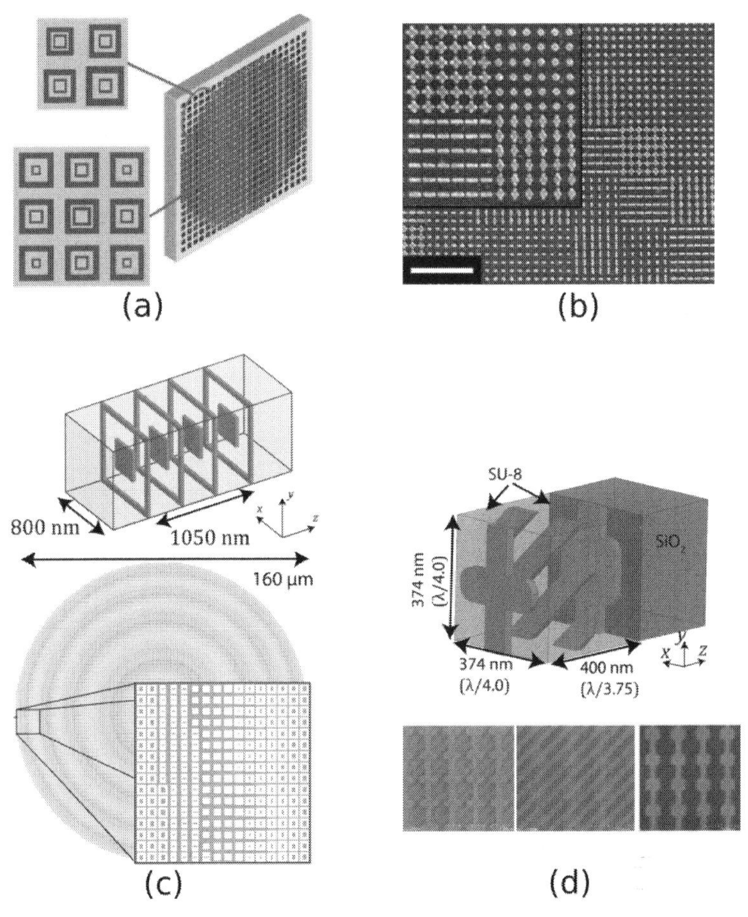

Figure 9.6: (a) Light concentrator build with loop antennas. Reprinted with permission from Ref. [34]. © 2011 The Optical Society. (b) A hologram mask built with nanorods, nanocrosses, and nanodisks. Reprinted with permission from Ref. [35]. © 2014 American Chemical Society. (c) Multilayer metasurface lens. Reprinted with permission from Ref. [36]. © 2013 AIP Publishing LLC. (d) A three-layer polarization converter. Reprinted with permission from Ref. [37]. © 2014 American Physical Society.

9.3 Summary

Although the field of optical metasurfaces is new, tremendous progress has been made in both theory and experimentation. Some of the metasurface devices such as the all-dielectric lens and axicon already have achieved performances close to those of conventional optical elements. The unique advantages of metasurfaces such as the sub-wavelength thickness and great design freedom make them very attractive in applications like space missions and integrated optics.

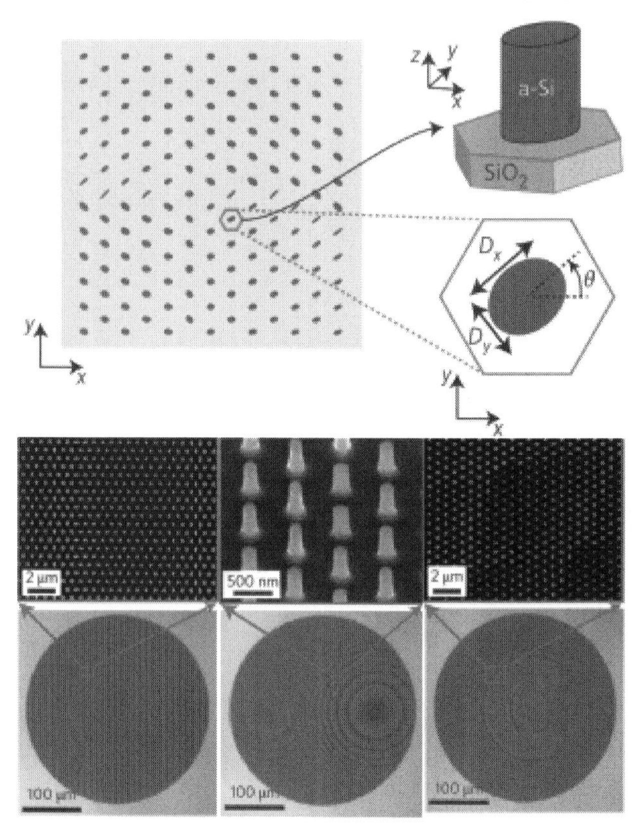

Figure 9.7: Dielectric metasurfaces with elliptical nanodisks give full control over light amplitude and phase. They are used to make beam splitters, polarization-dependent holograms, and optical vortex. Reprinted with permission from Ref. [40] © 2015 Nature Publishing Group.

A wide range of materials and shapes have been explored as the building blocks (meta-atoms) for metasurfaces. Metallic rods and V-shapes have been systematically studied, and various devices based on them have been realized. However, the metallic metasurfaces generally suffer from high dissipative loss, and thus demonstrate relatively low efficiencies. Dielectric metasurfaces have been explored only very recently, but have already demonstrated good performances. Among them the elliptical nanodisks are particularly promising as they provide great freedom for design while retaining high efficiency.

The functionalities of metasurfaces exceed those of the traditional optics. For example, active and flexible metasurfaces are being explored. Metasurfaces for sensing are also of interests to many scholars. In the future, we expect to see many more practical applications, as the full potential of metasurfaces for light control has not been realized yet.

Bibliography

[1] J. B. Pendry, "Negative refraction makes a perfect lens," *Physical Review Letters*, vol. 85, no. 18, pp. 3966–3969, 2000.

[2] R. A. Shelby, D. R. Smith, and S. Schultz, "Experimental verification of a negative index of refraction," *Science*, vol. 292, no. 5514, pp. 77–79, 2001.

[3] S. Zhang, W. Fan, K. J. Malloy, S. J. Brueck, N. C. Panoiu, and R. M. Osgood, "Near-infrared double negative metamaterials," *Optics Express*, vol. 13, no. 13, p. 4922, 2005.

[4] V. M. Shalaev, "Optical negative-index metamaterials," *Nature Photonics*, vol. 1, no. 1, pp. 41–48, 2007.

[5] C. M. Soukoulis and M. Wegener, "Past achievements and future challenges in the development of three-dimensional photonic metamaterials," *Nature Photonics*, vol. 5, pp. 1–8, 2011.

[6] S. Xiao, V. P. Drachev, A. V. Kildishev, X. Ni, U. K. Chettiar, H.-k. Yuan, and V. M. Shalaev, "Loss-free and active optical negative-index metamaterials," *Nature*, vol. 466, no. 7307, pp. 735–738, 2010.

[7] U. Guler, V. M. Shalaev, and A. Boltasseva, "Nanoparticle plasmonics: going practical with transition metal nitrides," *Materials Today*, vol. 18, no. 4, pp. 227–237, 2015.

[8] N. Kinsey, M. Ferrera, V. M. Shalaev, and A. Boltasseva, "Examining nanophotonics for integrated hybrid systems: a review of plasmonic interconnects and modulators using traditional and alternative materials," *Journal of the Optical Society of America B*, vol. 32, no. 1, p. 121, 2015.

[9] N. Yu, P. Genevet, M. A. Kats, F. Aieta, J.-P. Tetienne, F. Capasso, and Z. Gaburro, "Light propagation with phase discontinuities: generalized laws of reflection and refraction," *Science*, vol. 334, no. 6054, pp. 333–337, 2011.

[10] Z. Bomzon, V. Kleiner, and E. Hasman, "Pancharatnam-Berry phase in space-variant polarization-state manipulations with subwavelength gratings," *Optics Letters*, vol. 26, no. 18, p. 1424, 2001.

[11] Z. Bomzon, V. Kleiner, and E. Hasman, "Formation of radially and azimuthally polarized light using space-variant subwavelength metal stripe gratings," *Applied Physics Letters*, vol. 79, no. 11, p. 1587, 2001.

[12] G. Biener, A. Niv, V. Kleiner, and E. Hasman, "Formation of helical beams by use of Pancharatnam-Berry phase optical elements," *Optics Letters*, vol. 27, no. 21, p. 1875, 2002.

[13] E. Hasman, V. Kleiner, G. Biener, and A. Niv, "Polarization dependent focusing lens by use of quantized Pancharatnam-Berry phase diffractive optics," *Applied*

Physics Letters, vol. 82, no. 3, p. 328, 2003.

[14] A. Niv, Y. Gorodetski, V. Kleiner, and E. Hasman, "Topological spin-orbit interaction of light in anisotropic inhomogeneous subwavelength structures," *Optics Letters*, vol. 33, no. 24, p. 2910, 2008.

[15] F. Aieta, P. Genevet, N. Yu, M. A. Kats, Z. Gaburro, and F. Capasso, "Out-of-plane reflection and refraction of light by anisotropic optical antenna metasurfaces with phase discontinuities," *Nano Letters*, vol. 12, no. 3, pp. 1702–1706, 2012.

[16] Nanfang Yu, P. Genevet, F. Aieta, M. A. Kats, R. Blanchard, G. Aoust, J.-P. Tetienne, Z. Gaburro, and F. Capasso, "Flat optics: controlling wavefronts with optical antenna Mmetasurfaces," *IEEE Journal of Selected Topics in Quantum Electronics*, vol. 19, no. 3, pp. 4700423–4700423, 2013.

[17] X. Chen, L. Huang, H. Mühlenbernd, G. Li, B. Bai, Q. Tan, G. Jin, C.-W. Qiu, S. Zhang, and T. Zentgraf, "Dual-polarity plasmonic metalens for visible light," *Nature Communications*, vol. 3, p. 1198, 2012.

[18] L. Huang, X. Chen, H. Mühlenbernd, G. Li, B. Bai, Q. Tan, G. Jin, T. Zentgraf, and S. Zhang, "Dispersionless phase discontinuities for controlling light propagation," *Nano Letters*, vol. 12, no. 11, pp. 5750–5755, 2012.

[19] L. Huang, X. Chen, H. Mühlenbernd, H. Zhang, S. Chen, B. Bai, Q. Tan, G. Jin, K.-W. Cheah, C.-W. Qiu, J. Li, T. Zentgraf, and S. Zhang, "Three-dimensional optical holography using a plasmonic metasurface," *Nature Communications*, vol. 4, p. 2808, 2013.

[20] S. Sun, K.-Y. Yang, C.-M. Wang, T.-K. Juan, W. T. Chen, C. Y. Liao, Q. He, S. Xiao, W.-T. Kung, G.-Y. Guo, L. Zhou, and D. P. Tsai, "High-efficiency broadband anomalous reflection by gradient meta-surfaces," *Nano Letters*, vol. 12, no. 12, pp. 6223–6229, 2012.

[21] A. Pors, O. Albrektsen, I. P. Radko, and S. I. Bozhevolnyi, "Gap plasmon-based metasurfaces for total control of reflected light," *Scientific Reports*, vol. 3, p. 2155, 2013.

[22] Z. H. Jiang, L. Lin, D. Ma, S. Yun, D. H. Werner, Z. Liu, and T. S. Mayer, "Broadband and wide field-of-view plasmonic metasurface-enabled waveplates," *Scientific Reports*, vol. 4, p. 7511, 2014.

[23] Y. Yang, W. Wang, P. Moitra, I. I. Kravchenko, D. P. Briggs, and J. Valentine, "Dielectric meta-reflectarray for broadband linear polarization conversion and optical vortex generation," *Nano Letters*, vol. 14, no. 3, pp. 1394–1399, 2014.

[24] R. Blanchard, G. Aoust, P. Genevet, N. Yu, M. A. Kats, Z. Gaburro, and F. Capasso, "Modeling nanoscale V-shaped antennas for the design of optical phased arrays," *Physical Review B*, vol. 85, no. 15, p. 155457, 2012.

[25] P. Genevet, N. Yu, F. Aieta, J. Lin, M. A. Kats, R. Blanchard, M. O. Scully, Z. Gaburro, and F. Capasso, "Ultra-thin plasmonic optical vortex plate based on phase discontinuities," *Applied Physics Letters*, vol. 100, no. 1, p. 013101, 2012.

[26] N. F. Yu, F. Aieta, P. Genevet, M. A. Kats, Z. Gaburro, and F. Capasso, "A broadband, background-free quarter-wave plate based on plasmonic metasurfaces," *Nano Letters*, vol. 12, no. 12, pp. 6328–6333, 2012.

[27] F. Aieta, P. Genevet, M. A. Kats, N. Yu, R. Blanchard, Z. Gaburro, and F. Capasso, "Aberration-free ultrathin flat lenses and axicons at telecom wavelengths based on plasmonic metasurfaces," *Nano Letters*, vol. 12, no. 9, pp. 4932–4936, 2012.

[28] X. Ni, A. V. Kildishev, and V. M. Shalaev, "Metasurface holograms for visible light," *Nature Communications*, vol. 4, p. 2807, 2013.

[29] F. Monticone, N. M. Estakhri, and A. Alù, "Full control of nanoscale optical transmission with a composite metascreen," *Physical Review Letters*, vol. 110, no. 20, p. 203903, 2013.

[30] Y. Zhao and A. Alù, "Manipulating light polarization with ultrathin plasmonic metasurfaces," *Physical Review B*, vol. 84, no. 20, p. 205428, 2011.

[31] Y. Zhao and A. Alù, "Tailoring the dispersion of plasmonic nanorods to realize broadband optical meta-waveplates," *Nano Letters*, vol. 13, no. 3, pp. 1086–1091, 2013.

[32] A. B. Khanikaev, S. H. Mousavi, C. Wu, N. Dabidian, K. B. Alici, and G. Shvets, "Electromagnetically induced polarization conversion," *Optics Communications*, vol. 285, no. 16, pp. 3423–3427, 2012.

[33] W. Wan, J. Gao, and X. Yang, "Full-color plasmonic metasurface holograms," *ACS Nano*, vol. 10, no. 12, pp. 10671–10680, 2016.

[34] B. Memarzadeh and H. Mosallaei, "Array of planar plasmonic scatterers functioning as light concentrator," *Optics Letters*, vol. 36, no. 13, p. 2569, 2011.

[35] W. T. Chen, K. Y. Yang, C. M. Wang, Y. W. Huang, G. Sun, I. D. Chiang, C. Y. Liao, W. L. Hsu, H. T. Lin, S. Sun, L. Zhou, A. Q. Liu, and D. P. Tsai, "High-efficiency broadband meta-hologram with polarization-controlled dual images," *Nano Letters*, vol. 14, no. 1, pp. 225–230, 2014.

[36] C. Pfeiffer and A. Grbic, "Cascaded metasurfaces for complete phase and polarization control," *Applied Physics Letters*, vol. 102, no. 23, p. 231116, 2013.

[37] C. Pfeiffer, C. Zhang, V. Ray, L. J. Guo, and A. Grbic, "High performance bianisotropic metasurfaces: asymmetric transmission of light," *Physical Review Letters*, vol. 113, no. 2, p. 023902, 2014.

[38] M. Decker, I. Staude, M. Falkner, J. Dominguez, D. N. Neshev, I. Brener, T. Pertsch, and Y. S. Kivshar, "High-efficiency dielectric Huygens' surfaces," *Advanced Optical Materials*, vol. 3, no. 6, pp. 813–820, 2015.

[39] Y. F. Yu, A. Y. Zhu, R. Paniagua-Dominguez, Y. H. Fu, B. Luk'yanchuk, and A. I. Kuznetsov, "High-transmission dielectric metasurface with 2π phase control at visible wavelengths," *Laser & Photonics Reviews*, vol. 9, no. 4, pp. 412–418, 2015.

[40] A. Arbabi, Y. Horie, M. Bagheri, and A. Faraon, "Dielectric metasurfaces for

complete control of phase and polarization with subwavelength spatial resolution and high transmission," *Nature Nanotechnology*, vol. 10, no. 11, pp. 937–943, 2015.

[41] D. Lin, P. Fan, E. Hasman, and M. L. Brongersma, "Dielectric gradient metasurface optical elements," *Science*, vol. 345, no. 6194, pp. 298–302, 2014.

Part III

Applications of propagating eigenmodes

Chapter 10

Guiding light with resonant nanoparticles

Hong-Son Chu
Institute of High Performance Computing, Singapore

Thomas Y.L. Ang
Institute of High Performance Computing, Singapore

This chapter will discuss the use of polariton coupling in chains of nanoparticles (NPs) for resonant transmission of light for next-generation integrated circuits. We begin by exploring inter-particle coupling of simple two-dimensional structures to get physical insight into light transmission with resonantly coupled plasmonic particles. in Section 10.2, we will look into more complex three-dimensional particles on the silicon-on-insulator platform for on-chip integration. In Sections 10.3 and 10.4, we will study couplers and nanoparticle chain bends, respectively. The chapter will conclude with a summary in Section 10.5.

10.1 Introduction and background

Confinement and guiding of light with inter-particle resonant coupling have generated enormous interest in the optical research community owing to their unique optical phenomena. In particular, metal chain waveguides were shown be able to guide light beyond the diffraction limit [1, 2]. The optical antennas based on a single chain of NPs were exploited for ultra-sensitive chemical and biomolecular detection [3] and for fluorescence enhancement [4]. However, a great numbert of studies focused on the fundamental ability of waveguides based on nanostructures to guide optical waves at the diffraction limit and below. The polaritons of single nanostructures such as nanowires or nanoparticles have been studied in detail [5, 6]. They have been shown to depend strongly on the nanostructure's geometrical parameters, material properties, and the surrounding medium [7, 8]. However, when the nanostructures form a chain system, the polariton interaction between the individual parts create new hybridized polariton states enabling resonant energy transfer within the chain [5, 6, 9, 10]. This phenomenon makes the optical properties of coupled nanoparticle systems considerably different from those of their individual parts.

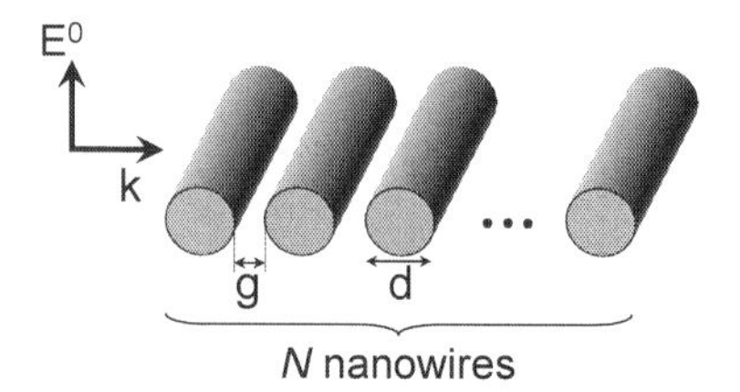

Figure 10.1: Schematic illustration of the single-chain array with N nanowires where $N = 5$, 10, 50, and 100. The diameter d of each nanowire is 50 nm; the gap distance g between the adjacent nanowires is 5 nm.

10.1.1 Guiding light with coupled nanoparticles

In order to explain the resonant light transmission occurring in inter-particle-coupled chains, we consider a two-dimensional case of coupled metal nanowires, that allows light to propagate across them. This is the simplest case that allows us to understand main effects of resonant inter-particle coupling on light transmission: the two-dimensional case simplifies the description, while the choice of metal for the nanowire material reduces the number of observed polaritonic resonances to electric ones. More complicated cases of dielectric nanoparticle chains with three-dimensional coupling and both electric and magnetic responses will be discussed in Sections. 10.2 through 10.4 for on-chip applications.

In this study, we focus on the investigation of the optical transmission along a chain of vertically aligned silver nanowires as a function of the chain length. We will show that chain length is one of the main parameters that define the polariton inter-particle coupling. To characterize the coupling, we will use surface charge distributions with near and far fields generated under illumination of the nanowire chain. The field and charge distributions are computed with the scattering matrix method (SMM) [11, 12], which is an efficient and attractive approach that minimizes computational resources required for accurate calculation of large coupled systems.

The single-chain system under consideration consists of N vertically aligned cylindrical silver nanowires, as depicted in Fig. 10.1, where N values are 5, 10, 50, and 100. Throughout this study, the nanowires are assumed to be of $d = 50$ nm diameter with the distance between two adjacent cylinder surfaces of $g = 5$. Note that g is chosen to be small enough so that polariton couplings between the nanowires are substantial. The chain is illuminated with a plane wave propagating in the k direction with the electric field E_0 polarized in the same plane and perpendicular to the nanowire chain. The material properties of silver used in the computation are obtained from experimental data [13] showing permittivity is a function of wavelength.

The evolution of the scattering cross-section (SCS) of a finite coupled silver nanowire chain as a function of the number N of nanowires is computed. The SCS data are normalized to their maximum values. Figure 10.2(a) illustrates the multiple polariton resonances. Note that the main resonance is always located on the right side

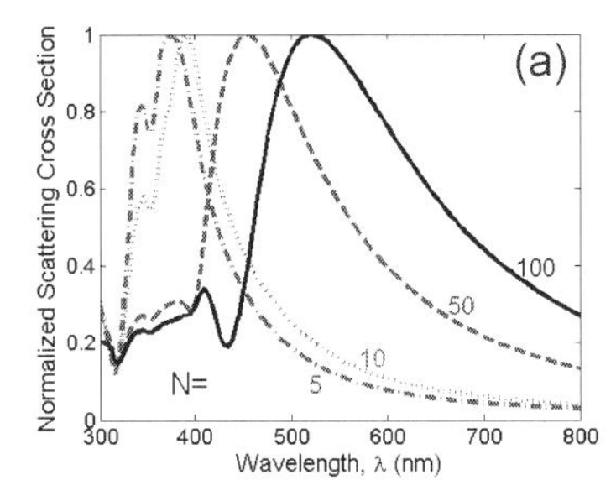

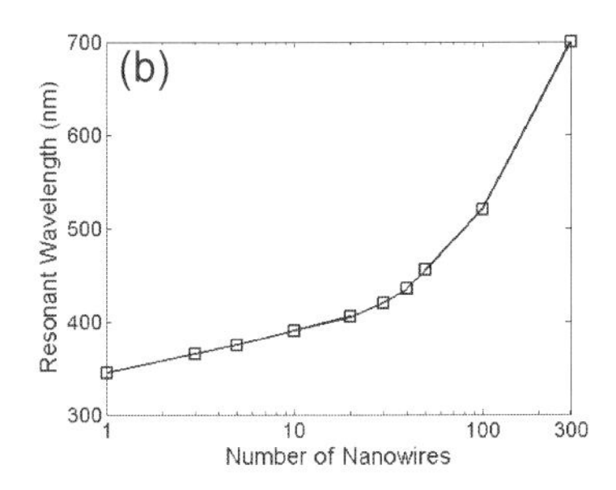

Figure 10.2: Evolution of the normalized scattering cross section of a single chain as a function of wavelength λ for different numbers of silver nanowires N in the chain. Reprinted with permission from Reference [2].

of the secondary resonances. The main resonance is attributed to the global coupling in the entire chain. It can be seen that a chain length increase causes a larger shift toward higher wavelengths, implying that the nanowire interaction extends beyond the first nearest neighbor. On the other hand, the secondary resonances are attributed to polaritons of a single nanowire mutually coupled to its close neighbors. The principal resonance broadens and redshifts from violet to blue when N increases.

Also, we can observe that depending on the number of the coupled nanowires, the frequency-dependent SCS displays two or more singularities. These singularities correspond to the polariton resonances of the nanowire chain, whose polarization characteristics influenced by the chain length. The main polariton resonance wavelength is plotted in Fig. 10.2(b) as a function of N. It shows that the main resonance monotonically increases the wavelength with the number of nanowires included in the chain. An increment in N affects the surface charge density distribution and also increases SCS of the chain, leading to a shift of the main resonance.

In Figs. 10.3(a) and (b), we plot the near-field intensity profile for the chains with $N = 5$ (a short-length chain) and with 100 nanowires (a long-length chain) corresponding to their respective main resonances and normalized to the incident field E_0. The field is strongest in the gap between two adjacent nanowires for the short chain, i.e., $N = 5$ nanowires, which corresponds to the dominant mode with the polarization along the chain axis. However, when N increases to 100 nanowires, the resonant mode with the polarization perpendicular to the chain axis becomes dominant with the field intensity locally concentrated at the top and bottom halves of the nanowires perpendicular to the direction of the wave propagation, as shown in Fig. 10.3(b). In addition, unexpected squeezing of the optical near field due to the eigenmode coupling is obtained in chains of different lengths.

The difference in the scattering and near fields for different chain lengths can be understood with an illustration of the surface charge distribution. Figures 10.3(c) and (d) represent the surface charge profiles (obtained with the electric field divergence at the main resonance) along the nanowire chains. Only the last five nanowires are plotted in 10.3(d). In Fig. 10.3(c) with $N = 5$, both dipole-like and quadrupole-like

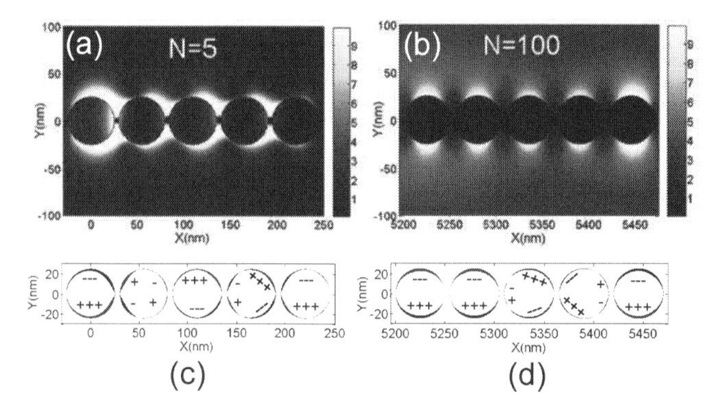

Figure 10.3: Normalized E-field intensity map [(a) and (b)] and their corresponding surface charge distributions [(c) and (d)], at the last five nanowires, observed at the polariton resonances of a single silver chain with $N = 5$ and $N = 100$, respectively. The $\pm$ signs denote the positive and negative surface charges. Reprinted with permission from Reference [2].

polarization surface charge distributions are observed and the opposite-signed surface charges are concentrated at the surfaces of the nanowires near the gap. From Fig 10.3(d) with $N = 100$, strong dipole-like surface charge polarizations are clearly observed. Contrary to the case of $N = 5$, the surface charge concentration is very weak in the gap and strong at the tops and bottoms of the nanowires. In addition, the negative surface charges generally concentrate on the nanowire surfaces in the upper halves of the cylinders, whereas the positive surface charges are found on the lower halves. As surface charges of the same sign repel each other, the coupling cannot act constructively, and we do not observe the same coupling in the gap, as compared to the case of $N = 5$. The surface charge distributions of the coupled nanowires act cooperatively to enhance the repulsive action in all coupled nanowires, and hence the main resonant blueshifts.

From the field intensity plots in Figs. 10.4(a) and (b), we observe transmission of light in the 50-nanowire chain at different polariton resonances. The chain is illuminated from the left at two different resonances: $\lambda = 455$ nm (the main resonance) and $\lambda = 380$ nm (the secondary resonance), as shown in Fig. 10.2(a). At the main resonance, the wave propagation to the end of the long chain is shown. However, the field intensity vanishes at 1000 nm, when the chain is illuminated at the secondary resonance. In addition, the transverse polarization mode dominates at the main resonance, whereas the longitudinal polarization mode is dominant at the secondary resonance. These different behaviors can be understood with an illustration of the surface charge distribution in Figs. 10.4(c) and (d), where only five nanowires from the middle of the chain are plotted. The results clearly demonstrate the dipolar and quadrupolar modes. At the resonance $\lambda = 380$ nm, the surface charge density is weak (about 25 times of that at $\lambda = 455$ nm) so that the field intensity rapidly vanishes after 1000 nm. However, at the main resonance with $\lambda = 455$ nm, the intensity of surface

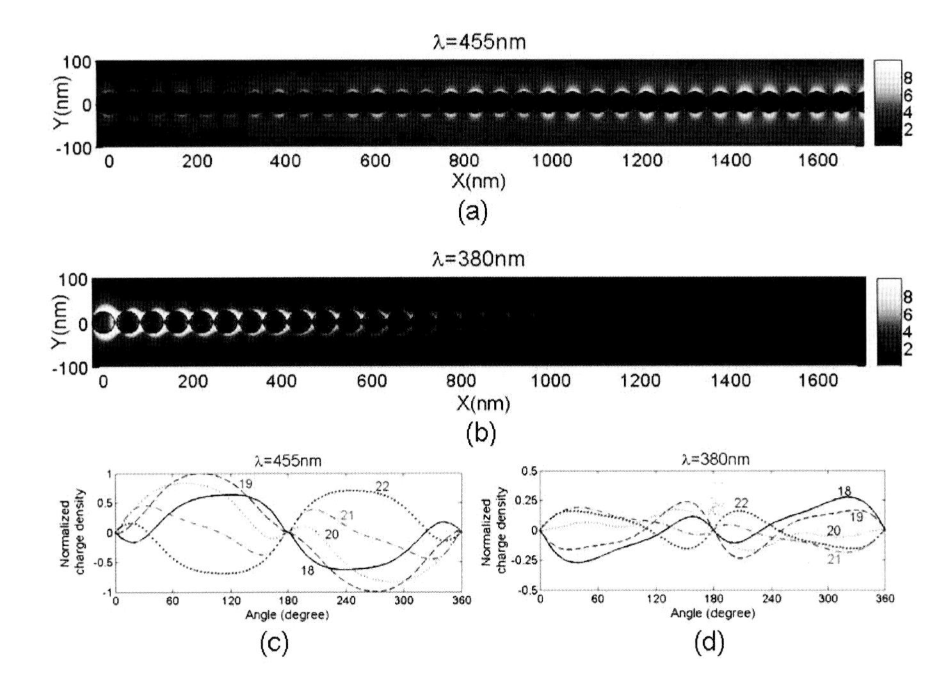

Figure 10.4: Distribution of the total electric field in the cross section of a 50 Ag nanowire chain for two different resonant wavelengths: (a) the main resonance with $\lambda = 455$ nm and (b) the secondary resonance with $\lambda = 380$ nm, as well as the corresponding surface charge distributions [(c) and (d)] observed at the center of chain in (a) and (b). The longitudinal polarization mode observed at $\lambda = 380$ nm vanishes after 1000 nm, whereas the transverse mode observed at $\lambda = 450$ nm can propagate for a longer distance. Reprinted with permission from Reference [2].

charge is strong and condensed in the both halves of the nanowires due to the dominant dipolar mode, as observed in Fig. 10.4(c). In fact, the surface charge density on the cylinder segments near the gap is even lower than on the rest of the circumference due to the repelling force. This is why the field minima are observed in the gap and the field maxima appear in the both halves of nanowires.

Thus, the transmission of coupled chain nanoparticles is strongly dependent on the number of particles included in a linear chain structure. For short chain length, only one longitudinal transmission mode exists and features strong field enhancements in the gap between two adjacent nanowires. However, when the chain length increases, the transverse transmission mode becomes dominant with the strong fields moved to the tops and bottoms of the nanowires. The maximum transmission length is obtained at the main resonance only.

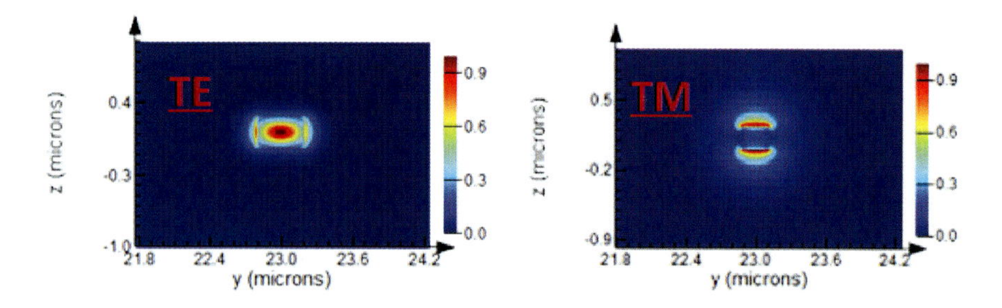

Figure 10.5: Cross-sectional profiles of electric field for quasi-TE and quasi-TM polaritons in submicron silicon-on-insulator channel waveguides for wavelength windows of $\lambda = 1550$ nm.

10.2 Design considerations for on-chip integration

We will now discuss more complicated case of three-dimensional chain waveguides with pronounced electric and magnetic responses. Compactness of such chains makes them suitable for on-chip integration, while electric and magnetic responses enable efficient control and routing of the light. Dielectric nanoparticles (DNPs) are good candidates because most of them are CMOS-compatible and flexible in terms of their polaritonic characteristics. In general, DNP chains have a broad functionality. They can be used as conventional waveguides to guide light, as power dividers to split and route light between different channels, and as waveguide crossings to guide light to different locations with minimal cross-stalk.

The basic building block of the DNP chain we consider is a silicon nanodisk with diameter d, thickness H, inter-particle gap distance g, and period p. For coupling in-plane laser light into the DNP chain, we consider a dielectric waveguide with width W and thickness H. The silicon-on-insulator (SOI) platform is used for both the DNP chain and the dielectric waveguide. Typically, $H = 220$ nm, while the thickness of the silicon oxide buffer is 2 μm, with a 725 μm silicon substrate. To facilitate back-end-of-line processing for metal interconnects, as well as for passivation to protect the device, the entire structure is covered by an upper cladding layer. These are the design rules usually dictated by the fabrication process; they will be used and fixed for the rest of chapter unless stated otherwise. The main design parameters that will be studied for the silicon DNP chain and dielectric waveguides are W for the waveguide and d, g for the DNP chain.

For most communication applications, the input optical fibers and silicon waveguides are designed for single polariton propagation. This means that only the low-order volume polariton (also known as the fundamental mode) is supported in the waveguide; all other higher-order polaritons are highly attenuating. This will help to avoid propagation loss associated with coupling to higher-order polaritons and inter-modal dispersion. To determine the single mode conditions for the silicon waveguide, we calculated the polariton complex wavenumbers at different waveguide width W for the two communication wavelength windows of $\lambda = 1310$ nm and $\lambda = 1550$ nm

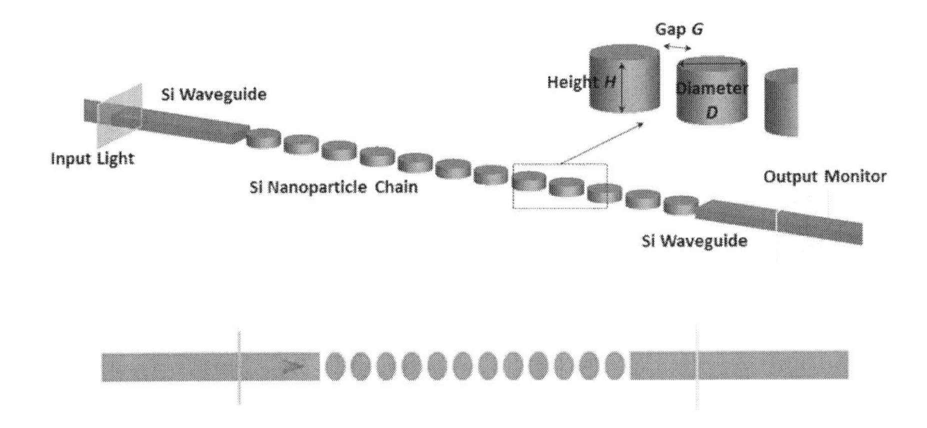

Figure 10.6: Coupling of input light from conventional silicon waveguide into a chain of silicon nanoparticles. The efficiency of the coupling can be quantified by the transmission at the output waveguide, which is measured by the output monitor in the simulation.

for both quasi-TE and and quasi-TM polarizations. We found that the range of W for single polariton condition in the SOI waveguide is $W = 200$ nm to 350 nm for $\lambda = 1310$ nm, and $W = 200$ nm to 450 nm for $\lambda = 1550$ nm. Thus, single mode condition is more stringent for smaller wavelength. These ranges of W will be used to design the silicon waveguides in the DNP chain system.

In silicon waveguides, quasi-TE polaritons is typically used, as this polarization is better confined within the waveguide compared to quasi-TM polaritons, as shown in Fig. 10.5. This nature of quasi-TE polaritons means lower bending loss in passive waveguides and higher modulation efficiency in active waveguide devices. However, input light with quasi-TM polarization might be required for transmission in the DNP. To achieve this, we should use a polarization rotator to convert the conventional quasi-TE light from the silicon waveguide into the quasi-TM light, just before it is coupled into the DNP chain.

In general, diameter D and period p of the DNPs affect the resonance wavelength and bandwidth, while gap distance g is mainly related to the coupling loss between DNPs. To facilitate fabrication based on deep UV lithography, DNP chain parameters are chosen to be $D = 340$ nm, $g = 150$ nm, with silicon thickness of $H = 220$. This dimensions also allow the light transmission with low propagation loss around $\lambda = 1310$ nm (for data communications applications) and $\lambda = 1550$ nm (for optical communications applications). These dimensions will be used for our discussion of passive devices based on DNP chain in the next few sections.

10.3 Nanocoupler for chain waveguide

Nanocoupler is one of the most important passive components, through which light from the waveguide can be coupled into the DNP chain and vice versa. This device can be used for two purposes: (i) to connect and route light between two types of photonic systems, namely the silicon photonics system and the DNP system, and (ii) to couple in-plane laser light source from the silicon waveguide into the DNP chain. This is illustrated in Fig. 10.6. An important figure-of-merit for the nanocoupler is the coupling efficiency and insertion loss. A nanocoupler is considered high performance, if the coupling of light from the silicon waveguide into the DNP chain occurs with high efficiency of > 0.9. The insertion loss includes the propagation loss, coupling loss due to modal mismatch, and back-reflections, when light travels between different photonic media.

The nanocoupler can also be viewed as a mode convertor. The polaritons in the silicon waveguide and DNP chain have different modal shapes and overlap integral. An ideal nanocoupler will help transform the polariton mode from the silicon waveguide into that of the DNP chain (and vice versa), such that there is a strong impedance matching with large overlap integral, resulting in coupling efficiency of close to 1. Note that in all our designs, the nanocoupler is composed of connected DNP chain consisting of 10 nanoparticles, with 2 waveguide couplers, one at the input waveguide, and another at the output waveguide. The final transmission T, and thus the coupling efficiency, which is measured at the output waveguide, can be used to assess the performance of the nanocoupler. This is illustrated in Fig. 10.6. We will now analyze and compare different types of nanocouplers connecting the silicon waveguide and the DNP chain.

10.3.1 Direct coupler

The simplest and most straightforward approach to couple light from the silicon waveguide into the DNP chain is to directly couple the edge of the straight waveguide with the chain, as illustrated in Fig. 10.6. The in-plane laser source, which is located at the photonics circuit, is used to excite the waveguide polaritons in the silicon waveguide. Either the quasi-TE or quasi-TM waveguide mode could be launched from the silicon waveguide into the DNP chain. A polarization rotator can be used, if the silicon waveguide and DNP chain support different types of polarized modes.

The main design parameters for the direct nanocoupler between the waveguide and DNP chain are the waveguide width W and particle diameter D. As illustrated in Fig. 10.7, varying the width W of the waveguide, while keeping all other parameters in the photonic system, affects the final transmission T. The general trend is that T starts to increase to > 0.6, when $W < D$. However, if we increase W to $W = D$, the final transmission T drops to < 0.5. These suggest that better mode coupling between the waveguide and DNP can be achieved with waveguide width smaller than the diameter of nanoparticles. With $W < D$, there is better phase (or impedance) matching with higher overlap between the polaritons of the two different structures. Using wide waveguide, i.e. with $W > D$, one can reduce the transmission, as the phase mismatch increases with W. However, we note that for ease of fabrication, it

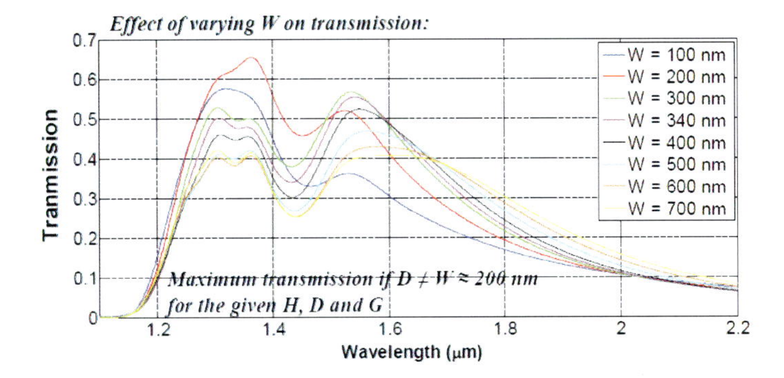

Figure 10.7: Effect of varying waveguide width W on the transmission of the dielectric nanoparticle chain.

might be easier to have $W = D$. Then, given a fixed D, there is a limit on how far we can reduce W to meet the condition of $W < D$ for higher transmission without resulting in polariton cut-off. For the silicon waveguide considered, the minimum size of W that supports guiding volume polaritons is 200 nm. As such, below, we will explore other designs of nanocoupler that utilize $W = D$. In particular, we will consider the case of $W = D$, as it is the most straightforward approach, without the need to chirp or taper the diameter of the nanoparticles at the entry and exit of the DNP chain, unlike the case of $W > D$.

10.3.2 Tapered coupler

To improve the modal mismatching of the straight waveguide with the DNP, one way is to redesign the edge of the straight waveguide such that both waveguides have similar cross-sectional shape. In conventional waveguide photonics, a taper is typically used to connect two waveguides of different widths W_1 and W_2. This can help to gradually convert the polariton from the wider waveguide to the narrower waveguide as it propagates down the taper. In this section, we will explore the use of a similar tapering concept to design an effective coupler between the waveguide and DNP chain.

The taper coupler between the waveguide and DNP chain is illustrated in Fig. 10.8. It consists of tapered periodic round edges along its sidewall that has the periodicity identical to that of the DNP. The taper coupler has a width W when it initially connects to the regular waveguide before tapering to a width of 100 nm at the termination point shaped like a DNP with dimensions identical to those of each nanoparticle in the adjacent DNP chain. The tapered periodic round edges, together with the tapered width, help gradually transform the waveguide polariton into a DNP polariton before it is coupled into the nanoparticle chain. In this way, the transmission loss due to impedance mismatch is reduced.

The key design parameter for the taper coupler is the number N of periodic structures on its sidewall, which affect the maximum transmission at the output port. It

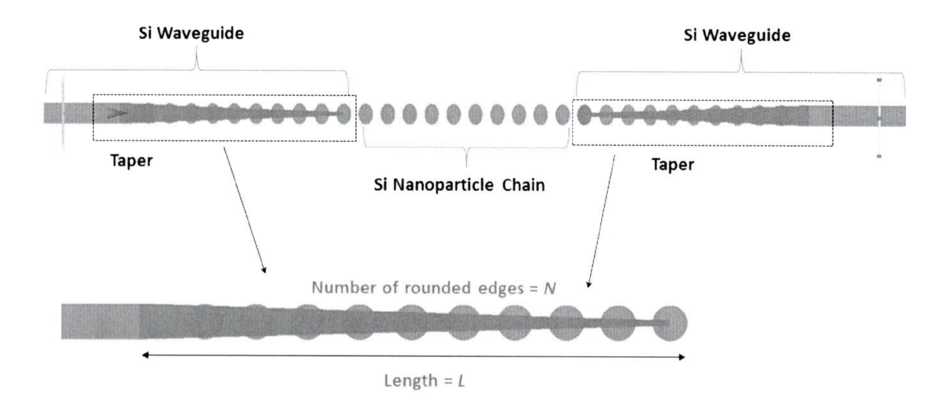

Figure 10.8: The tapered coupler that connects the conventional silicon waveguide with the chain of silicon nanoparticles. To gradually match the shape and effective index (or impedance) between the silicon waveguide and DNP chain, periodic round edges with tapered waveguide are used as couplers.

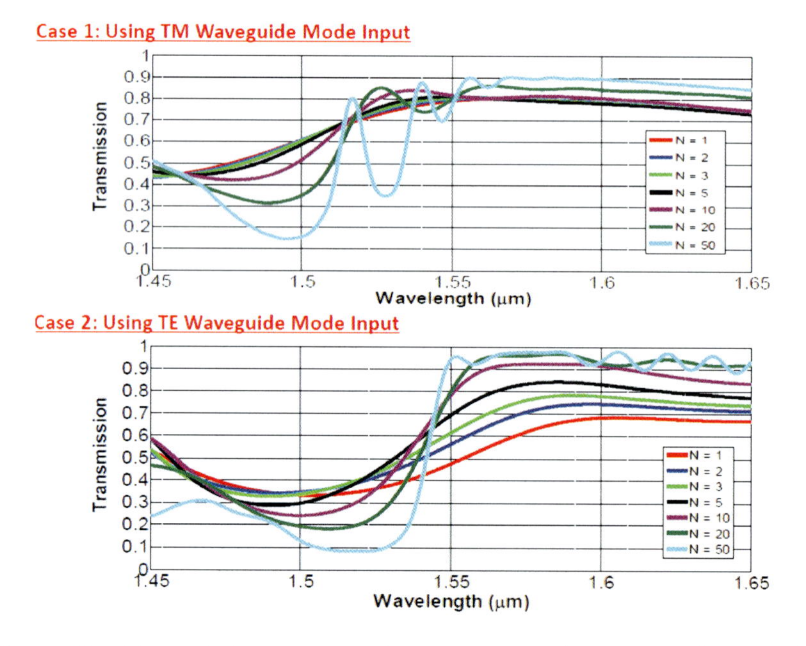

Figure 10.9: Transmission spectra at the output port, when a tapered coupler is employed to guide light from the waveguide into the nanoparticle chain.

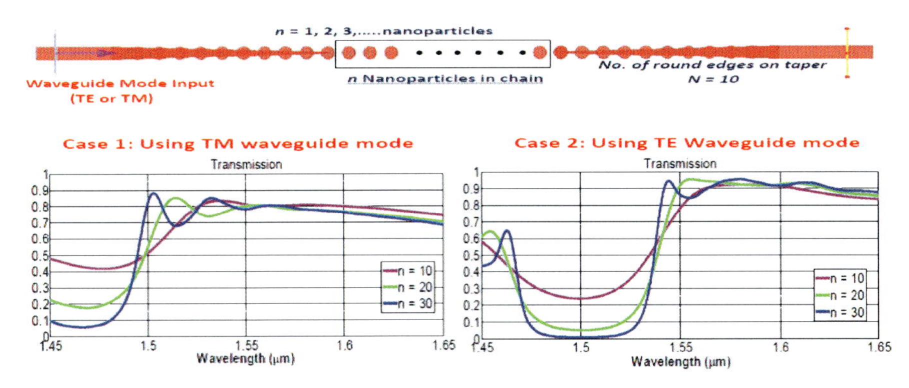

Figure 10.10: Effect of DNP number in the chain. Increasing N can be used to reshape the transmission spectra and to make photonic bandgaps more distinct.

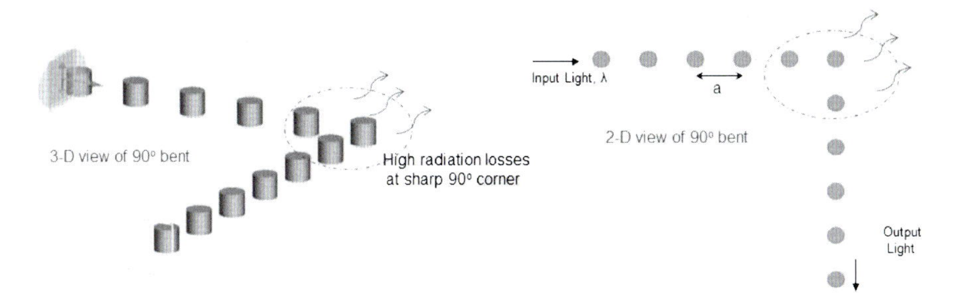

Figure 10.11: Conventional bend chain of nanoparticles with associated high radiation losses at a sharp 90 degree corner due to the impedance mismatch.

can be seen in Fig. 10.9 that increasing N results in higher transmission before it starts to decrease. As N approaches 50, bandgap and ripples start to appear in the transmission spectra. An optimized range of N for high transmission above 0.8 for both quasi-TE and quasi-TM waveguide polarizations would be < 50 for the considered DNP and waveguide. Thus far, the number of nanoparticles in the chain is fixed at 10 for simplicity. However, by changing N, we can alter the shape of the transmission spectra, as illustrated in Fig. 10.10. One main feature of increasing N is that the photonic bandgaps become more distinct.

10.4 Nanoparticle bend chain

Bend is one of the basic building blocks of photonic circuits. It changes the radial direction of the wave propagation, thereby allowing the input light to be routed easily to different parts of the photonic circuit. To achieve high density of photonic integration, compact and sharp 90 degree bends enabling light propagation with minimal loss are highly desired. However, light transmission through a bend DNP chain illustrated in Fig. 10.11 suffers from high transmission loss attributed to the strong

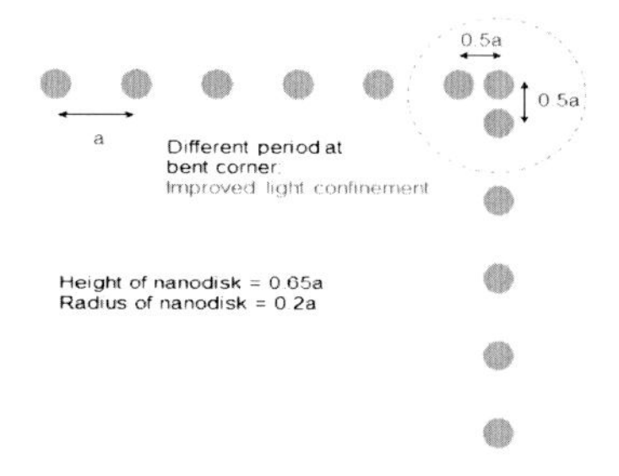

Figure 10.12: Design of bend chain of dielectric nanoparticles with different periodicities at the bend corner to improve the transmission.

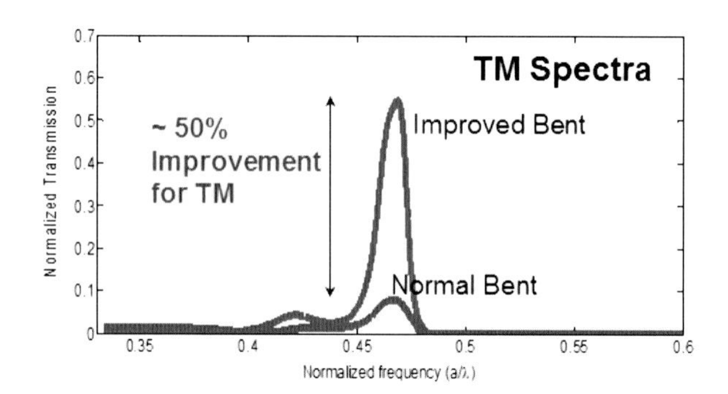

Figure 10.13: Transmission of the improved bend chain of nanoparticles, when quasi-TM waveguide mode is launched at the input port.

radiation caused by mode mismatch when light propagates around the bend. To improve the transmission of the bend DNP chain, we consider a configuration of bend DNP, as illustrated in Fig. 10.12. The main concept of this design is to use two types of periodicity for the bend chain structure: periodicity of a for the straight chains in the vertical and horizontal directions and periodicity of $0.5a$ at the edge of the chain where the two straight chains meet at the 90 degree bend. In this configuration, each nanoparticle has radius $0.2a$ and height $0.65a$. The air gap separation between the nanoparticles at the edge of the chain is $0.1a$, thus, giving us periodicity of $0.5a$ for the nanoparticle at the corner.

The use of different periodicities at the corner of the chain can better confine and guide light around the sharp bend, in particular, for the case, when the input light from the rectangular waveguide is quasi-TM polarized. The respective transmission spectra are shown in Fig. 10.13. They demonstrate that the considered design improves the transmission by 50%, as compared with the normal bend chain of DNP structure for the quasi-TM polarization.

10.5 Summary

The ability of resonantly coupled nanoparticles to transmit light allows us to use them as small-scale alternatives of conventional waveguides. In particular, they can be used in integrated photonic circuits. In this chapter, we have explored silicon nanodisk chains on SOI platform suitable for on-chip integration. We demonstrated that it can be integrated with conventional straight silicon waveguides to have transmission of above 90%. In addition, we have discussed and shown how the efficiency above 60% can be obtained for the bend chain by utilizing a different periodicity of nanoparticles at the bend corner.

Bibliography

[1] M. Quinten, A. Leitner, J. R. Krenn, and F. R. Aussenegg, "Electromagnetic energy transport via linear chains of silver nanoparticles," *Opt. Lett.*, vol. 23, p. 1331, 1998.

[2] H. S. Chu, W. B. Ewe, W. S. Koh, and E. P. Li, "Remarkable influence of the number of nanowires on plasmonic behaviors of the coupled metallic nanowire chain," *Appl. Phys. Lett.*, vol. 92, p. 103103, 2008.

[3] N. Felidj, J. Aubard, G. Levia, J. R. Krenn, A. Hohenau, G. Schider, A. Leitner, and F. R. Aussenegg, "Optimized surface-enhanced Raman scattering on gold nanoparticle arrays," *Appl. Phys. Lett.*, vol. 82, p. 3095, 2003.

[4] Y. Chen, K. Munechika, and D. S. Ginger, "Dependence of fluorescence intensity on the spectral overlap between fluorophores and plasmon resonant single silver nanoparticles," *Nano Lett.*, vol. 7, no. 3, pp. 690–696, 2007.

[5] J.-C. Weeber, A. Dereux, C. Girard, J. R. Krenn, and J.-P. Goudonnet, "Plasmon polaritons of metallic nanowires for controlling submicron propagation of light," *Phys. Rev. B*, vol. 60, pp. 9061–9068, Sep 1999.

[6] J. P. Kottmann and O. J. F. Martin, "Retardation-induced plasmon resonances in coupled nanoparticles," *Opt. Lett.*, vol. 26, pp. 1096–1098, Jul 2001.

[7] T. Laroche and C. Girard, "Near-field optical properties of single plasmonic nanowires," *Appl. Phys. Lett.*, vol. 89, no. 23, 2006.

[8] Q. Xu, J. Bao, F. Capasso, and G. M. Whitesides, "Surface plasmon resonances of free-standing gold nanowires fabricated by nanoskiving," *Angew. Chem.*, vol. 118, no. 22, pp. 3713–3717, 2006.

[9] S. A. Maier, P. G. Kik, and H. A. Atwater, "Observation of coupled plasmon-polariton modes in au nanoparticle chain waveguides of different lengths: Estimation of waveguide loss," *Appl. Phys. Lett.*, vol. 81, p. 1714, 2002.

[10] H. S. Chu, W. B. Ewe, E. P. Li, and R. Vahldieck, "Analysis of sub-wavelength light propagation through long double-chain nanowires with funnel feeding," *Opt. Expr.*, vol. 15, p. 4216, 2007.

[11] E. P. Li, Q. X. Wang, Y. J. Zhang, and B. L. Ooi, "Analysis of finite-size coated electromagnetic bandgap structure by an efficient scattering matrix method,"

IEEE J. Select. Top. Quant. Electron., vol. 11, p. 485, 2005.

[12] T. Decoopman, G. Tayeb, S. Enoch, D. Maystre, and B. Gralak, "Photonic crystal lens: From negative refraction and negative index to negative permittivity and permeability," *Phys. Rev. Lett.*, vol. 97, p. 073905, 2006.

[13] E. D. Palik, *Handbook of Optical Constants of Solids.* Academic Press, San Diego, 1991.

Chapter 11

Sub-wavelength slot waveguides

Hong-Son Chu
Institute of High Performance Computing, Singapore

This chapter aims to give an overview of sub-wavelength photonic and plasmonic slot waveguides together and discuss their main properties. Section 11.2 will describe and summarize various nanophotonic and plasmonic waveguides based on dielectric, metal, and hybrid dielectric-metal material platforms. Next, we will discuss optical performance of coupled metallic nanoparticle double chains in Section 11.3. Given the focus on sub-wavelength slot waveguides for enhancement and confinement of light, three CMOS-compatible waveguide platforms including silicon slot, hybrid silicon-copper dielectric slot and copper slot will be studied in Section 11.4. Summary appears in Section 11.5.

11.1 Introduction

Nanophotonics and plasmonics are exciting and promising technologies based on the interaction of light with nanostructured materials. Their application in various fields such as communication and all-optical signal processing can make positive contributions to several societal challenges. First, they can provide faster data communication for supercomputers, data centers, and optical routers. Second, they can help reduce energy consumption because they require only small amounts of electrostatic energy to drive the optical transmitter devices. Finally, they will accelerate advances in the fields of healthcare and other sciences through their use in integrated bio-medical devices, gas sensors, and bio-analysis [1]. However, to become successful market technologies, it is still necessary to achieve more data bandwidths, less power consumption, and reduced cost margins of on-chip photonics [2, 3].

A solution to fulfill both size and power requirements for future nanoscale photonic integrated circuit (nPIC) technologies lies in the use of dielectric and metallic optical components scaled beyond the diffraction limit of light. The advantages of this sub-diffraction limit are threefold: small physical device sizes, faster operating speeds, and reduced optical power requirements arising through strong light-matter interaction [4]. More specifically, the significantly enhanced optical localization within nanophotonic and plasmonic components may help overcome the typically weak interaction between light and matter, which in turn induces the energy

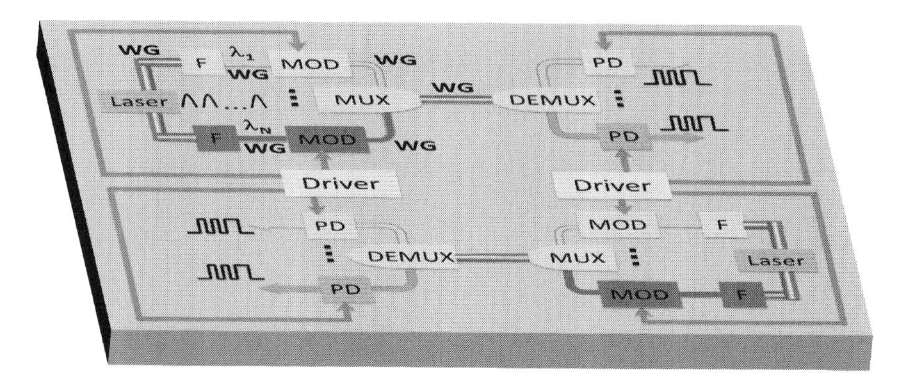

Figure 11.1: Schematic illustration of a typical optical link built from dielectric and/or metallic nanophotonic devices for on-chip data transfer. WG = waveguide, F = filter, MOD = modulator, PD = photodetector, MUX = multiplexer, and DEMUX = demultiplexer.

necessary to obtain a desired effect, for instance, electronic modulation of an optical signal or nonlinear optical frequency mixing [5–8]. In order to address these demands, optical components and circuits show a great promise for the scalability and performance of future nPICs.

By definition, an optical waveguide is a physical structure able to guide electromagnetic waves at optical wavelengths. Optical waveguides are used as building-block components in integrated optical circuits or as transmission media in local and long haul optical communication systems. In general, optical waveguides can be classified according to their geometry (planar, strip, or fiber waveguides), mode structure (single-mode, multi-mode), refractive index distribution (step or gradient index), and material (glass, polymer, semiconductor). Moreover, nanophotonic and plasmonic waveguides, as reported, help to confine and guide the light beyond the diffraction-limit in the guided-wave devices and circuits of integrated optics.

Figure 11.1 illustrates a typical nanophotonic and plasmonic link for on-chip data transfer. The nanolaser source generates the optical signal to be guided through straight or bent waveguides. Then, this signal is modulated by a nanophotonic modulator. For the practical reason of integration, the electro-optic modulator is preferred; it allows encoding of electrical signals into the optical data stream. The modulated signal then passes through the nanoscale photodetector to convert the propagating electromagnetic waves into electrical signals.

11.2 Overview of sub-wavelength waveguides

Although many successful achievements have been realized with the current photonic integrated circuits (PIC), they are still creating large footprints compared to their electronic counterparts. It is because the on-chip optical waveguides typically have a low index contrast, and therefore light there is weakly confined and requires a large waveguide core, which is typically of micrometer size. As a result, such waveguide has a large footprint and, thus consumes valuable chip area for interconnects. This is one of the reasons why current PICs integrate only a small number of components onto a single chip, which is often many square centimeters in size.

Recently, nanophotonics has been demonstrated as a vital technology to realize compact photonic integrated circuits. Compared to glass- or polymer-based waveguide technologies, the silicon technology has provided good waveguiding properties because the high refractive index contrast between silicon and its native oxide cladding allows optical confinement in submicron waveguide cores. Silicon nanophotonic wires can scale down the footprints of integrated optical components by a few orders of magnitude. As a result, compact optical components become possible [9].

Figure 11.2(a) shows a silicon wire on a silicon-on-insulator (SOI) waveguide platform. The high confinement of light in these waveguiding structures has two key benefits for developing nonlinear optical devices. First, the small mode effective area enhances the nonlinear coefficient of the waveguide. Second, the mode confinement allows waveguide dispersion engineering such that the key phase matching condition can be met for development of several nonlinear optical devices such as four-wave mixing and second or third harmonic generation.

As discussed in Section 2.3.3, the guiding mechanism of a dielectric slab waveguide is the *total internal reflection* (TIR) in a high index material (core) surrounded by a low-index material (cladding); the TIR can strongly confine light inside the high-index core. However in general, this waveguiding mechanism is still facing few challenges such as (i) the diffraction limit for the light bound inside high-index dielectric core, and (ii) lower flexibility to achieve high field enhancement for the light-matter interaction impact.

In order to overcome these challenges, a silicon *slot waveguide* platform was proposed and demonstrated. It allows for a large portion of the optical mode to propagate through a low index region of the waveguide. By definition a slot waveguide is an optical waveguide that guides strongly confined light at a sub-wavelength scale in low refractive index regions [6, 10, 11]. A slot waveguide consists of two strips or slabs of high refractive-index (n_H) materials separated by a sub-wavelength scale low refractive index (n_S) slot and surrounded by low refractive-index (n_C) cladding materials, as sketched in Fig. 11.2(b). Alternatively, we can use photonic crystals [12], which are periodic structures with high index contrast. An example consisting of air holes etched into a silicon layer is illustrated in Fig. 11.2(c). Because of their wavelength-scale periodicity, photonic crystals can have a photonic band gap, i.e., a wavelength region where no light can penetrate the crystal. A defect in such a photonic crystal made by changing or removing a row of holes can sustain a guided mode in the photonic crystal. Light in the photonic band gap is bound to the waveguide defect because it is not allowed to propagate through the areas of photonic crystal on both sides of the waveguide [13, 14].

The use of materials with negative dielectric permittivity is one of the most feasible ways of circumventing the diffraction limit and achieving nanoscale localization of the electromagnetic energy. The most readily available materials for this purpose are metals below the plasma frequency. As discussed in Section 2.3, metal structures

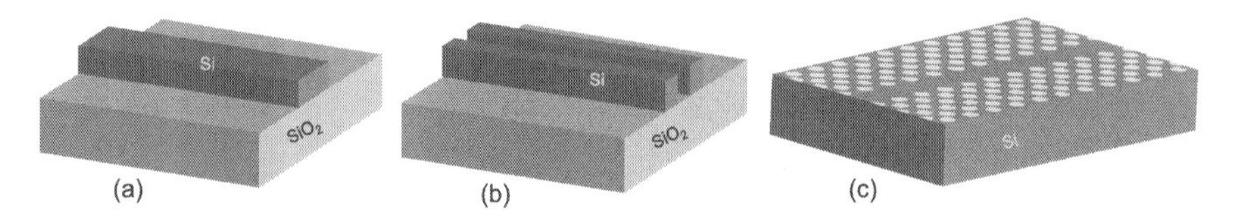

Figure 11.2: Schematic illustration of (a) typical silicon wire on SOI platform, (b) silicon slot waveguide, and (c) photonic crystal waveguide with holes in the silicon thin film.

and interfaces are known to guide *surface polaritons* (SPs) [15]. In recent years, it has been demonstrated theoretically and experimentally that metal-based waveguides taking the advantage of SP-enabled tight modal confinement promise to overcome the size mismatch between microscale photonics and nanoscale electronics. Thus, SP technology is a potential platform for the next generation of optical interconnects that enables the deployment of small-footprint and low-energy integrated circuitry and holds great promise for on-chip optical interconnects of high integration density at the chip scale [3, 16].

Recent advances in fabrication led to the realization of a wide variety of SP-based devices. Several types of SP waveguide platforms have been developed, which differ in topology, material composition and propagation mechanisms. Each waveguide platform possesses different waveguiding characteristics [16, 17]. Figure 11.3 is an illustration of most popular SP-based waveguides reported in the literature for both theoretical and experimental studies.

The first SP waveguide platform is the waveguide geometry consisting of a three-layer structure made of metal and dielectric and exhibiting extremely strong field localization in one dimension. The simplest plasmonic waveguide of this type is known as the insulator-metal-insulator (IMI) waveguide. It is a one-dimensional version of a cylindrical nanowire waveguide, which was one of the first structures identified for strong light confinement. Structures of coupled nanowires [18–21] also exhibit strong field confinement, but feature high dissipation, as shown in Fig. 11.3(a). Metallic slabs [see Fig. 11.3(b)] are relatively easy to fabricate, but expected to exhibit large bend losses, and may be sensitive to structural imperfections [22–24]. The metal-insulator-metal (MIM) structures shown in Fig. 11.3(c), have been intensively studied and proposed as suitable solutions for both mode confinement and propagation losses [25, 26].

Among the mentioned configurations, MIM geometry has received the most attention ease of fabrication and truly sub-wavelength confinement of the optical mode without cutoff [23]. Furthermore, by altering the thickness of the insulator layer, the planar MIM system can support both surface and volume polaritons [27]. In the MIM system, light can be coupled into and out of the waveguide structures using slit openings in the metal, enabling broadband propagation of electromagnetic energy over distances of several micrometers. In addition, metallic grooves, as sketched in Fig. 11.3(d), demonstrated different functional devices for on-chip integrated cir-

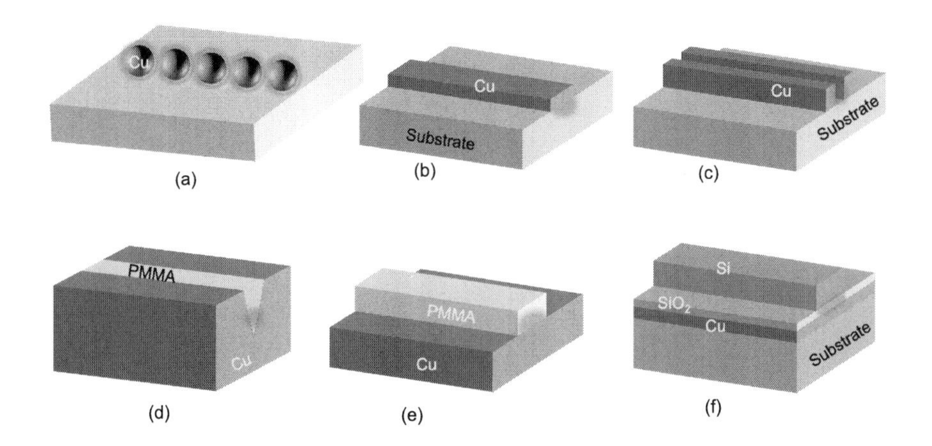

Figure 11.3: Schematic illustration of typical SP waveguides: (a) inter-particle-coupled single chain, (b) metal slab on a dielectric (or insulator-metal-insulator waveguide), (c) metal-insulator-metal on a substrate, (d) channel SP waveguide with PMMA dielectric filled in the groove, (e) dielectric loaded SP waveguide, and (f) hybrid plasmonic-photonic waveguide by means of metal-oxide-semiconductor platform. The glow areas represent general patterns of the electric-field confinement in different platforms.

cuit application [28, 29]. But this structure is still faces some challenges such as the control of groove angle for the respective confinement and a large footprint to be compatible with the current silicon nanophotonics.

Next, the combination of dielectric and plasmonic waveguiding principles, called dielectric-loaded waveguides, as sketched in Fig. 11.3(e), exhibit high confinement and better propagation loss than the SPs waveguide based on the metal-dielectric interface [30, 31]. However the waveguiding characteristics of this platform depend strongly on the dielectric materials; for example, if the dielectric is a high-index semiconductor, the propagation loss becomes excessively high. Recently, by using the hybrid metal-insulator-semiconductor material platform, hybrid plasmonic waveguides [see Fig. 11.3(f)] demonstrated a potential for CMOS-compatible silicon photonics and nanophotonics integrated circuits [32–34]. This waveguide platform demonstrated a deep-sub-wavelength mode down to about one-tenth of the diffraction limit of light in each lateral direction and reasonable propagation lengths.

The main focus of this chapter is sub-wavelength waveguide platforms that could provide ample optical mode to propagate and confine through its low-index region. Therefore, we will consider the coupled metallic interparticle double-chain and slit waveguides made of two slabs of high-index dielectrics or metals. These kinds of waveguide platforms exhibit few advantages including (i) they are able to guide light at the nanoscale dimensions, and (ii) they hugely enhance light-matter interaction. More importantly, the second point is a key property to achieve high confinement for integrated nonlinear optics, optical bio-sensing, hybrid electro-optical modulators and other applications.

11.3 Metal-nanoparticle double-chain waveguide

11.3.1 Waveguide specifications

The coupled metallic nanoparticle double chain is a type of the slot waveguide. This waveguide platform exhibits a great ability to guide and squeeze light in the nanometer size area. In this section, we present the optical properties of an SP inter-particle-coupled double chain, which consists of two parallel chains of silver nanowires fed by a V-shaped funnel at the wavelength of 600 nm. The main thrust of this study was to demonstrate the influence of the waveguide geometry including the funnel arm, center arm, and distance between the two chains on the propagation characteristics of the waveguide. Since the opening angle of the V-shaped funnel is important for optimum light capturing, this parameter was also included in the investigation. For this study, the structure was excited with a dipolar source operating at three different wavelengths to demonstrate the confinement and guidance of the excited waves along the core of a metallic inter-particle-coupled double chain. As an example, a waveguiding distance of 3.3 μm will be used for demonstration.

In order to increase the excitation efficiency of SPs in coupled nanowires and efficiently guide the lightwave in the desired direction, we introduced a funneling array to feed the waveguide. This funneling consists of a double chain of a silver nanowire array [35]. The electric dipole source is used to mimic the practical situation as closely as possible. The system under study is shown in Fig. 11.4. It consists of a V-shaped feed configuration (called funnel arm) and double chain of silver nanowires. The V-shaped feed opens with an aperture of α (default value set to 90°) and consists of 10 silver nanowires, while the two parallel chains consist of 120 silver nanowires. The center-center separations of adjacent cylinders in the funnel-arm and in the center arm may be different and are represented by d_{fn} and d, respectively. The two parallel chains are separated by a distance of h (center-center separation of adjacent cylinders). For simplicity, the medium surrounding the waveguide is assumed to be air ($\varepsilon_r = 1$). The cylindrical silver nanowires have a radius of $r = 25$ nm. The dipole source is located between the first cylinders in the funnel region and in the center plane of the double-chain waveguide, as shown in Fig. 11.4. It illuminates the TM-polarized light, which means that E is perpendicular to the z direction [36], at 600 nm wavelength. From experimental data [37], the relative permittivity for silver is taken as $\varepsilon_r(\omega) = -13.98 + i0.95$. The computed field intensity is extracted along the horizontal symmetry plane (xz plane or line AB), that is, along the axis of propagation, as shown in Fig. 11.4.

11.3.2 Operation characteristics

The performance of the waveguide was computed with the surface integral equation (SIE) methodthat allows the study on SPs of arbitrarily shaped homogeneous objects.

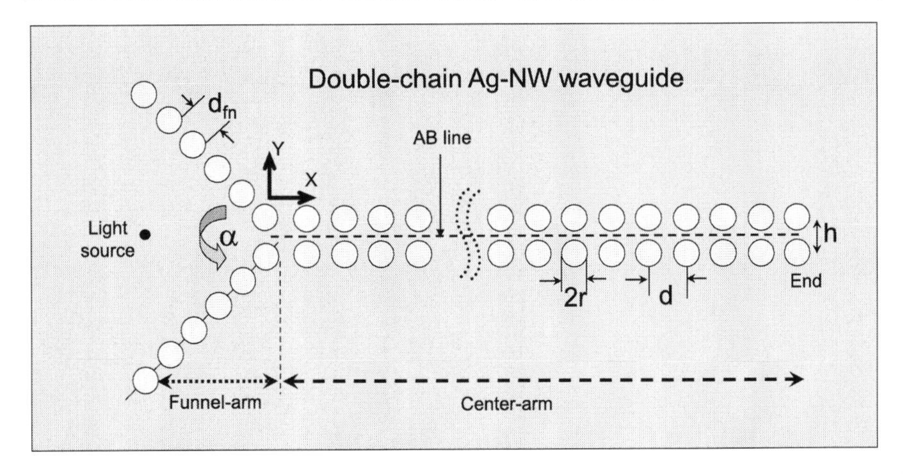

Figure 11.4: SP waveguide using 130 cylindrical silver nanowires. The double-chain waveguide consists of a funnel arm and a center arm. The dipole source at the left is used to excite the structure with TM-polarized light. Reprinted with permission from Reference. [20].

In contrast to time domain methods, the SIE method does not require medium modeling such as the Drude or Lorentz models to simulate scattering effects in metallic nanostructures. The SIE method can directly use experimental permittivity data for characterizing wave propagation in such structures. The method is formulated by considering the total fields and the boundary conditions at the surface of the object. By expanding the surface currents using pulse basis functions and applying point matching, the SIE method can be converted into a matrix equation solved by a matrix solver [36, 38, 39].

It is expected that for a given wavelength, the confinement of light and its propagation between the parallel double chains can be significantly enhanced depending on the distance between the parallel chains and the center-center separation of adjacent nanowires. It is also expected that the opening angle of the funnel region will have a direct impact on the amount of light captured. The H_z-field distribution along the center arm is investigated for different center-center spacings between the silver nanowires in the funnel region (AgNW), $d_{\mathrm{fn}} = 2.04r$, $2.2r$, $2.6r$, $3r$, and $4r$, which correspond to actual values of $d_{fn} = 51$, 55, 65, 75, and 100 nm. Figure 11.5 illustrates the relative field intensity along the AB line (defined in Fig. 11.4). It is evident that for $d_{fn} = 2.2r$, $2.6r$, $3r$, and $4r$, there is little effect on the field distribution along the double-chain region. This changes, however, when two nanowires in the funnel region become too close (or too far apart), such as for $d_{fn} = 2.04r$. As observed, a value of $d_{fn} = 2.6r$ can be considered as the optimum center-center distance for two adjacent nanowires to obtain maximum energy transport along the double-chain waveguide.

Figure 11.6 shows the field amplitude of excited waves along the waveguide for different distances h separating the parallel chains. The separation between the

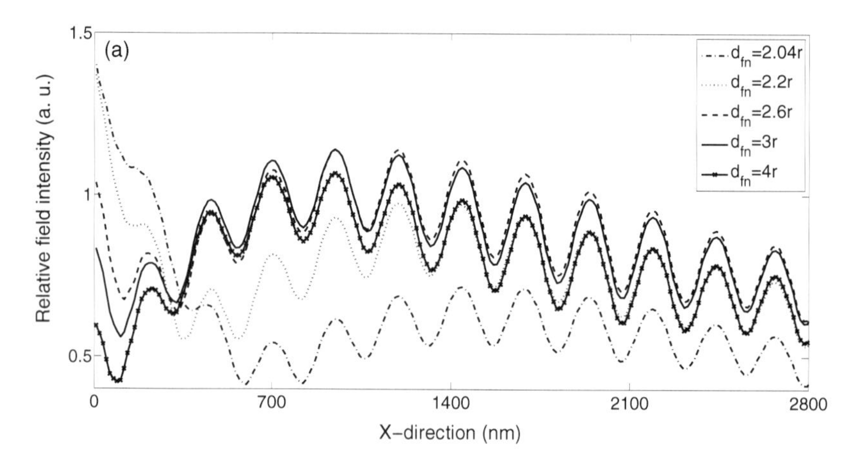

Figure 11.5: H_z-field intensity distribution along the propagation direction (AB) in the double-chain region for different center-center separation d_{fn} in the funnel region with V-shaped aperture $\alpha = 90°$. Reprinted with the permission from Reference [20].

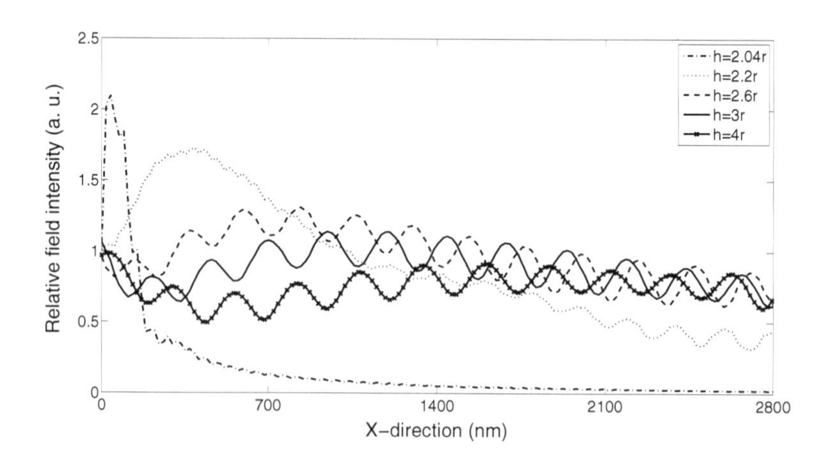

Figure 11.6: H_z-field intensity distribution along the propagation direction (AB) in the double-chain region with respect to different gaps h between the two parallel chains in the center-arm, with $d_{fn} = 2.6r$, $d = 2.2r$, and V-shaped aperture $\alpha = 90°$. Reprinted with the permission from Reference [20].

nanowires in the funnel region and the double-chain region are kept constant at $d_{fn} = 2.6r = 65$ nm and $d = 2.2r = 55$ nm, respectively. It can be seen that for values of $h = 2.04r$ and $2.2r$, the field intensity decays rapidly along the propagation direction and the intensity at the end of the double-chain region drops to zero for $h = 2.04r$, which corresponds to 5 nm. For gap widths ranging from $h = 2.2r$ to

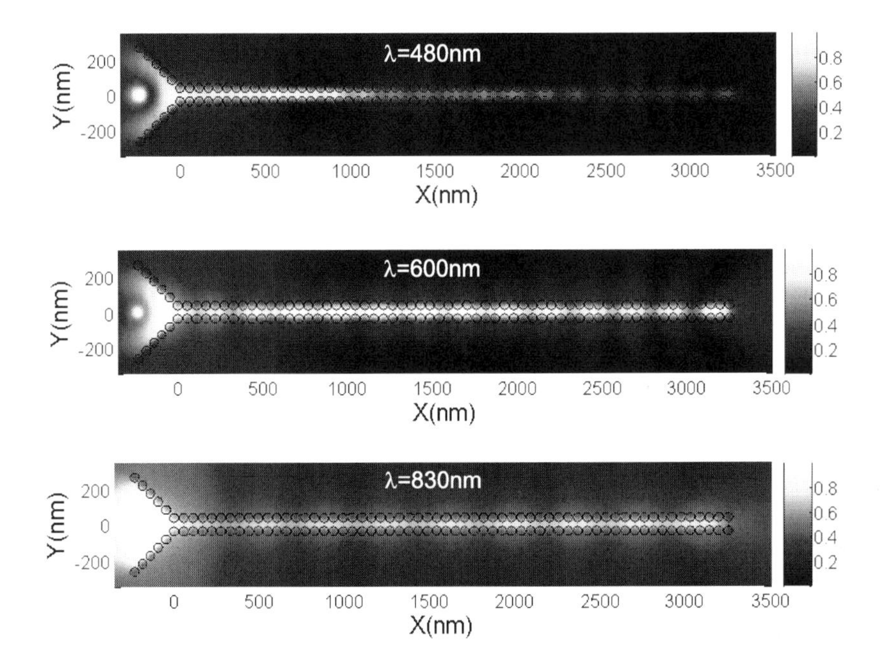

Figure 11.7: Normalized H_z-field intensity distribution in the waveguide for different wavelengths: (a) $\lambda = 480$ nm, (b) $\lambda = 600$ nm, and (c) $\lambda = 830$ nm. The field-intensity is strongly confined and well guided within the double-chain waveguide at $\lambda = 600$ nm. Reprinted with the permission from Reference [20].

$h = 3r$, the field intensities change only slightly, reaching a maximum for $h = 3r = 75$ nm. This value can be considered the optimum distance between the chains to obtain the maximum light energy propagating along the structure.

The fields excited at different wavelengths λ = 480, 600, and 830 nm are plotted in Fig. 11.7. Note that the light source is located at the left side of the waveguide. The field intensity distribution along the propagation direction in the center of the waveguide (AB) is plotted in Fig. 11.8. The dimensions in the funnel region and the double-chain waveguide sections are taken at $d_{fn} = 2.6r = 65$ nm, $d = 2.2r = 55$ nm, and V-shaped aperture $\alpha = 90^{\circ}$. The results in Fig. 11.8 confirms that at $\lambda = 600$ nm the wave propagates with the highest H_z-field intensity in the region between the chains, while at $\lambda = 480$ nm the intensity rapidly decays further as the wave propagates away from the source. At $\lambda = 830$ nm, the light intensity at the end of the waveguide is significantly smaller than at $\lambda = 600$ nm.

Figure 11.9 represents the effect of the opening angle of the V-shaped aperture on the H_z-field intensity along and at the end of the system. It is evident that the opening angle of $\alpha = 90^{\circ}$ in our previous analysis is not the optimum one. In fact, a smaller opening angle improves the light coupling into the structure significantly, as

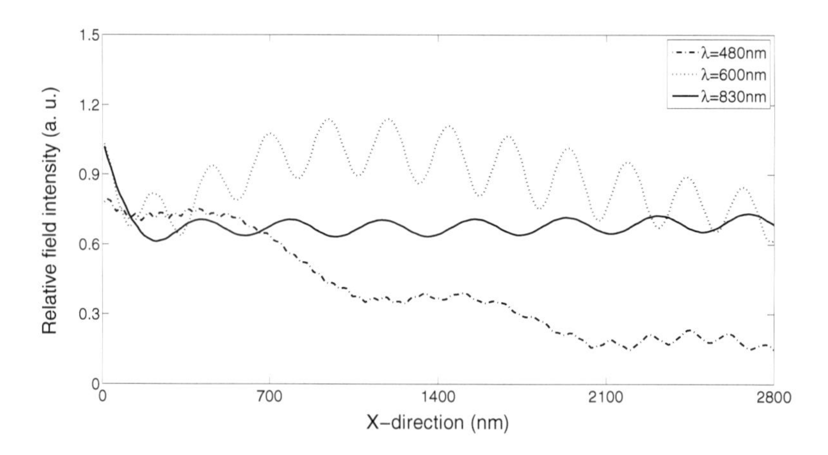

Figure 11.8: Magnetic field intensity in the center of the waveguide along the propagation direction for different excitation wavelengths: λ = 470, 600 and 830 nm. Reprinted with the permission from Reference [20].

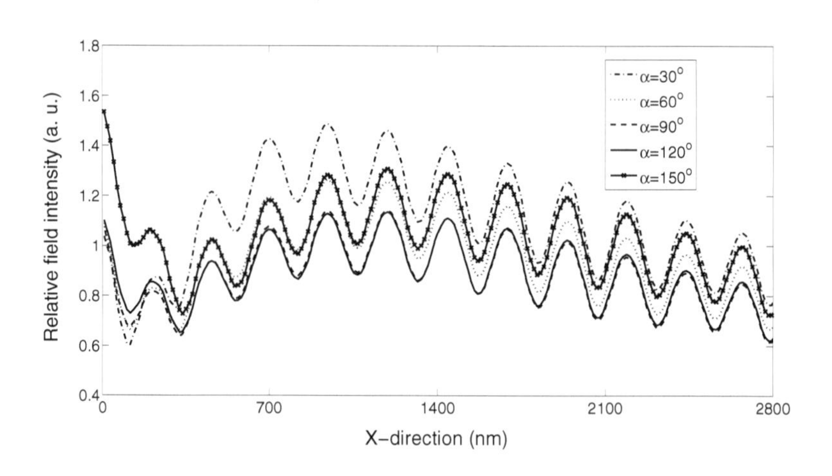

Figure 11.9: H_z-field intensity distribution along the propagation direction (AB) in the double-chain region for different V-shaped aperture α with $d_{\text{fn}} = 2.6r$, $d = 2.2r$, and $h = 3r$. Reprinted with the permission from Reference [20].

it captures more light from the source due to less abrupt transition from the funnel to the double-chain region and hence reduced back reflection.

Thus, the efficiency of SP excitation and the resultant confinement of light along the coupled nanoparticle double chain depend on the waveguide geometry and on structural dimensions in the funnel region for launching the light into the structure. Another important conclusion is that with the separation distance of 25 nm, the double-chain waveguide can support SPs at the wavelengths at which the respective single-chain waveguides (see Section 10.1.1) are unable to do.

11.4 Sub-wavelength slab-slot waveguides

The silicon technology in high performance photonic devices and optical integrated circuits has been demonstrated exciting prospects for a great number of applications, including datacom and sensing. For example, silicon integrated lab-on-a-chip systems, in which photonic sensing functions are integrated with electronic intelligence and wireless communications, can be employed in environmental monitoring and medical diagnostics [40]. Silicon-on-insulator (SOI) sub-micrometer silicon wire waveguides have demonstrated promise for integrated photonics [9, 41]. Due to the successful development of Si technology, sub-wavelength dielectric slot waveguides have been introduced. This waveguide platform consists of two closely placed silicon wires, with ability to realize SOI nanometer guiding structures, usually known as SOI slot waveguides [42]. A great variety of optical devices have been proposed or fabricated by using silicon slot waveguides, including micro-ring resonators [43], switches [8], optical modulator switches [44], electrically pumped light emitting devices [45], directional couplers [46], and beam splitters [47]. However, sub-wavelength slot waveguides in general can be built from any high-index dielectric, semiconductor, metal, or hybrid dielectric-metal materials. These kinds of waveguide platforms serve as promising building blocks to develop novel functional linear and nonlinear devices.

In this section, we will compare the performance characteristics of three fully CMOS-compatible sub-wavelength slot waveguides: (i) silicon-silica-silicon (Si-based), (ii) copper-silica-copper (Cu-based), and (iii) hybrid silicon-silica-copper (hybrid) slot waveguides as aforementioned.

11.4.1 Waveguide specifications

To facilitate the fabrication and integration process, in-plane or vertical platforms for three slot waveguides can be used. The waveguides are shown in Fig. 11.10. The refractive index at the telecommunication wavelengths ($\lambda = 1.55\mu$m) of the Si slab and the SiO_2 cladding medium are taken as $n_{\text{Si}} = 3.48$ and $n_{\text{SiO}_2} = 1.45$, while the complex index of Cu is taken as $n_{\text{Cu}} = 0.28 + \text{i}11.05$ [48]. The thickness of all three waveguides is chosen to be 220 nm to complete the standard CMOS fabrication process [49]. The devices were fabricated on 200 mm SOI wafers with 220 nm of lightly p-doped Si on a 2 μm buried oxide (BOX) layer. The thick oxide serves

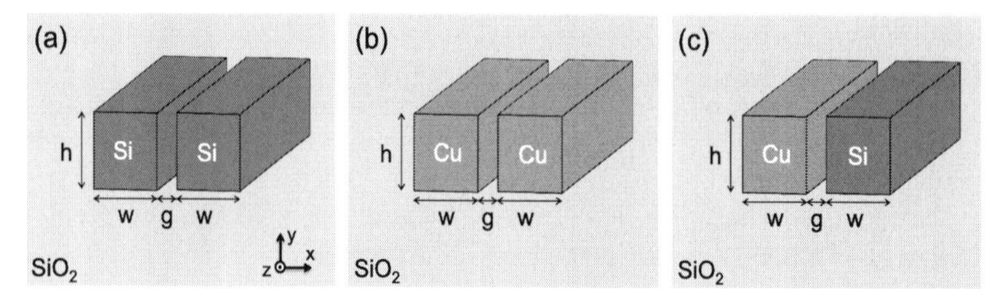

Figure 11.10: Illustration of sub-wavelength slot waveguides: (a) Si-SiO_2-Si, (b) Cu-SiO_2-Cu, and (c) hybrid Si-SiO_2-Cu.

to optically isolate the circuit from the substrate, reducing losses due to substrate leakage.

The principle of operation of this waveguide configuration is based on the discontinuity of an electric field at the high-index/low-index or metal/low-index interfaces. As an example, when an electromagnetic wave is propagating in the z direction, as shown in Fig. 11.10(a), the electric field component of the quasi-TE mode (which is aligned in the x direction) undergoes a discontinuity proportional to the square of the ratio between the refractive indices of the silicon and the low-refractive-index SiO_2 slot.

By using a full-vectorial mode solver approach [50], we computed the propagation properties of the three slot waveguides. For polaritons propagating in the z direction, the electric fields are given by

$$E(x,y)\,e^{i(k_z z-\omega t)}, \tag{11.1}$$

where ω is the angular frequency, and k_z is a complex propagation constant. The complex modal effective index is then defined as

$$N_{\text{eff}} = ck_z/\omega. \tag{11.2}$$

The eigensolver finds these polaritons by solving Maxwell's equations on a cross-sectional mesh of the waveguide. The finite difference algorithm used for meshing the waveguide geometry has the ability to accommodate arbitrary waveguide structures. After meshing the structure, Maxwell's equations are written in the matrix form and solved using sparse matrix techniques for complex modal effective index and field profiles of the eigenmodes. The real part of the obtained complex modal index is assigned to the effective index value,

$$n_{\text{eff}} = \text{Re}(N_{\text{eff}}), \tag{11.3}$$

which means the phase constant. In the meantime, the imaginary part gives the polariton propagation length, which is calculated as

$$L_{\text{p}} = \frac{\lambda_0}{4\pi\,\text{Im}(N_{\text{eff}})}. \tag{11.4}$$

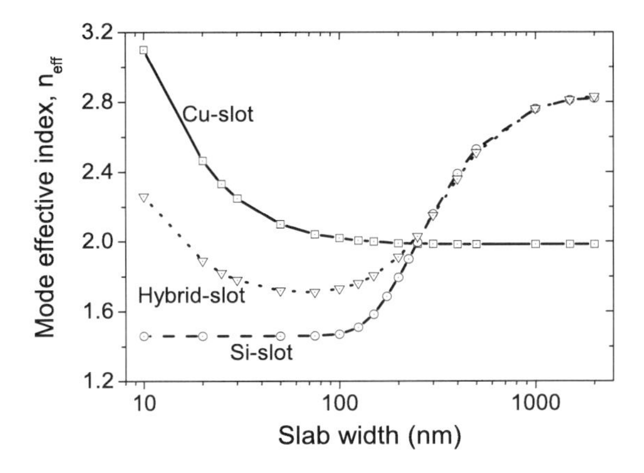

Figure 11.11: Effective index as a function of slab widths for Si-SiO_2-Si, Cu-SiO_2-Cu, and Si-SiO_2-Cu slot waveguides.

The confinement is quantified with the power confinement factor P_c defined as the ratio of the optical power focused inside the SiO_2 slot to the total waveguide optical power [10],

$$P_c = \frac{\iint\limits_{(g\times h)} S_z \, \mathrm{d}x \, \mathrm{d}y}{\iint\limits_{-\infty}^{\infty} S_z \, \mathrm{d}x \, \mathrm{d}y}, \tag{11.5}$$

where S_z is the z component of the time-average Poynting vector.

11.4.2 Operation characteristics

We studied the effective index and power confinement of the first-order modes as a function of the Si-slab or Cu-slab widths w ranging from 10 nm to 2000 nm for three different waveguides (Si-slot, Cu-slot, and hybrid-slot based) at a fixed SiO_2 slot width (gap) of 50 nm and a wavelength of $\lambda = 1.55\mu$m. Figure 11.11 shows the variation of effective indices n_{eff} of the three waveguides as a function of w. It demonstrates that n_{eff} of the Cu-slot waveguide decreases rapidly as the Cu-slab width increases from 10 nm to 80 nm. For a value w larger than 80 nm, this variation stabilizes, as the SPs become fully confined inside the SiO_2-slot. The results also reveal a cut-off of w for the Si-slot waveguide. In particular, if the Si-slab width is smaller than 100 nm, the effective index is close to the SiO_2 refractive index. The hybrid Si-Cu slot waveguide is a best trade-off between truly confined Cu-slots and extremely-low-loss Si-slot waveguides. As observed, if the slab width is equal to or greater than 250 nm, the bounded slab polariton is dominant. This is the case when the effective index is the same for both Si-slot and hybrid-slot waveguides.

For deeper understand of waveguide confinement properties in Si- and Cu-based

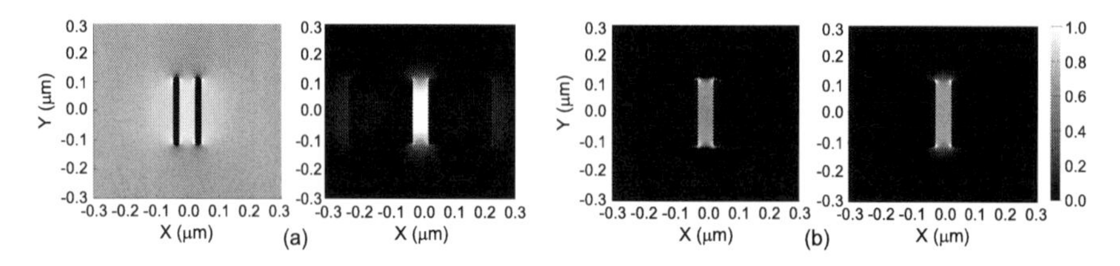

Figure 11.12: Electric-field intensity for (a) Si-SiO_2-Si and (b) Cu-SiO_2-Cu slot waveguides with the slab widths of 20 and 200 nm, while the slot width is fixed at 50 nm.

slot waveguides, we plotted their electric-field intensity in Fig. 11.12 for slab widths of 20 and 200 nm. They demonstrate that the waveguides with small slab widths are a hybrids of MIM (Cu-SiO_2-Cu) and IMI (SiO_2-Cu-SiO_2) modes, where the IMI mode significantly contributes when the Cu-slab is very thin. This phenomenon cannot be seen when the Cu-slab is too thick as only the SP slot mode (or MIM mode) is totally dominant. For the Si-slot waveguide with s small slab width (below 100 nm), the effective index is close to the SiO_2 index as the Si-slot waveguide encounters the diffraction limit. This results in a very week power confinement inside the slot area. For thicker Si-slabs below 500 nm, the slot mode becomes stronger. In contrast, if the Si-slab width increases and exceeds than 500 nm, the bound modes appear inside the Si-slab and become dominant.

In Fig. 11.13 we plot the power confinement factors as functions of the waveguide slab widths w for three aforementioned waveguide structures with the slot widths fixed at 50 nm. One can see that the power confinement of the Cu-slot waveguide is nearly constant and less dependent on the width of Cu-slab. Its value is more than twice those of the two other waveguides and makes the metallic slot waveguide more attractive, especially if better confinement is needed. In contrast, for the two other cases, the confinement factor features the pronounced maximum. In both cases, the optimal maximum power confinement can be achieved at $w \approx 200$ nm.

Next, we consider the effect of the slot width on the propagation constant and the power confinement. We maintain the waveguide thickness of 220 nm following the standard of SOI wafer. The slab width for both Si and Cu is fixed at 200 nm for simplicity. Note that an increase in the slot width g decreases the effective index n_{eff}, as shown in Fig. 11.14. In fact, the effective index dependence on g is more pronounced for for Cu-slot waveguide. For example, the effective index for Cu-SiO_2-Cu waveguide with $g = 10$ nm is approximately at 4.7, while for $g = 500$ nm it decreases to ~ 1.5. In contrast to this, the variation range of the effective index is small, approximately [2.6 to 1.5] for the hybrid Si-SiO_2-Cu waveguides and [2.1 to 1.5] for the Si-SiO_2-Si slot waveguide.

Figure 11.15 shows the electric-field intensity for the three waveguides for the case of a very small slot width $g = 10$ nm. It confirms the large discontinuity and high electric-field intensity confinement inside the slot. The Cu slot presents the largest contrast ratio between the intensity in the SiO_2 slot and the exterior ones. Also, we

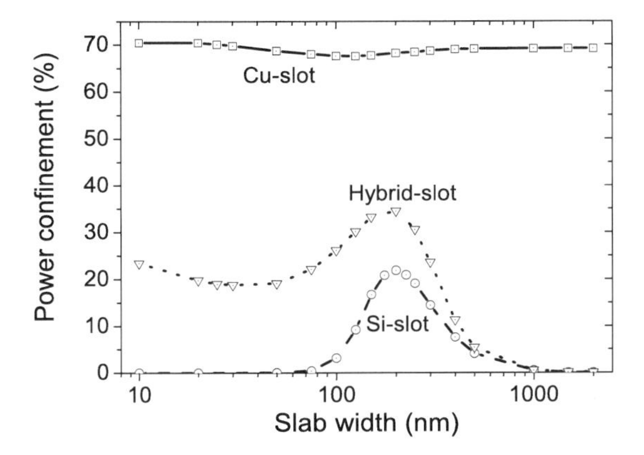

Figure 11.13: Power confinement factor as a function slab widths for for Si-SiO_2-Si, Cu-SiO_2-Cu, and Si-SiO_2-Cu slot waveguides with a fixed slot width of 50 nm.

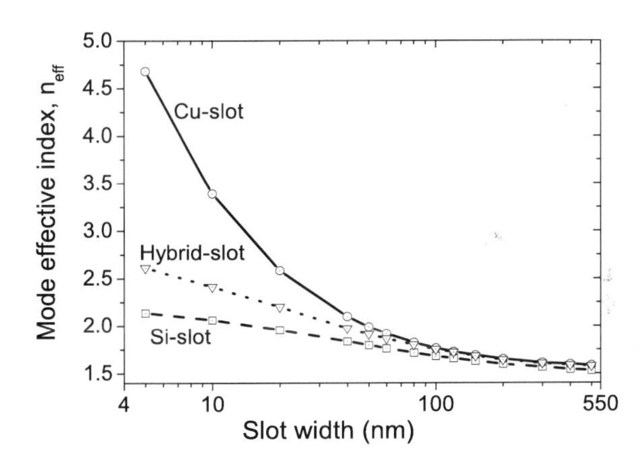

Figure 11.14: Effective index as a function of slot width for Si-SiO_2-Si, Cu-SiO_2-Cu, and Si-SiO_2-Cu slot waveguides. The slab width is fixed at 200 nm for all waveguide types.

find that the hybrid slot waveguide achieves a balanced confinement value, which is well calibrated between the pure Si slot and metallic slot waveguides. When the slot width increases to a large value, for example, $g = 500$ nm, the field intensity is less confined in the slot, as shown in Fig. 11.16. The discrepancy of the field intensity between the slot area and the outer areas such as Si-SiO_2 and Cu-SiO_2 interfaces is quite small and makes the waveguide platforms very weak in light-matter interaction compared to the smaller slot width cases.

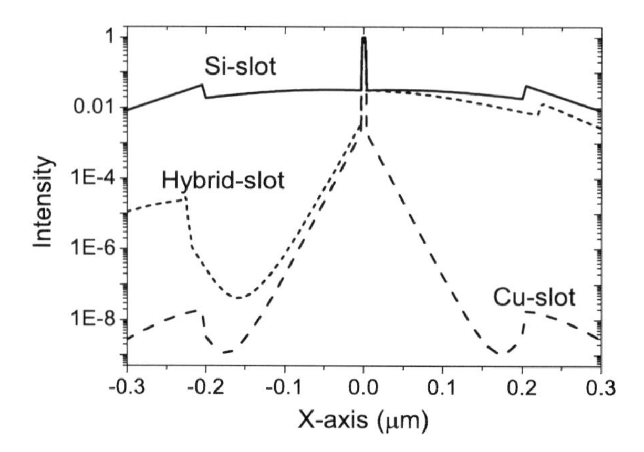

Figure 11.15: Electric-field intensity along the cross central line for Si-SiO_2-Si, Cu-SiO_2-Cu, and Si-SiO_2-Cu slot waveguides with a slot width of 10 nm.

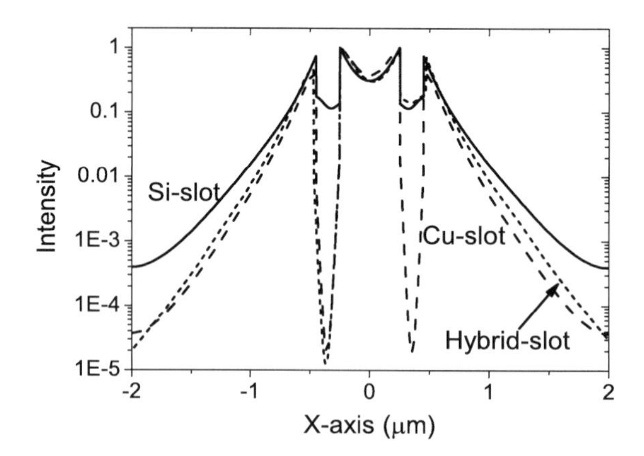

Figure 11.16: Electric-field intensity along the cross central line for Si-SiO_2-Si, Cu-SiO_2-Cu, and Si-SiO_2-Cu slot waveguides with a slot width of 500 nm.

For comparison, the power confinements in the three studied waveguides are plotted in Fig. 11.17. We can see that the Cu slot waveguide exhibits far stronger confinement than its counterparts for small slot widths. In this case, light propagation in the Cu slot waveguide shows a much higher intensity than levels achievable in two other waveguides. In contrast, the Si slot has the lowest power confinement, below 20%. Although the Cu slot waveguide is a best candidate for achieving high power confinement, it evidently encounters high propagation loss. To overcome this challenge, the hybrid-slot waveguide can be used; it achieves a best trade-off between waveguide

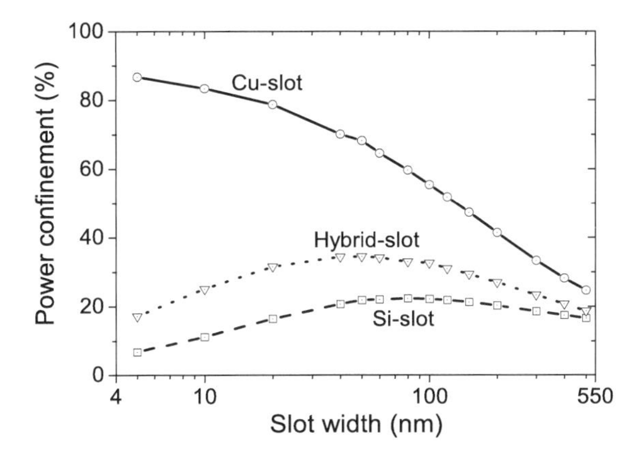

Figure 11.17: Power confinement factor as a function of slot width g for Cu-SiO_2-Cu (Cu-slot), hybrid Si-SiO_2-Cu (hybrid-slot), and Si-SiO_2-Si (Si-slot) slot waveguides. The slab width g is taken at 50 nm.

loss and confinement. The compromised performance of the hybrid waveguide can be explained by two factors. First the low propagation loss is due to the presence of the Si-SiO_2 interface, which has much lower absorption loss than Cu-SiO_2, and the electric-field is mainly spread over the oxide layer with relatively less contact with the metal surface. Second, the dielectric discontinuity at the semiconductor oxide interface generates strong surface polarization charges that form an effective optical capacitor, leading to a high confinement of the guided polariton.

In order to clarify the dependence of the propagation characteristics on the operating wavelength, we studied the propagation length and confinement factor of the slot polaritons in the near-infrared spectrum range from 1 μm to 2 μm. For the sake of simplicity, the dimensions of the Si nanowire are taken at w = 200 nm and h = 220 nm, as they provide reasonable propagation length and confinement. For comparison, the effective indices in the three different studied waveguides are plotted in Fig. 11.18. We can see that the Cu slot waveguide is more stable than its counterparts. The Cu slot waveguide can be considered as nearly dispersionless in the considered wavelength range while the Si-slot waveguide has a variation of 37%, and the hybrid provides a compromise between two waveguides.

Thus, the sub-wavelength slot waveguides discussed above possess two unique properties. First, they support excitation of high electric fields at levels that cannot be achieved with conventional slab waveguides. By utilizing these fields, we can realize highly efficient interaction between fields and active materials useful for all-optical switching and parametric amplification in integrated nanophotonics. Second, a strong E-field confinement is localized in a nanometer-sized low-index region; therefore, slot waveguides can be used to enhance the sensitivity of compact optical sensing devices or the efficiency of near-field optical probes. These findings are particularly

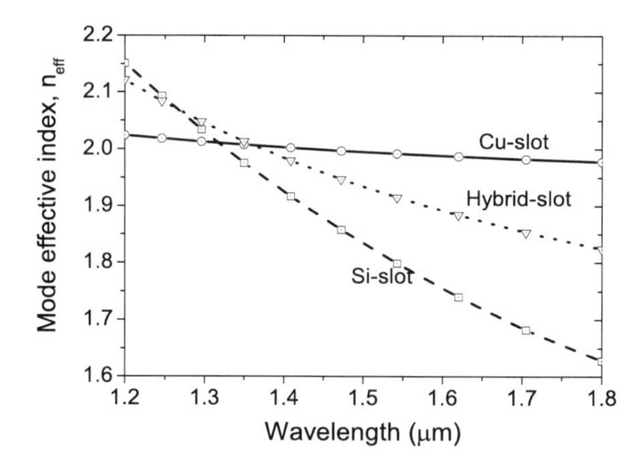

Figure 11.18: Effective index as a function of wavelength for Cu-SiO_2-Cu (Cu-slot), hybrid Si-SiO_2-Cu (hybrid-slot), and Si-SiO_2-Si (Si-slot) slot waveguides. The slab width g and slot width w are taken, respectively, at 200 and 50 nm.

useful for development of novel sensors able to detect single molecules at low concentration levels and for development of nonlinear optical devices for datacom and quantum communication.

11.5 Summary

In this chapter, we reviewed various slot waveguides ranging from nanoparticle double chain to slab slot waveguides. These platforms are useful not only for waveguiding, but also for sensing and nonlinear optics, as they provide strong electric field enhancement. For example, by optimizing the dimensions of metal nanoparticle-double-chain waveguides and excitation wavelengths we can efficiently couple light into the waveguide and transmit it for the longest distance with strong field enhancement inside the gap between adjacent particles. Similar effects are observed for slab-slot waveguides that allows us to excite extremely high fields within a nanometer-sized slot. Among the three studied CMOS-compatible platforms for slab slot waveguides, the Si-slot exhibits the lowest loss for guiding light, while the Cu-slot provides the highest field confinement enabling the best light-matter interaction properties. More importantly, the hybrid-slot is feasible to achieve the best trade-off between confinement and propagation loss, which makes it a potential platform for future waveguide-based devices for datacom and sensing applications.

Bibliography

[1] Committee on Harnessing Light: Capitalizing on Optical Science Trends and Challenges for Future Research and National Materials and Manufacturing Board and Division on Engineering and Physical Sciences and National Re-

search Council, *Optics and Photonics: Essential Technologies for Our Nation*. National Academies Press, 2013.

[2] Y. A. Vlasov, "Silicon CMOS-integrated nano-photonics for computer and data communications beyond 100 G," *IEEE Communications Magazine*, p. S67, Feb 2012.

[3] E. Ozbay, "Plasmonics: merging photonics and electronics at nanoscale dimensions," *Science*, vol. 311, p. 189, 2006.

[4] R. Kirchain and L. Kimerling, "A roadmap for nanophotonics," *Nature Photonics*, 2007.

[5] T. Baehr-Jones, B. Penkov, J. Huang, P. Sullivan, J. Davies, J. Takayesu, J. Luo, T.-D. Kim, L. Dalton, A. Jen, M. Hochberg, and A. Scherer, "Nonlinear polymer-clad silicon slot waveguide modulator with a half wave voltage of 0.25 V," *Applied Physics Letters*, vol. 92, no. 16, 2008.

[6] C. Koos, P. Vorreau, T. Vallaitis, P. Dumon, W. Bogaerts, R. Baets, B. Esembeson, I. Biaggio, T. Michinobu, F. Diederich, W. Freude, and J. Leuthold, "All-optical high-speed signal processing with siliconorganic hybrid slot waveguides," *Nature Photonics*, vol. 3, p. 216, 2009.

[7] J. A. Dionne, K. Diest, L. A. Sweatlock, and H. A. Atwater, "Plasmostor: a metal-oxide-Si field effect plasmonic modulator," *Nano Letters*, 2009.

[8] A. Martinez, J. Blasco, P. Sanchis, J. V. Galan, J. Garcia-Ruperez, E. Jordana, P. Gautier, Y. Lebour, S. Hernandez, R. Guider, N. Daldosso, B. Garrido, J. M. Fedeli, L. Pavesi, J. Marti, and R. Spano, "Ultrafast All-Optical Switching in a Silicon-Nanocrystal-Based Silicon Slot Waveguide at Telecom Wavelengths," *Nano Letters*, vol. 10, 2010.

[9] G. T. Reed, *Silicon Photonics: The State of the Art*. Wiley, 2008.

[10] V. R. Almeida, Q. Xu, C. A. Barrios, and M. Lipson, "Guiding and confining light in void nanostructure," *Optics Letters*, vol. 29, p. 1209, 2004.

[11] Q. Xu, V. R. Almeida, R. R. Panepucci, and M. Lipson, "Experimental demonstration of guiding and confining light in nanometer-size low-refractive-index material," *Optics Letters*, vol. 29, p. 1626, 2004.

[12] J. D. Joannopolous, S. G. Johnson, J. N. Winn, and R. D. Meade, *Photonic Crystals: Molding the Flow of Light*. Princeton University Press, 1995.

[13] M. Loncar, T. Doll, J. Vuckovic, and A. Scherer, "Design and fabrication of silicon photonic crystal optical waveguides," *IEEE Journal of Lightwave Technology*, vol. 18, p. 1402, 2000.

[14] W. Bogaerts, V. Wiaux, D. Taillaert, S. Beckx, B. Luyssaert, P. Bienstman, and R. Baets, "Fabrication of Photonic Crystals in Silicon-on-Insulator using 248 nm Deep UV Lithography," *IEEE Journal of Selected Topics in Quantum Electronics*, vol. 8, p. 928, 2002.

[15] H. Raether, *Surface Plasmons on Smooth and Rough Surfaces and on Gratings*. Springer, 1988.

[16] D. K. Gramotnev and S. I. Bozhevolnyi, "Plasmonics beyond the diffraction limit," *Nature Photonics*, vol. 4, p. 83, 2010.

[17] V. J. Sorger, R. F. Oultona, R.-M. Ma, and X. Zhang, "Toward integrated plasmonic circuits," *MRS Bulletin*, vol. 37, p. 728, 2012.

[18] M. Quinten, A. Leitner, J. R. Krenn, and F. R. Aussenegg, "Electromagnetic energy transport via linear chains of silver nanoparticles," *Optics Letters*, vol. 23, p. 1331, 1998.

[19] S. A. Maier and et al., "Local detection of electromagnetic energy transport below the diffraction limit in metal nanoparticle plasmon waveguides," *Nature Materials*, vol. 2, p. 229, 2003.

[20] H. S. Chu, W. B. Ewe, E. P. Li, and R. Vahldieck, "Analysis of sub-wavelength light propagation through long double-chain nanowires with funnel feeding," *Optics Express*, vol. 15, p. 4216, 2007.

[21] H. S. Chu, W. B. Ewe, W. S. Koh, and E. P. Li, "Remarkable influence of the number of nanowires on plasmonic behaviors of the coupled metallic nanowire chain," *Applied Physics Letters*, vol. 92, p. 103103, 2008.

[22] W. L. Barnes, A. Dereux, and T. Ebbesen, "Surface plasmon subwavelength optics," *Nature*, vol. 424, p. 824, 2003.

[23] R. Zia, M. D. Selker, P. B. Catrysse, and M. I. Brongersma, "Geometries and materials for subwavelength surface plasmon modes," *Journal of the Optical Society of America A*, vol. 21, p. 2442, 2004.

[24] R. Zia, J. A. Schuller, A. Chandran, and M. L. Brongersma, "Plasmonics: The next chip-scale technology," *Material Today*, vol. 9, p. 20, 2006.

[25] L. Liu, Z. Han, and S. He, "Novel surface plasmon waveguide for high integration," *Optics Express*, vol. 13, p. 6645, 2005.

[26] K. C. Y. Huang, M.-K. Seo, T. Sarmiento, Y. Huo, J. S. Harris, and M. L. Brongersma, "Electrically driven subwavelength optical nanocircuits," *Nature Photonics*, vol. 8, p. 244, 2014.

[27] J. A. Dionne, H. Lezec, and H. A. Atwater, "Highly confined photon transport in subwavelength," *Nano Letters*, vol. 6, p. 1928, 2006.

[28] S. I. Bozhevolnyi, V. S. Volkov, E. Devaux, J.-Y. Laluet, and T. W. Ebbesen, "Channel plasmon subwavelength waveguide components including interferometers and ring resonators," *Nature*, vol. 440, p. 508, 2006.

[29] E. Moreno, F. J. Garcia-Vidal, S. G. Rodrigo, L. Martin-Moreno, and S. I. Bozhevolnyi, "Channel plasmon-polaritons: modal shape, dispersion, and losses," *Optics Letters*, vol. 31, p. 3447, 2006.

[30] B. S. et al., "Dielectric strips on gold as surface plasmon waveguides," *Applied Physics Letters*, vol. 88, p. 094104, 2006.

[31] H. S. Chu, W. Ewe, and E. Li, "Tunable propagation of light through a coupled-bent dielectric-loaded plasmonic waveguides," *Journal of Applied Physics*, vol. 106, p. 106101, 2009.

[32] R. F. Oulton, V. J. Sorger, D. A. Genov, D. F. P. Pile, and X. Zhang, "A hybrid plasmonic waveguide for subwavelength confinement and long-range propagation," *Nature Photonics*, vol. 2, p. 496, 2008.

[33] H. S. Chu, E. Li, P. Bai, and R. Hegde, "Optical performance of single-mode hybrid dielectric-loaded plasmonic waveguide-based components," *Applied Physics Letters*, vol. 96, p. 221103, 2010.

[34] M. Z. Alam, J. Meier, J. S. Aitchison, and M. Mojahedi, "Propagation characteristics of hybrid modes supported by metal-low-high index waveguides and bends," *Optics Express*, vol. 18, p. 12971, 2010.

[35] S. K. Gray and T. Kupka, "Propagation of light in metallic nanowire arrays: finite-difference time-domain studies of silver cylinders," *Physical Review B*, vol. 68, p. 045415, Jul 2003.

[36] A. F. Peterson, S. L. Ray, and R. Mittra, *Computational Methods for Electromagnetics*. Wiley-IEEE Press, 1997.

[37] E. D. Palik, *Handbook of Optical Constants of Solids*. Academic Press, San Diego, 1991.

[38] T.-K. Wu and L. Tsai, "Scattering by arbitrarily cross-sectioned layered, lossy dielectric cylinders," *IEEE Transactions on Antennas and Propagation*, vol. 25, pp. 518–524, Jul 1977.

[39] Y. Chang and R. Harrington, "A surface formulation for characteristic modes of material bodies," *IEEE Transactions on Antennas and Propagation*, vol. 25, pp. 789–795, Nov 1977.

[40] B. Jalali and S. Fathpour, "Silicon photonics," *Journal of Lightwave Technology*, vol. 24, 2006.

[41] R. Soref, "The past, present, and future of silicon photonics," *IEEE Journal of Selected Topics in Quantum Electronics*, vol. 12, p. 1678, 2006.

[42] L. Chen, J. Shakya, and M. Lipson, "Subwavelength confinement in integrated metal slot waveguide on silicon," *Optics Letters*, 2006.

[43] T. Baehr-Jones, M. Hochberg, C. Walker, and A. Scherer, "High-Q optical resonators in silicon-on-insulator based slot waveguides," *Applied Physics Letters*, vol. 86, p. 081101, 2005.

[44] T. Baehr-Jones, M. Hochberg, G. Wang, R. Lawson, Y. Liao, P. A. Sullivan, L. Dalton, A. K.-Y. Jen, and A. Scherer, "Optical modulation and detection in slotted silicon waveguides," *Optics Express*, vol. 13, p. 5216, 2005.

[45] C. A. Barrios and M. Lipson, "Electrically driven silicon resonant light emitting device based on slotwaveguide," *Optics Express*, vol. 13, p. 10092, 2005.

[46] C.-H. Chen, L. Pang, C.-H. Tsai, U. Levy, and Y. Fainman, "Compact and integrated TM-pass waveguide polarizer," *Optics Express*, vol. 13, p. 5347, 2005.

[47] T. Fujisawa and M. Koshiba, "Polarization-independent optical directional coupler based on slot waveguide," *Optics Letters*, vol. 31, p. 56, 2006.

[48] S. Roberts, "Optical properties of copper," *Physical Review*, vol. 118, pp. 1509–

1518, Jun 1960.

[49] Y. A. Vlasov and S. J. McNab, "Losses in single-mode silicon-on-insulator strip waveguides and bends," *Optics Express*, vol. 12, pp. 1622–1631, Apr 2004.

[50] Z. Zhu and T. G. Brown, "Full-vectorial finite-difference analysis of microstructured optical fibers," *Optics Express*, vol. 10, pp. 853–864, Aug 2002.

Chapter 12

Photodetectors

Ching Eng Png
Institute of High Performance Computing, Singapore

Song Sun
Institute of High Performance Computing, Singapore

Ping Bai
Institute of High Performance Computing, Singapore

In this chapter, we will review photodetection technologies based on three material systems: semiconductors, low-dimensional materials, and metals. In Section 12.2, the photodetection technologies based on InGaAs, Ge, and all-Si will be reviewed. In Section 12.3, the development of photodetectors based on low-dimensional materials will be introduced. Different metallic photodetectors including quantum-tunneling based types will be discussed in Section 12.4. The chapter will conclude with future outlook presented in Section 12.5.

12.1 Introduction and background

The concepts of photonics and optoelectronic devices were first raised in the late 1980s. However, only in the past decade have they become intensively research fields, driven primarily by the rapid development of communications technologies. As transistor sizes shrink, the traditional technologies based on copper have been pushed to their limits and can no longer satisfy growing demands for high-speed and large-bandwidth applications. Under this circumstance, photonics is destined to have a central role, owing to high-speed transmission and outstanding low-noise properties. Recent advances and breakthroughs have proven that photonics can deliver low-cost and high-performance solutions not only for the communication industries, but also for other applications such as bio-detection and imaging [1].

So far, various photonic devices have been developed including light emission sources, optical modulators, and other passive components [2]. Among these devices, photodetectors (PDs) are placed at the ends of optical paths to convert optical signals into electrical ones. As a result, their conversion efficiency greatly affects the overall performance of entire systems. In general, an ideal PD should possess high responsivity or sensitivity, high detection speed, large bandwidth, high quantum efficiency

(QE), low dark currents (small stand-by power consumption and low noise level), and low applied voltage. Besides the above characteristics, another important criterion for a photodetector is its compatibility with the complementary-metal-oxide-semiconductor (CMOS) technology to take advantage of: i) good noise immunity, low power dissipation, and reliability of the CMOS fabrication process; ii) direct integration with other circuit components without changing foundry processes; and iii) perhaps most importantly, high packing density and scalability that are essential in construction of very-large-scale-integration (VLSI) systems.

To achieve high performance and CMOS-compatible photodetection, an appropriate material system is necessary and several have been aggressively investigated in the past 30 years. The commonly adopted photodetection mechanism utilizes interband transitions in semiconductors to absorb photons, where the operation wavelength can be adjusted by proper selection of existing semiconductor materials or creation of the required bandgaps with band-engineering techniques. A number of semiconductor systems have been verified with photodetection capability including Si, Ge, and InGaAs [3, 4]. Recently, low-dimensional materials such as graphene, carbon nanotubes and noble metals have been explored for next generation PDs. Graphene is an atomically thin film with a unique band structure, possessing ultra-high carrier mobility and a wide optical absorption spectrum [5]. A carbon nanotube is a two-dimensional structure with a diameter-dependent tunable bandgap from ultraviolet to infrared [6]. Ultra-high responsivity and large bandwidth have been demonstrated, but the full potential of these low-dimensional materials has not been realized yet. Meanwhile, noble metals are also of great technological interest as a bridge between photonics and electronics to enhance light-matter interaction. Metallic PDs exhibit small footprint, high responsivity per volume, and potentially high-speed [7].

12.2 Semiconductor-based photodetectors

12.2.1 $In_xGa_{1-x}As$ photodetectors

$In_xGa_{1-x}As$ alloy is currently the most mature material system for photodetection, whereby the bandgap of $In_xGa_{1-x}As$ can be tuned for wavelengths between 0.85 and 3.6 μm, making it ideal for broadband photodetection. By simply changing the alloy composition, the photodetection responsivity can be maximized at the desired wavelength to enhance the signal-to-noise ratio. PD based on $In_xGa_{1-x}As$ currently outperforms Si and Ge owing to its direct bandgap properties and relatively large absorption coefficient, making it able to cover a broader spectra range (see Fig. 12.1) [2].

The development of $In_xGa_{1-x}As$ fabrication technology involves various PD structures, including *p*-doped-intrinsic-*n*-doped (*p-i-n*), metal-semiconductor-metal (MSM), and an avalanche photodetector (APD). The classic type is the *p-i-n* that relies on both carrier drift and diffusion to generate photocurrents. Through evanescent coupling of the *p-i-n* detector with the Si waveguide, a high responsivity of 1.1 A/W at 1550 nm has been achieved by Sheng et al. with a low dark current of 10 pA and a high QE over 90% [8]. An alternative Si-waveguide butt-coupled PD with a 9

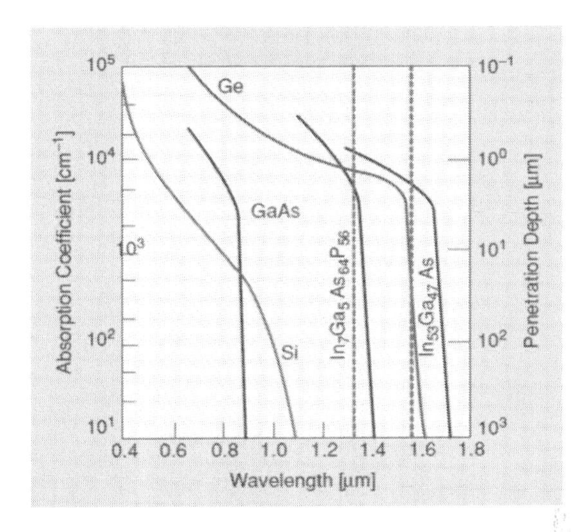

Figure 12.1: Absorption coefficients and penetration depths of various bulk materials as a function of wavelength. Reprinted with permission. Reference [2]. © 2006 IEEE.

GHz 3dB-bandwidth at −4 V bias and a high detection speed of 10 Gb/s has been demonstrated [9].

MSM PDs consist of interdigitated Schottky metal contacts on top of absorption layers that are compatible with the current field-effect transistor technology. The MSM PDs generally have higher speed and lower dark currents than *p-i-n* PDs due to their low capacitance, but suffer from smaller responsivity. Kim et al. reported a back-illuminated InGaAs MSM PD with a record responsivity of 0.96 A/W and 92% QE at 5 V bias, although the bandwidth is only 4 GHz [10]. Cheng et al. demonstrated an InGaAs MSM PD that is monolithically integrated with an InP waveguide that fully covers the *C* and *L* bands with a nearly-flat wavelength dependent responsivity [11].

APDs are *p-n* junction photodiodes operated at high electric fields to achieve photocurrent gains >1, where an incident photon can generate more than one electron-hole pair. In high speed long distance optical systems, APD is normally the best choice owing to its large internal gain, which provides sensitivity of 5 to 10 dB higher than the *p-i-n* photodiodes. For instance, using narrow InGaAs multiplication layer widths, a very low dark current of 10 to 300 nA with a 3 dB bandwidth of 80 GHz was achieved [12].

12.2.2 Ge-on-Si photodetectors

Today, fully CMOS-integrated Ge PDs are commercially available and share the market with InGaAs PDs. In terms of material properties, Ge is more suitable for photodetection than InGaAs, because the lattice mismatch between Ge and Si is 4.2%,

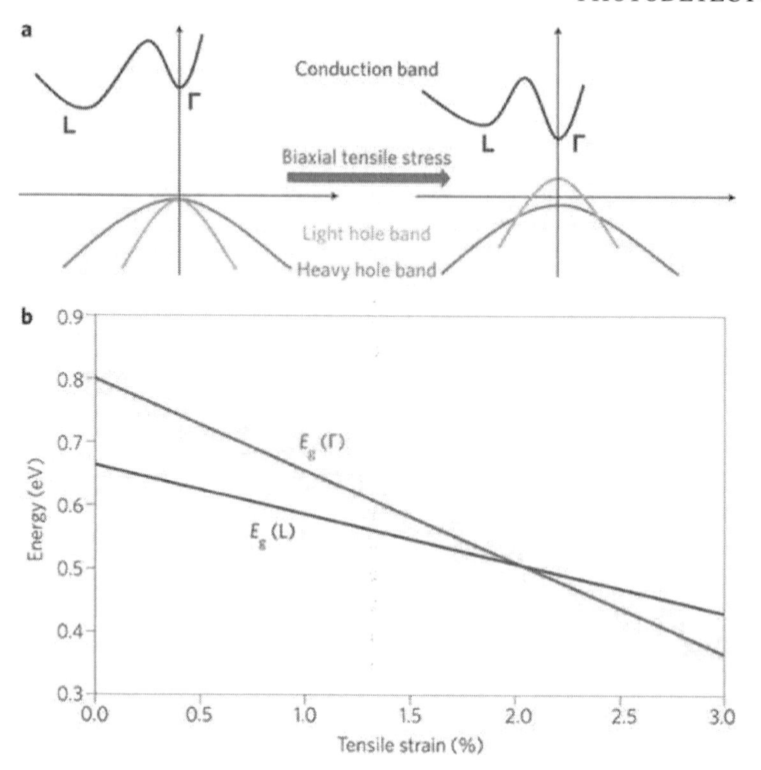

Figure 12.2: The effect of tensile strain on the band structure of Ge. (a) Diagram showing how the band diagram changes as biaxial tensile strain is applied. (b) Plot of the bandgap energies for the Γ and L bands as a function of tensile strain. Reprinted with permission. Reference [13]. © 2010 Nature publishing group.

relatively smaller than that of InGaAs/Si (∼8.1% lattice mismatch), which makes Ge a better choice to be integrated with a CMOS circuit. Furthermore, the epitaxial growth technique has also been developed to grow Ge directly on Si without using the buffer layer, and more importantly, introduce tensile strain to the Ge layer during fabrication, which could transform Ge from an indirect to a direct bandgap material (see Fig. 12.2), greatly improving optoelectronic properties [13].

Ge PDs have similar structures to those of InGaAs PDs, including the *p-i-n*, MSM, and APD. The early Ge *p-i-n* PDs used a SiGe buffer layer that suffered from high dark currents due to large thermionic emission even from high-quality Ge crystals. Later, it was found that coupling of the photodiode to a waveguide can significantly improve responsivity while maintaining a small dark current; as a result, a responsivity larger than 1 A/W was achieved [14].

The performance of Ge MSM PDs was limited initially by the high dark currents due to low hole Schottky barrier height caused by Fermi level pinning near the valence band edge. To overcome this problem, Ang et al. utilized a Si:C layer to manipulate the Schottky barrier and reported an extremely low dark current of 11.5 nA with a high responsivity of 0.76 A/W and a moderate QE ∼60% [15]. Alternatively, implanting and segregating the boron and sulfur dopants can suppress the total dark currents by 3 to 4 orders of magnitude, so the responsivity of Ge MSM PDs can be boosted to 1.76 A/W at 5 V bias [16].

Since Si has a much smaller ionization ratio than the III-V materials, the Ge-on-Si APD is believed to be better than III-V APDs in terms of higher gain-bandwidth product and better sensitivity. Kang et al. achieved a gain-bandwidth product of 340 GHz with a small ionization k-value of 0.09 and a sensitivity of −28 dBm at 10 Gb/s [17]. Assefa et al. manipulated the optical and electrical fields within the Ge layer and achieved 70% noise reduction with a detection speed up to 43 Gb/s with bit-error rate of 10^{12} [18].

12.2.3 All-Si photodetectors

Si crystal has a relatively large indirect bandgap ∼1.1 eV, corresponding to a cut-off wavelength below 1100 nm. Si is not the most suitable material for PDs compared to direct bandgap III-V alloys such as InGaAs or Ge, which have smaller bandgaps. Nevertheless, all-Si PDs are advantageous in terms of monolithic integration with the standard CMOS technology and possibility to significantly reduce fabrication costs. In general, all-Si PD operation is based on three common physical mechanisms: mid-bandgap absorption (MBA), surface-state absorption (SSA), and internal photon emission (IPE).

MBA PDs are developed based on the ability of high energy particles to introduce defect states located within the bandgap of the intrinsic Si crystal, hence enabling detection of sub-bandgap optical radiation. Knight et al. first proposed an all-Si MBA *p-i-n* photodiode with the defect states introduced by ion implantation [19]. With modern fabrication technology, Geis et al. was able to reduce the transversal cross section of the waveguide to increase absorption length and reduce carrier recombination. They demonstrated a responsivity of 0.5 to 10 A/W with a 60 V bias, and a bandwidth >35 GHz [20].

SSA PDs are based on a principle similar to those of MBA PDs, where surface states are introduced into the bandgap of the intrinsic Si. By increasing the overlap between the polaritonic mode and the waveguide surface, Baehr-Jones et al. built an SSA PD in an ultrahigh vacuum environment. They reported the responsivity of 36 mA/W under 11 V bias and dark current of 12 A [21]. Ideally, surface states should have larger carrier mobilities and fewer recombination sites compared to the bulk defect states in the MBA PDs, resulting in higher detection speed and larger QE. However, SSA is very sensitive to Si surface conditions and external temperature, which requires a delicate fabrication process that hinders mass production.

IPE PDs rely on the principle that photo-excited electrons in metal can gain energy above the Schottky barrier and subsequently move into the conduction band of the semiconductor. Elabd et al. first proposed ultra-thin metal films (∼2 nm) to increase the carrier escape probability of the photo-excited electron [22]. The responsivity of 250 mA/W at 1500 nm was achieved with a trade-off in low total optical absorption. Recently, it has been demonstrated that the QE can be significantly improved by coupling the IPE PD with a Si waveguide. The responsivity of 4.6 mA/W at 1550 nm was reported with −1 V bias [23].

12.3 Photodetectors based on low-dimensional materials

Low-dimensional materials provide a fascinating ground between molecular and bulk materials. Interesting properties such as *excitons*,[1] which are often negligible in bulk materials, are greatly accentuated in low-dimensional materials. These unique electronic and optical properties make photodetection promising even in extremely small nanostructures of one atomic layer (graphene) or a few nanometers in two (carbon nanotubes) or three dimensions (quantum dots). In this section, we briefly review photodetection technologies based on graphene and carbon nanotubes [1].

12.3.1 Graphene-based photodetectors

Since its first isolation in 2004, graphene is rapidly becoming an appealing material for photonics and optoelectronics due to its ultra-high bandwidth photodetection capability. Graphene's ultrafast carrier mobility, tunable optical properties via electrostatic doping, low dissipation, and good stability make it an ideal platform for development of high speed PDs. In addition, graphene is compatible with CMOS technology, which is preferable for low-cost and large-scale integration into the CMOS circuits.

The band structure of a single-layer graphene can be calculated with the tight-binding description, where the conduction and valence bands meet at the six Dirac points of the Brillouin zone. Near the Dirac point, the band diagram possesses linear energy-momentum dispersion. Due to this peculiar dispersion property, electrons in graphene behave like massless Dirac fermions with high mobility at $\sim$1/300 the speed of light. Besides the excellent carrier dynamics, the gapless band structure makes graphene capable of broadband photodetection from the UV to even the terahertz region. High optical absorption coefficient of 7×10^5 cm^{-1} has been found from 300 to 2500 nm, which is much higher than those of conventional semiconductors [5].

Generally, the physical mechanisms of photodetection in graphene can be characterized into three major categories: photoelectric, photo-thermoelectric, and photo-bolometric effects. For the photoelectric effect, graphene typically aligns longitudinally between two electrodes and the photo-excited electrons form excitons separated and propelled under external bias to create photocurrent. For photo-thermoelectric effect, a photo-excited electron-hole pair can cause ultrafast heating of the carriers in graphene due to strong electron-electron interaction. By forming inhomogeneous thermoelectric power across the graphene channel, these hot carriers can produce a built-in bias based on the Seebeck effect for a graphene interface junction or a

[1]In this chapter, the term 'exciton' refers to coupled electron-hole pairs, which is different from the description of elementary polarization excitations in Section 1.3.3. We refer the readers to Ref. [5] for more detailed explanation of excitons in solid-state electronics.

dual-gate graphene to generate photocurrent. Photo-bolometric effect is associated with the light-induced change in conductance via optical heating instead of direct photocurrent generation. The temperature-dependent conductance can be attributed to changes in both carrier mobility and carrier density. Note that besides the three major mechanisms, other processes may be responsible for graphene photodetection including photogating, tunneling, and plasma-wave-assisted mechanisms. The readers are referred to Refs. [1, 5] for more detailed information.

The first graphene PD was reported by Xia et al. It was a transistor-based photodetector made of single or double graphene layers, demonstrating an ultrafast photodetection of up to 40 GHz [24]. Konstantatos et al. covered the graphene layer with a thin film of colloidal quantum dots to improve the device gain to 10^8 with a responsivity of 10^7 A/W, but with a limited response speed [25]. Later, Gan et al. developed an evanescently coupled graphene waveguide structure and achieved a responsivity of >0.1 A/W with a uniform response between 1450 and 1590 nm and high speed of 12 Gb/s [26]. Recently, Kim et al. further improved the responsivity to 0.4 to 1.0 A/W in a broad spectral range from ultraviolet to near-infrared using an all-graphene *p-n* vertical junction structure, as shown in Fig. 12.3 [27].

12.3.2 Carbon nanotube-based photodetectors

Carbon nanotubes (CNTs) are nearly ideal one-dimensional (1D) systems with diameters of a few nanometers and lengths that be scaled up to centimeters. By controlling the arrangement of a carbon-atom structure with respect to its axis, CNTs can be direct bandgap semiconductors or even metals with nearly ballistic conduction. The bandgap of the CNT is inversely proportional to its diameter (the bandgap ranges from 0.2 to 1.5 eV for CNT diameters from 0.8 to 3 nm), which allows CNT photodetection from ultraviolet to infrared. Besides the tunable bandgap, CNT intrinsically has high absorption coefficient from 10^4 to 10^5 cm^{-1} in the infrared region, which is one order of magnitude larger than those of conventional semiconductors [6].

Photoexcitation creates an electron in the conduction band, leaving a hole in the valence band. For the bulk semiconductors, the electron-hole interaction is negligible, and they behave as free carriers. On the other hand, for low-dimensional materials, especially CNTs, substantial coulombic interaction due to the strong quantum confinement between the electrons and holes binds them into excitons, offering an alternative valley for charge transfer in the low-dimensional materials. Excitons can be formed efficiently in CNTs under photoexcitation due to the 1D confinement and subsequently decay into quasi-free electrons and holes. In addition, carrier multiplication or robust exciton is found to be effective in the CNT due to impact excitation with picosecond response, therein promising improved photodetection [1, 6].

High performance infrared photodetectors based on CNTs are usually fabricated with molecular beam epitaxial (MBE) or chemical vapor deposition (CVD), where the photodetection is attributed to thermal or photoeffect. For the thermal effect, the current is generated as a result of the change in conductance due to the temperature change under light illumination. St-Antoine et al. demonstrated a thermopile using single-walled CNT thin film with a detectivity of 2×10^6 cm Hz$^{1/2}$/W in the visible

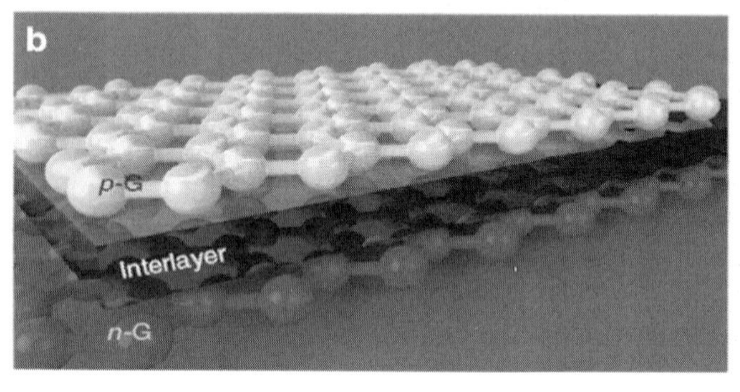

Figure 12.3: (a) A typical graphene *p*-*n* diode for photodetection and (b) its magnified version showing an interlayer between metallic *p*- and *n*-graphene (*p*-G and *n*-G) layers. Ag was used for the electrodes. Reprinted with permission. Reference [27]. © 2014 Macmillan Publishers Limited.

and near-infrared region [28]. Lu et al. showed a bolometer with a higher detectivity of 3.3×10^6 cm Hz$^{1/2}$/W using a multiwall CNT [29]. Alternatively for the photoeffect, Arnold et al. reported a detectivity $>$1010 cm Hz$^{1/2}$/W at 400 to 1500 nm with a 3 dB bandwidth of 31 MHz using a CNT/C60 heterojunction (see Fig. 12.4) [30]. More details can be found in Ref. [6].

Besides the graphene and CNT, other low-dimensional materials also exhibit photodetection capabilities. They include transition metal chalcogenides (TMDCs), semiconducting nanowires, and quantum dots (QDs). The readers are referred to a review [1] for more details.

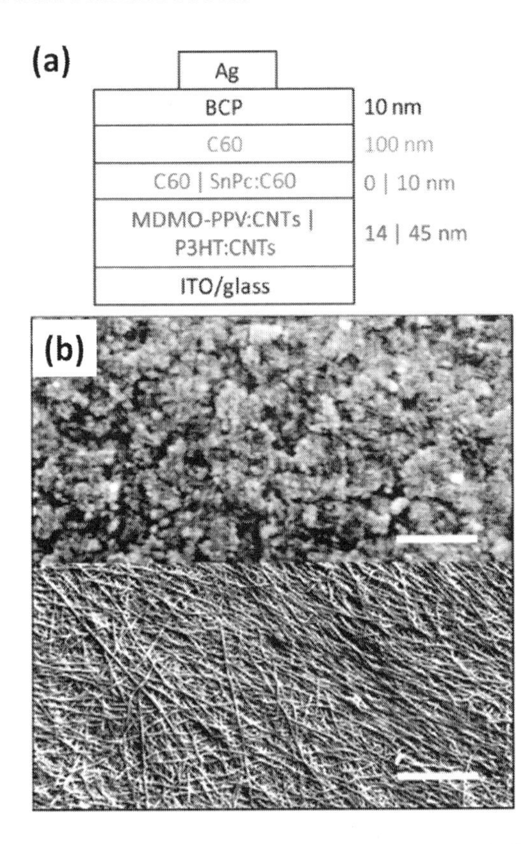

Figure 12.4: (a) Schematic of CNT/organic heterojunction. (b) Film morphology. Reprinted with permission. Reference [30]. © 2009 American Chemical Society.

12.4 Metal-based photodetectors

One of the challenges for the integration of optics and electronics is that the dimensions of optics (in micrometers) are significantly larger than the sizes of electronics (in nanometers). This incompatibility often leads to substantial penalties in power dissipation, area, and latency in photodetectors. Smaller active regions can be designed for PDs. However, this will result in low responsivity because of the short effective absorption length. Here, surface polaritons would be able to enhance and focus electromagnetic waves to improve the performance of the PDs with small active regions.

12.4.1 Electrode surface polaritons

Waveguide PDs are common devices for converting optical to electronic signals by coupling light from waveguides to semiconductors. To increase speed and reduce footprint, their sensitivity needs to be enhanced. Polaritonic resonances in metals

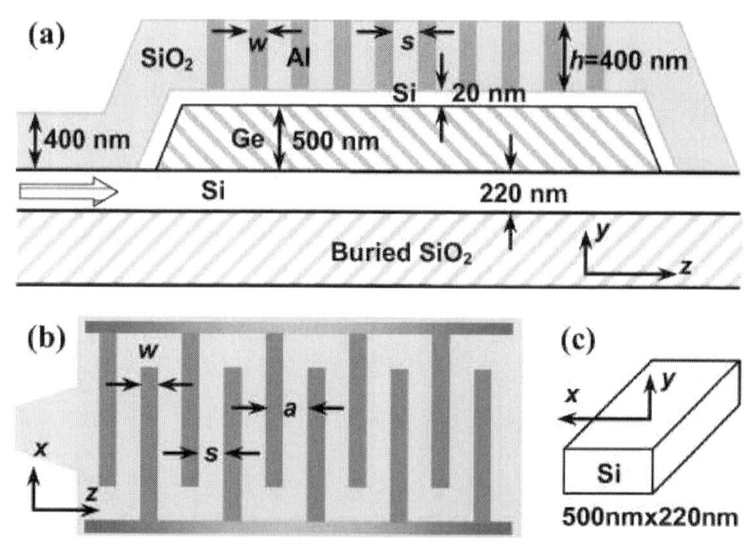

Figure 12.5: Scheme of a Ge-on-SOI MSM photodetector with interdigitated electrodes. (a) Cross-sectional view of the device. (b) Top view of the interdigitated fingers. (c) Perspective view of the Si core layer with thickness of 220 nm and width of 500 nm, where light propagates in the z direction. Reprinted with permission. Reference [31]. © 2010 American Institute of Physics.

have been studied to improve the sensitivity of waveguide PDs. Ren et al. introduced surface polaritons by adding thin Al interdigitated electrodes on top of a Ge-based PD (see Fig. 12.5) [31]. The surface polaritons supported by the Al electrodes considerably enhanced the coupling of light from the silicon waveguide into the PD, enabling the use of smaller devices and higher speed. The resultant strong field intensity permeates the Ge active region, enabling high absorption under TM light illumination. Measured results showed the TM-wave photocurrent is three times higher than the TE one. The responsivity measured was 1.081 A/W at 1550 nm.

Surface polaritons have also been used to enhance the performance of Si waveguide internal photon emission (IPE) PDs (see Sec. 12.2.3). They can dramatically enhance the electromagnetic fields near the metal interface and increase the total absorption, resulting in increased photocurrent. Levy et al. demonstrated an on-chip nanoscale silicon surface-polariton IPE PD, as shown in Fig. 12.6 [32]. The fabricated PDs showed enhanced detection capability attributed to the increased probability of the internal photoemission enhanced by the surface polaritons. The responsivity of 13.3 mA/W was measured at 1310 nm. Halas et al. also reported an active optical PD consisting of an array of independent rectangular gold nanorods, which support two types of resonances on surface polaritons [33], creating hot electron-hole pairs for photocurrents. The experiment measured photocurrent responsivities up to 9 A/W at the telecommunications wavelength with nanoantenna dimensions of only 50 nm (wide) $\times$ 30 nm (high) $\times$ 158 nm (long).

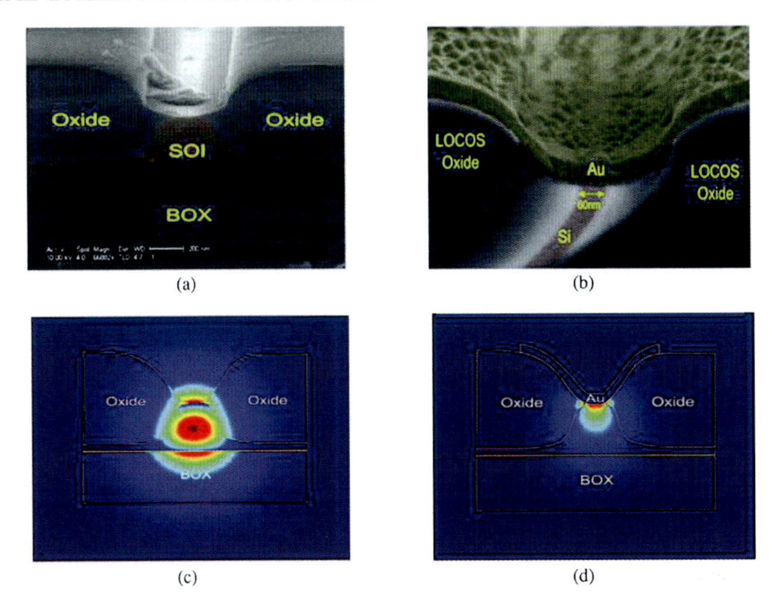

Figure 12.6: (a) SEM micrograph of the photonic waveguide after local oxidation process before the metallization step. (b) SEM micrograph of the Schottky contact. (c) Volume polariton intensity profile of the photonic waveguide. (d) Surface polariton intensity profile of the plasmonic waveguide (Schottky contact). Reprinted with permission. Reference [32]. © 2011 American Chemical Society.

12.4.2 Metal antenna-based photodetectors

Metal antennas have recently been used to enhance the performance of the PDs, which are typically placed close to the active region of the PDs to confine optical near fields into a subwavelength volume. Subwavelength MSM PDs using metal dipole antennas have been demonstrated, as shown in Fig. 12.7 [34]. The detector active region has an ultra-small volume of 150 nm $\times$ 60 nm $\times$ 80 nm, on the order of 10^{-4} of the operating wavelength at 1310 to 1480 nm. An enhancement by a factor of 20 in the photocurrent has been achieved due to the polariton resonance in the metal antenna.

This concept was also extended to metal waveguide PDs to effectively couple the optical power from the waveguide to the antenna by placing a dipole antenna at the end of a metal-insulator-metal (MIM) waveguide. Bai et al. showed a metal MIM waveguide and an MSM photodetector consisting of a dipole antenna formed by two L-shaped metallic nanorods, whereby the electromagnetic fields in the gap between two antenna arms can be enhanced more than a hundred times [35]. The L-shaped nanorods operate as both a dipole nanoantenna and a half-wavelength nanocavity to efficiently couple and confine the electromagnetic fields from the metal waveguide. This detector has a very small active volume (50 nm $\times$ 50 nm $\times$ 50 nm) and potential terahertz detection speed.

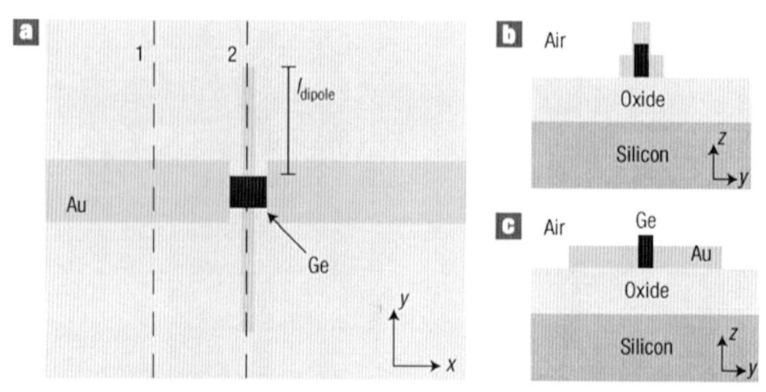

Figure 12.7: (a) Top view of the open-sleeve dipole antenna oriented in the y direction and two line electrodes in the x direction. (b) Cross section of the germanium nanowire lying under the two line electrodes. (c) Cross section showing germanium in the gap region between the two antenna arms. Reprinted with permission. Reference [34]. © 2010 Nature Publishing Group.

Among all metal waveguides, hybrid waveguides are the most suitable for intrachip optical data transmission due to their excellent confinement and relatively low propagation loss. Hybrid waveguides represent a metal-insulator-semiconductor structure, which can act as a monopole antenna with more efficient operation than the dipole antenna configuration. The monopole is essentially a half-dipole mounted on a ground plane. If the ground plane is large enough, the electromagnetic waves reflected from it will seem to come from an image antenna that forms the missing half of the dipole. Figure 12.8 shows a hybrid waveguide PD consisting of a monopole antenna [36]. The waveguide is formed by sandwiching a thin layer of silica between an aluminum slab and a silicon nanowire. The metallic part of the waveguide serves as the conducting ground of the monopole antenna. The antenna feed can thus be placed at the terminal and the monopole antenna on top of it to form a nanocavity or the active volume of the PD. Simulation results show that 42% of optical power is absorbed in the active volume, which is well above the 27% power absorption with a dipole antenna [35]. The optical coupling was further improved using coupled cavities, and 78% absorption efficiency was achieved [36].

12.4.3 Photodetectors without semiconductors

Unlike most PDs relying on semiconductors, polaritons supported by metals provide a direct conversion between photons and electrons, and thus metal PDs could be built without using semiconductors. Melosh et al. designed a purely metal PD by using polariton excitations in a simple MIM structure similar to a rectenna, as shown in Fig. 12.9(a) [37]. When light illuminates from the top of the MIM structure, polaritons excited on the top metal can create a high concentration of hot electrons that can tunnel through the thin insulating barrier to produce photocurrents. Without po-

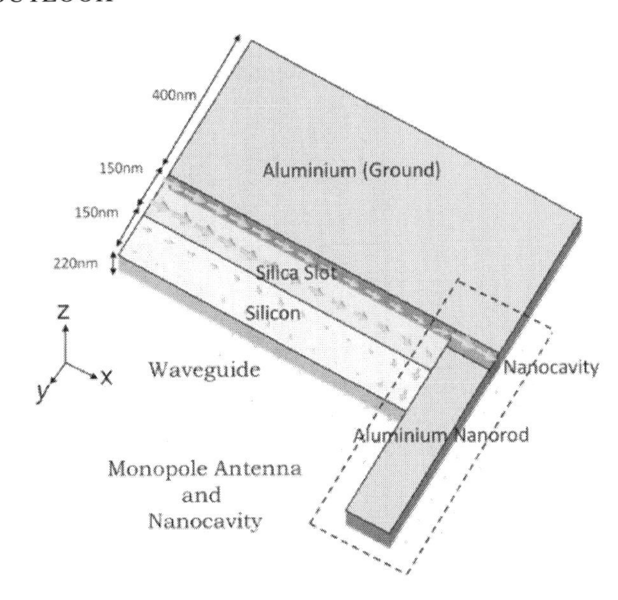

Figure 12.8: Schematic of a monopole antenna waveguide photodetector whose nanocavity is formed by coupling a hybrid waveguide with a monopole antenna. Reprinted with permission. Reference [36]. © 2009 OSA publishing.

lariton excitation, the efficiency of this hot carrier tunneling is low and cannot be considered for photodetection.

When the MIM structure is replaced by a nanoparticle dimer separated by a sub-nanometer gap, charge-transfer polaritons (CTPs) can be excited [38]. This discovery opened new opportunities in nanoscale optoelectronics, single-molecule sensing, and nonlinear optics. Quantum mechanical effects become important when two nanoparticles are placed so closely that electrons can tunnel across the gap. Theoretically, the CTP can be excited only when the vacuum gap is in the scale of 0.3 to 0.5 nm. Recently the CTP was experimentally observed at a large gap up to 1.3 nm, with a self-assembled molecular layer filling the gap, as shown in Fig. 12.9(b) [38]. The measured and simulated results showed that the frequency of the tunneling currents is approximately 200 THz and can be tuned by changing the molecules. This is revolutionary considering that traditional semiconductor optoelectronic detectors can only operate at frequencies limited to tens of GHz. As these PDs also have nanometer footprints, they are promising PDs for next generation optoelectronic circuits, with operation speeds over tens of thousands times higher than the familiar microprocessors.

12.5 Future outlook

Among semiconductors, InGaAs and Ge are currently the most mature platforms that will continue dominating for a time. All-Si PD research will continue improving the

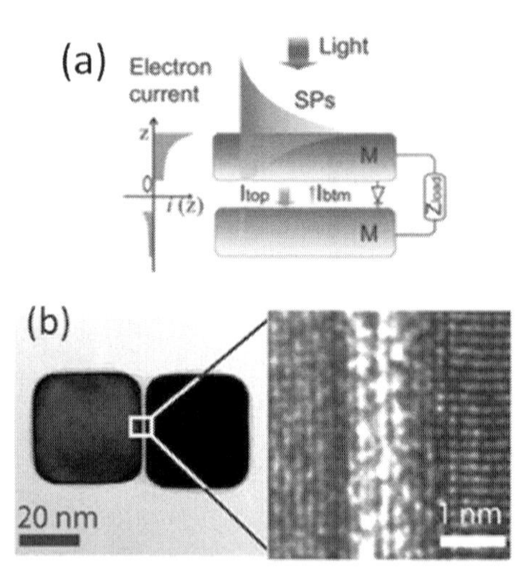

Figure 12.9: (a) Schematic of an MIM optical energy converter. Polariton-enhanced hot electrons in the top electrodes can largely increase the forward current. (b) TEM image of quantum tunnel junctions made of two silver nanoparticles bridged by a self-assembled monolayer in which CTP has been detected. Reprinted with permission. (a) Reference [37]. © 2011 American Chemical Society. (b) Reference [38]. © 2014 American Association for the Advancement of Science.

device performance (sensitivity, speed, bandwidth etc.) to enable suitability for high-performance applications. PDs based on low-dimensional materials (e.g., graphene, CNT, QD etc.) and metals are still in their infancy, but have already shown their potential for nanoscale high-performance PDs. We foresee that low-cost and high-performance PDs may eventually materialize with the rapid progress in the research of low-dimensional materials and metals.

Bibliography

[1] C. Png, S. Sun, and P. Bai, "State-of-the-art photodetectors for optoelectronic integration at telecommunication wavelength," *Nanophotonics*, vol. 4, pp. 277–302, 2015.

[2] B. Jalali, M. Paniccia, and G. Reed, "Silicon photonics," *IEEE Microw. Mag.*, vol. 7, pp. 58–68, 2006.

[3] J. Wang and S. Lee, "Ge-photodetectors for Si-based optoelectronic integration," *Sensors*, vol. 11, pp. 696–718, 2011.

[4] A. Rogalski, "Infrared detectors: status and trends," *Prog. Quant. Electron.*, vol. 27, pp. 59–210, 2003.

[5] F. Bonaccorso, Z. Sun, T. Hasan, and A. Ferrari, "Graphene photonics and optoelectronics," *Nat. Photon.*, vol. 4, pp. 611–622, 2010.

[6] P. Avouris, M. Freitag, and V. Perebeinos, "Carbon-nanotube photonics and optoelectronics," *Nat. Photon.*, vol. 2, pp. 341–350, 2008.

[7] L. Novotny and N. Hulst, "Antennas for light," *Nat. Photon.*, vol. 5, p. 8390, 2011.

[8] Z. Sheng, L. Liu, J. Brouckaert, S. He, and D. Throurhout, "InGaAs PIN photodetectors integrated on silicon-on-insulator waveguide," *Opt. Express*, vol. 18, pp. 1756–1761, 2010.

[9] S. Feng, Y. Geng, K. Lau, and A. Poon, "Epitaxial III-V-on-silicon waveguide butt-coupled photodetectors," *Opt. Lett.*, vol. 37, pp. 4035–4037, 2002.

[10] J. Kim, H. Griem, R. Friedman, E. Chan, and S. Ray, "High-performance back-illuminated InGaAs/InAlAs MSM photodetector with a record responsivity of 0.96 A/W," *IEEE Photon. Technol. Lett.*, vol. 4, pp. 1241–1244, 1992.

[11] Y. Cheng, Y. Ikku, M. Takenaka, and S. Takagi, "InGaAs MSM photodetector monolithically integrated with InP photonic-wire waveguide on III-V CMOS," *IEICE Electron. Expr.*, vol. 11, pp. 1–8, 2014.

[12] L. Tarof, D. Knight, K. Fox, C. Miner, N. Puetz, and H. Kim, "Planar InP/InGaAs avalanche photodetectors with partial charge sheet in device periphery," *Appl. Phys. Lett.*, vol. 57, pp. 670–672, 1990.

[13] J. Michel, J. Liu, and L. Kimerling, "High-performance Ge-on-Si photodetectors," *Nat. Photon.*, vol. 4, pp. 527–534, 2010.

[14] D. Ahn, C. Hong, J. Liu, W. Giziewicz, M. Beals, L. Kimerling, J. Michel, J. Chen, and F. Kartner, "High performance, waveguide integrated Ge photodetectors," *Opt. Express*, vol. 15, pp. 3916–3921, 2007.

[15] K. Ang, S. Zhu, J. Wang, K. Chua, M. Yu, G. Lo, and D. Kwong, "Novel silicon-carbon (Si:C) Schottky barrier enhancement layer for dark-current suppression in Ge-on-SOI MSM photodetectors," *IEEE Electron Dev. Lett.*, vol. 7, pp. 704–707, 2008.

[16] N. Harris, J. Baehr, A. Lim, T. Liow, G. Lo, and M. Hochberg, "Noise characterization of a waveguide-coupled MSM photodetector exceeding unity quantum efficiency," *IEEE J. Lighw. Technol.*, vol. 31, pp. 23–27, 2013.

[17] Y. Kang, H. Liu, M. Morse, M. Paniccia, M. Zadka, S. Litski, G. Sarid, A. Pauchard, Y. Kuo, H. Chen, W. Zaoui, J. Bowers, A. Beling, D. McIntosh, X. Zheng, and J. Campell, "Monolithic germanium/silicon avalanche photodiodes with 340 GHz gain-bandwidth product," *Nat. Photon.*, vol. 3, pp. 59–63, 2009.

[18] I. Kim, K. Jang, J. Joo, S. Kim, S. Kim, K. Choi, J. Oh, S. Kim, and G. Lim, "High-performance photoreceivers based on vertical-illumination type Ge-on-Si photodetectors operating up to 43 Gb/s at $\sim$1550 nm," *Opt. Express*, vol. 21, pp. 30716–30723, 2013.

[19] A. Knights, A. House, R. MacNaughton, and F. Hopper, "Optical power monitoring function compatible with single chip integration on silicon-on-insulator," *Proceedings of Conference on Optical Fiber Communication, Technical Digest Series*, vol. 2, pp. 705–706, 2003.

[20] M. Geis, S. Spector, M. Grein, J. Yoon, D. Lennon, and T. Lyszczarz, "Silicon waveguide infrared photodiodes with >35 GHz bandwidth and phototransistors with 50 A/W response," *Opt. Express*, vol. 17, pp. 5193–5204, 2009.

[21] T. Baehr-Jones, M. Hochberg, and A. Scherer, "Photodetection in silicon beyond the band edge with surface states," *Opt. Express*, vol. 16, pp. 1659–1668, 2008.

[22] H. Elabd, T. Villani, and W. Kosonocky, "Palladium-silicide Schottky-Barrier IR-CCD for SWIR applications at intermediate temperatures," *IEEE Trans. Electron. Dev. Lett.*, vol. EDL-3, pp. 89–90, 1982.

[23] M. Casalino, L. Sirleto, M. Iodice, N. Saffioti, M. Gioffre, I. Rendina, and G. Coppola, "Cu/p-Si Schottky barrier-based near infrared photodetector integrated with a silicon-on-insulator waveguide," *Appl. Phys. Lett.*, vol. 96, pp. 241112–241114, 2010.

[24] F. Xia, T. Mueller, Y. Lin, A. Garcia, and P. Avouris, "Ultrafast graphene photodetector," *Nat. Nanotechnol.*, vol. 4, pp. 839–843, 2009.

[25] G. Konstantatos, M. Badioli, L. Gaudreau, J. Osmond, M. Bernechea, F. Arquer, F. Gatti, and F. Koppens, "Hybrid graphene-quantum dot phototransistors with ultrahigh gain," *Nat. Nanotechol.*, vol. 7, pp. 363–368, 2012.

[26] X. Gan, R. Shiue, Y. Gao, I. Meric, T. Heinz, K. Shepard, J. Hone, S. Assefa, and D. Englund, "Chip-integrated ultrafast graphene photodetector with high responsivity," *Nat. Photon.*, vol. 7, pp. 883–887, 2013.

[27] C. Kim, S. Kum, D. Shin, S. Kang, J. Kim, C. Jang, S. Joo, L. Lee, J. Kim, S. Choi, and E. Hwang, "High photoresponsivity in an all-graphene $p-n$ vertical junction photodetector," *Nat. Commun.*, vol. 5, pp. 3249:1–7, 2014.

[28] St-Antoine, D. Menard, and R. Martel, "Single-walled carbon nanotube thermopile for broadband light detection," *Nano Lett.*, vol. 11, pp. 609–613, 2011.

[29] R. Lu, J. Shi, F. Baca, and J. Wu, "High performance multiwall carbon nanotube bolometers," *J. Appl. Phys.*, vol. 108, p. 084305, 2010.

[30] M. Arnold, J. Zimmerman, C. Renshaw, X. Xu, R. Lunt, C. Austin, and S. Forrest, "Broad spectral response using carbon nanotube/organic semiconductor/C_{60} photodetectors," *Nano Lett.*, vol. 9, pp. 3354–3358, 2009.

[31] F. Ren, K. Ang, J. Song, Q. Fang, M. Yu, G. Lo, and D. Kwong, "Surface plasmon enhanced responsivity in a waveguided germanium metal-semiconductor-metal photodetector," *Appl. Phys. Lett.*, vol. 97, p. 091102, 2010.

[32] I. Goykhman, B. Desiatov, J. Khurgin, J. Shappir, and U. Levy, "Locally oxidized silicon surface-plasmon Schottky detector for telecom regime," *Nano Lett.*, vol. 11, pp. 2219 – 2224, 2011.

[33] M. Knight, H. Sobhani, P. Nordlander, and N. Halas, "Photodetection with active optical antennas," *Science*, vol. 332, pp. 702–704, 2011.

[34] S. Tang, L. Kocabas, S. Latif, A. Okyay, D. Ly-Gagnon, K. Saraswat, and D. Miller, "Nanometre-scale germanium photodetector enhanced by a near-infrared dipole antenna," *Nat. Photon.*, vol. 2, pp. 226 – 229, 2008.

[35] P. Bai, M. Gu, X. Wei, and E. Li, "Electrical detection of plasmonic waves using an ultra-compact structure via a nanocavity," *Opt. Express*, vol. 17, pp. 24349–24357, 2009.

[36] K. Ooi, P. Bai, M. Gu, and L. Ang, "Design of a monopole-antenna-based resonant nanocavity for detection of optical power from hybrid plasmonic waveguides," *Opt. Express*, vol. 19, pp. 17075–17085, 2011.

[37] F. Wang and N. Melosh, "Plasmonic energy collection through hot carrier extraction," *Nano Lett.*, vol. 11, pp. 5426–5430, 2011.

[38] S. Tan, L. Wu, J. Yang, P. Bai, M. Bosman, and C. Nijhuis, "Quantum plasmon resonances controlled by molecular tunnel junctions," *Science*, vol. 343, pp. 1496–1499, 2014.

Chapter 13

Integrated nonlinear photonics

Jun Rong Ong
Institute of High Performance Computing, Singapore

In this chapter, we will review developments in the field of integrated nonlinear photonics. Section 13.1 will introduce the field with exposure to nonlinear optical effects and materials. Section 13.2 will discuss the use of silicon for integrated nonlinear photonics. This will be followed by discussion of integrated nonlinear quantum photonics in Section 13.3. Finally, the chapter will conclude with future outlook for integrated nonlinear photonics in Section 13.4.

13.1 Introduction and background

The study of the nonlinear response of optical media under illumination by high intensity light has its origins in the 1960s [1]. At that time, the invention of the laser provided the intense coherent light required to observe the first nonlinear optical processes. Within a few years, experimentalists were able to measure a number of wave mixing products (second harmonic, third harmonic, Raman scattering etc.) in a plethora of nonlinear optical crystals [2, 3]. The theoretical framework of nonlinear optics was then formulated shortly after the first experimental observations [4].

Subsequently, the invention of the low-loss silica optical fiber in the 1970s added a new dimension to nonlinear optics. These low-loss fiber optics designed to transmit optical data over large distances, also allowed light waves to interact coherently within the waveguides. Despite the low nonlinearity of silica fibers, the long interaction length greatly enhances nonlinear optical effects, such as stimulated Raman and Brillouin scattering, self-phase modulation and parametric four-wave mixing [5–7]. Such optical processes were initially viewed as a nuisance causing optical cross-talk between communication channels. However, the ability to control these nonlinear optical phenomena eventually led to a number of useful applications in the rich field of *nonlinear fiber optics* [8].

The field of *integrated photonics* grew in parallel with the emergence of optical fiber communications and micro-electronics fabrication. This led to the development of active and passive devices such as diode lasers, optical modulators, and planar lightwave circuits consisting of filters, couplers, and multiplexers [9]. The guided wave configuration of planar integrated photonics and the resultant high field confinement together with long interaction length were soon identified and exploited to demonstrate a variety of intensity-dependent functionalities [10]. Advancements in fabrication technology and utilization of high confinement geometries later pushed the size scales of integrated nonlinear optics to the sub-micron regime. In general, nanoscale nonlinear optics and photonics involve the use of metallic nanostructures and/or high-index dielectrics [11, 12]. In this chapter, we will cover the basics of nonlinear optical phenomena in such material platforms. In particular, we will discuss in some detail the field of integrated nonlinear photonics in silicon, which has been and continues to be an active field of research [13, 14].

Integrated photonics technology has recently begun to impact the quantum optics community, forming a distinct area of research termed *quantum photonics*. With its capacity for complex large-scale on-chip photonic circuits, integrated quantum photonics has been suggested as a solution to the challenge of scaling increasingly complex quantum optical experiments and devices [15]. The field of nonlinear optics has been intimately linked to quantum optics since its inception and was integral in the earliest demonstrations of entangled photon states via spontaneous parametric down-conversion [16]. Of particular interest in the context of our discussion on nonlinear photonics is the ability of such photonic circuits to simultaneously exhibit certain quantum optical functionalities such as acting as a source of correlated photon pairs and serving as a single photon frequency converter.

13.1.1 Physical basis

In a linear isotropic material, the polarization $\mathbf{P}(\omega)$ is given by the linear polarization [see Chapter 1, Eq. (1.40)],

$$\mathbf{P}_L(\omega) = \varepsilon_0 \chi(\omega) \mathbf{E}(\omega), \tag{13.1}$$

where $\varepsilon_0 = 8.85418782 \cdot 10^{-12}$ F/m is the electric constant, and $\chi(\omega)$ is the linear susceptibility. The wave equation in such a medium is commonly written as

$$\nabla^2 \mathbf{E}(\omega) + [1 + \chi(\omega)] \varepsilon_0 \mu_0 \omega^2 \mathbf{E}(\omega) = 0, \tag{13.2}$$

where $\mu_0 = 4\pi \cdot 10^{-7}$ H/m is the magnetic constant. For a nonlinear optical material, the polarization has additional higher-order components $\mathbf{P}(\omega) = \mathbf{P}_L(\omega) + \mathbf{P}_{NL}(\omega)$ such that

$$\nabla^2 \mathbf{E}(\omega) + [1 + \chi(\omega)] \varepsilon_0 \mu_0 \omega^2 \mathbf{E} = -\mu_0 \omega^2 \mathbf{P}_{NL}(\omega). \tag{13.3}$$

This is an inhomogeneous wave equation with the nonlinear polarization, where the nonlinear response of the material to an applied electric field acts as an additional source term. For a general anisotropic material, the i-th component of the polarization is given by

$$\begin{aligned} P_i(\omega) &= \varepsilon_0 \sum_j \int \chi_{ij}^{(1)}(\omega_1) E_j(\omega_1)\, \mathrm{d}^1\Omega \\ &+ \varepsilon_0 \sum_{jk} \int \chi_{ijk}^{(2)}(\omega_1, \omega_2) E_j(\omega_1) E_k(\omega_2)\, \mathrm{d}^2\Omega \\ &+ \varepsilon_0 \sum_{jkl} \int \chi_{ijkl}^{(3)}(\omega_1, \omega_2, \omega_3) E_j(\omega_1) E_k(\omega_2) E_l(\omega_3)\, \mathrm{d}^3\Omega \\ &+ \ldots, \end{aligned} \tag{13.4}$$

where nonlinear susceptibilities $\chi^{(i)}$ are tensors of the $(i+1)$-th rank, and

$$\mathrm{d}^n\Omega = \delta\left(\sum_{i=1}^{n} \omega_i - \omega\right) \prod_{i=1}^{n} \mathrm{d}\omega_i$$

with $\delta(\omega)$ being the Dirac delta function. Note that the integrations in every term of $P_i(\omega)$ are performed over all frequencies ω_i, where only those give non-zero contributions, which satisfy the condition

$$\sum_{i=1}^{n} \omega_i = \omega. \tag{13.5}$$

Following this condition, the nonlinear origin of the higher harmonics and frequency mixing is clearer. For example, given that the applied electric field $\mathbf{E}$ oscillates at frequency ω_0, the resulting polarization $\mathbf{P}$ will have a linear term at $\omega = \omega_0$, and nonlinear ones at $\omega = 2\omega_0,\ 3\omega_0,\ 4\omega_0, \ldots$.

In most materials, the second- and third-order nonlinearities are of the greatest interest. Like linear optical effects, they are contributed by both the bound and conduction charges. In systems with strong contributions from conduction electrons, second-order nonlinearity is very common due to the nonlinear dynamics of conduction electrons [17, 18]. In material systems where there is negligible contribution of conduction charges, the nonlinear response is defined by the lattice symmetry. For example, in *centrosymmetric* materials, where the properties at a point (x, y, z) are indistinguishable from those at the point $(-x, -y, -z)$, second-order effects are generally absent, while third-order effects are present. It follows directly from Eq. (13.4), where

$$P_i(2\omega) = \varepsilon_0 \chi_{ijk}^{(2)} E_j(\omega) E_k(\omega). \tag{13.6}$$

Reversing the sign of the applied electric field will also reverse the sign of the nonlinear polarization due to the centrosymmetry,

$$-P_i(2\omega) = \varepsilon_0 \chi_{ijk}^{(2)} [-E_j(\omega)][-E_k(\omega)] = \varepsilon_0 \chi_{ijk}^{(2)} E_j(\omega) E_k(\omega). \tag{13.7}$$

Note that $-P(2\omega) = P(2\omega)$ is satisfied only when $P(2\omega) = 0$. This illustrates that

$\chi^{(2)}_{ijk} = 0$ in non-conductive centrosymmetric systems. The same applies for other even-order nonlinear susceptibilities,

$$\chi^{(2m)}_{ijk\ldots} = 0, \; m \in \mathbb{N}. \tag{13.8}$$

Strictly speaking, this conclusion is valid only for unbounded pure centrosymmetric systems with no conduction electrons. In real crystals with defects and finite dimensions, it deviates from zero due to symmetry breaking. However, for most cases the even-order nonlinear effects remain negligible compared to the odd-order ones.

In traditional nonlinear optics, the nonlinear optical material in question is generally an optical crystal placed at the focus of an intense laser beam. In order to maximize the nonlinear optical effects, the temporal and spatial walk-offs of the interacting beams have to be minimized. The beams have to overlap in space and also have to be phase matched in order to add coherently at the output of the crystal.

If we assume the beams are aligned in space with the z axis and are highly collimated, we can take the linear solutions of Eq. (13.2) to be monochromatic plane waves of the form $A(\omega)\,\mathrm{e}^{ikz}$. In addition, if the nonlinear polarization source term is small, as is generally the case, it can be taken into account as a perturbation to the linear solution. In this case, the total solutions of Eq. (13.3) will still be of the same form except for the amplitude coefficients, which become weak functions of space, $A(\omega,z)$, where $|\partial A(\omega,z)/\partial z| \ll |kA(\omega,z)|$. In this way, the beams propagating at different frequencies are coupled.

Substitution of the plane wave solutions into Eq. (13.3) and taking the slowly varying amplitude approximation [1] yields a set of coupled amplitude equations. For example, difference frequency generation (DFG) is a three-wave interaction described with three coupled equations:

$$\frac{\mathrm{d}A_1}{\mathrm{d}z} = i\kappa_1 A_3 A_2^* \,\mathrm{e}^{\mathrm{i}\Delta kz}, \tag{13.9}$$

$$\frac{\mathrm{d}A_2}{\mathrm{d}z} = i\kappa_2 A_3 A_1^* \,\mathrm{e}^{\mathrm{i}\Delta kz}, \tag{13.10}$$

$$\frac{\mathrm{d}A_3}{\mathrm{d}z} = i\kappa_3 A_1 A_2 \,\mathrm{e}^{-\mathrm{i}\Delta kz}, \tag{13.11}$$

where $\omega_1 = \omega_3 - \omega_2$, the raised asterisk (*) represents the complex conjugate, $\kappa_{1,2,3}$ are the coupling constants, and $\Delta k = k_3 - k_1 - k_2$ is the so-called phase mismatch. The generated frequency ω_1 is the difference in frequencies of the input waves ω_2 and ω_3. In low efficiency DFG, the input waves A_2 and A_3 do not change much in amplitude and hence can be taken to be constants as the first approximation. Equation (13.10) can then be integrated over the length L of the nonlinear crystal to give

$$|A_1|^2 \propto |A_2|^2 |A_3|^2 L^2 \left[\frac{\sin(\Delta kL/2)}{\Delta kL/2}\right]^2. \tag{13.12}$$

As can be seen, the generated output intensity depends on the input intensities, the length of the interaction and the phase mismatch factor. It is useful to define a phase matched coherence length as $L_{\mathrm{coh}} = 2\pi/\Delta k$, at which Eq. (13.12) is equal to zero. At approximately half the coherent length, energy starts to flow from the output wave back to the input pump wave, reducing output wave power. Such coupled amplitude equations are common in nonlinear optics and also in integrated nonlinear photonics, especially in guided waves. However, at the nanophotonics regime, particularly in metallic nanostructures with high-field regions, phase matching is not necessary due to the wavelength-scale interaction volume. In this case, the coupled-wave picture is not suitable.

13.1.2 Material platforms and properties

A common goal of nonlinear optics is to maximize the generally weak output of nonlinear optical wave-mixing phenomena. To reiterate, from Eq. (13.12), the strengh of the generated difference frequency is dependent on the intensities of the mixing frequencies, the length of the interaction, and the phase mismatch. In free-space nonlinear optics, the laser beam divergence limits the maximum intensity and interaction length of the mixing waves. Waveguiding is the ideal platform to enhance both the field intensity and length of interaction. Using waveguides, beams are prevented from expanding outward by confining them in one or two dimensions. In ideal transparent materials, no power is lost, while the optical wave propagates inside the waveguide and hence maximum intensity is maintained throughout the interaction length. Moreover, the waves are localized in the same region of space inside the waveguide, eliminating the need to carefully align beams and ensuring that the optical interaction is maintained for the full waveguide length. In fact, in long silica fibers, the nonlinear phenomena can be undesirably enhanced; this is the case with spontaneous Raman scattering, which causes inter-channel cross-talk via frequency mixing. The challenge of waveguided nonlinear photonics is to make use of the strong dependence of the nonlinear interaction on the transverse dimensions of the waveguides, in order to maximize desired nonlinear optical effects and minimize undesired side effects.

In integrated optics, optical devices of different functionalities are combined to form an optical circuit, which is able to perform more complex functions. Such optical devices composing a circuit are connected via waveguides laid out on a planar substrate. The idea of integrated nonlinear photonics is to take advantage of the nonlinear properties of optical devices to serve useful functions in the overall circuit.

To analyze optical waveguides, we begin with the assumption that they are long enough to be well approximated by an infinitely long waveguide with uniform cross-section. We can then solve the eigenvalue problem to obtain eigenfields of guided waves,

$$\mathbf{F}_n = \mathbf{f}_n(x,y)\, \mathrm{e}^{\mathrm{i}k_n z}. \tag{13.13}$$

Fields of the n-th guided eigenwave has two parts, $\mathbf{f}_n(x,y)$ being the cross-section profile (a three-component vector field) and $\mathrm{e}^{\mathrm{i}k_n z}$ being the variation in the propagation direction z. The eigenvalue $k_n = (2\pi/\lambda)n_{\mathrm{eff}}(\lambda)$ is used to define the mode effective index n_{eff}. The variation of k_n with wavelength gives the dispersion of guided waves.

Except for the simple case of a slab waveguide, solving for guided eigenwaves is a challenging problem. Generally it can be done numerically only with the use of different methods [19]. In the context of nonlinear optics, the eigenwaves are the solutions to Maxwell's equation in the optical waveguide. Following the prescription described above, we can substitute them in Eq. (13.3) to obtain coupled amplitude equations similar to Eq. (13.10) [20]. We can then use these equations to design and optimize nonlinear optical interactions in waveguides.

Material	Refractive Index (1.55 μm)	Centrosymmetry	Band Gap (eV)	Ref.
Ge	4.04 (3 μm)	yes	0.67	[21]
SiGe	3.6 (4 μm)	yes	0.95	[22]
Si	3.47	yes	1.11	[13, 14]
AlGaAs	3.3	no	1.74	[12, 23]
InGaP	3.1	no	1.9	[24]
As_2S_3	2.8	yes	2.6	[25]
SiC	2.6	yes	3.0	[26]
SiN	2.0	yes	3.0	[27]
GaN	2.32	no	3.4	[28]
PP-$LiNbO_3$	2.2	no	4.0	[29]
Diamond	2.4	yes	5.5	[30]
AlN	2.0	no	6.0	[31]
Silica (doped)	1.7	yes	6.0 (> 6.0)	[27, 32]

Table 13.1: Nonlinear photonic waveguide materials

Table 13.1 is a non-exhaustive list of photonic materials used to make nonlinear optical waveguides. The refractive index reported is at 1.55 μm, unless otherwise stated. This is mainly due to the inherent interest for compatibility with telecommunications technology, which is centered around the optical fiber transparency window at 1.55 μm. The presence of centrosymmetry indicates the lack of second order nonlinear effects; the band gap gives the material transparency range and hence feasible operation wavelengths. In general, a thin layer of material is patterned into waveguides on top of a low-index cladding layer such as silicon dioxide. The refractive index contrast between the core and cladding allows for confinement of the guided wave. High-index materials allow for tighter mode confinement and thus higher intensities, which is beneficial for nonlinear phenomena. Wide band gap materials are of interest due to the lack of multi-photon absorption. For example, when the photon energy at the operating wavelength is less than half the band gap energy, two-photon absorption vanishes.

Current microfabrication technology provides the ability to precisely control the sub-micrometer scale properties of the waveguides, enabling techniques to significantly increase overall nonlinear optical process efficiencies. One example is waveguide dispersion engineering, where the waveguide cross-section dimensions are tuned to give a desired waveguide dispersion, allowing the reduction of the phase mismatch of nonlinear optical processes. Another well established technique called quasi-phase matching, as in periodically-poled lithium niobate (PPLN) waveguides, is achieved through the creation of artificial periodic media. This allows the flow of energy from the pump wave to the output wave beyond the coherence length by periodic reversal of the sign of the nonlinear coupling coefficient.

Apart from the dielectric and semiconductor materials in Table 13.1, there is an interest in using metals to define so-called plasmonic waveguides that utilize surface polaritons (SPs) [33]. SPs have deep subwavelength confinement and high intensity enhancement. Hence, plasmonic waveguides are very interesting as nonlinear optical waveguides. However, due to the high absorption losses of the metals, plasmonic waveguides are necessarily much shorter than dielectric counterparts, which limits the interaction length. Recent demonstrations of compact low-loss high-speed plasmonic optical modulators have shown that the short interaction length can be overcome by the gain in field enhancement over conventional dielectric waveguides [34, 35].

Another way that nonlinear plasmonics evades this interaction length limit is by shifting outside of the guided wave paradigm and confining the interaction to nanostructured surfaces or nanoparticles [36, 37]. Second-order nonlinear processes are inherently present in all metals due to a strong contribution from the nonlinear dynamics of the conduction electrons. A variety of nonlinear optical processes have been observed from plasmonic nanostructures, such as second and third harmonics, four-wave mixing and saturable absorption. Research into nonlinear plasmonic phenomena is of scientific as well as practical interest with a variety of suggested applications such as measuring nanostructured surface or nanoparticle size uniformity and high sensitivity chemical and biological detection. Since these nanostructures generally have sub-wavelength dimensions, phase matching in the sense of guided wave photonics is not relevant. The current challenge is to design nanostructures with good mode matching between the nanostructure resonances at the intended coupled harmonics and enhance far-field emission by tailoring nanostructure shape and geometry [38–41].

13.2 Integrated nonlinear silicon photonics

13.2.1 Overview and challenges

Integrated silicon photonics on the silicon-on-insulator platform is considered a promising technology for electronics and photonics integration. The primary goal is to enable high-speed and low-cost opto-electronics chips packaged in a compact form factor, while maintaining compatibility with current CMOS fabrication facilities. Owing to the high index of silicon, patterned waveguides are usually of sub-micrometer size at telecommunications wavelengths (1.55 μm). The high field confinement of these waveguides results in high intensities at low input powers, which is suited for observation of nonlinear optical phenomena. As a result, nonlinear silicon photonics has seen a great amount of research over the past decade.

Silicon has a centrosymmetric crystal structure (diamond cubic) and hence it is expected that the lowest order nonlinear polarization response in silicon waveguides is of the third order. Several demonstrations of second-order nonlinear response have been shown by applying a strain layer of silicon nitride on top of the silicon waveguide [42–44]. However, the majority of research has been conducted on various third-order effects due to the electronic nonlinear polarization (Kerr intensity-dependent refraction, two-photon absorption), as well as from photon-phonon coupling (Raman, Brillouin scattering).

Acronym	Process	Description	Frequency notion
THG	Third Harmonic Generation	Generation of tripled frequency	$\omega_2 = 3\omega_1$
FWM	Four Wave Mixing	Two frequencies mix to generate two new frequencies	$\omega_1 + \omega_2 = \omega_3 + \omega_4$
SPM	Self Phase Modulation	Intensity dependent refractive index change	-
XPM	Cross Phase Modulation	Optical phase change imparted by other wave	-
TPA	Two Photon Absorption	Simultaneous absorption of two photons	-
FCA	Free Carrier Absorption	Intraband absorption by free carrier	-
FCD	Free Carrier Dispersion	Free carrier induced refraction	-
SRS	Stimulated Raman Scattering	Inelastic scattering by optical phonon	$\omega_1 - \omega_2 = \omega_R$
SBS	Stimulated Brillouin Scattering	Inelastic scattering by acoustic phonon	$\omega_1 - \omega_2 = \omega_B$

Table 13.2: List of nonlinear processes in silicon waveguides

The electronic third-order nonlinear susceptibility of silicon, after considerations of symmetry, have two independent components, which are of the ratio

$$\chi^{(3)}_{1111} \approx 2.36\chi^{(3)}_{1122}.$$

The effective nonlinear susceptibility is dependent on the polarization directions of the interacting fields [13]. Generally, $\chi^{(3)}$ is not measured directly, but is inferred from the nonlinear refractive index n_2 and the two-photon absorption (TPA) coefficient β_{TPA}:

$$n_2 = \frac{3}{4}\frac{Z_0}{n_0^2}\text{Re}[\chi^{(3)}], \tag{13.14}$$

$$\beta_{\text{TPA}} = -\frac{3}{2}\frac{k_0 Z_0}{n_0^2}\text{Im}[\chi^{(3)}], \tag{13.15}$$

where n_0 is the linear refractive index, Z_0 is the impedance of free space. Measurements of the dispersion of n_2 and β_{TPA} have been carried out by several groups (see Fig. 13.1) [45, 46]. A common way to compare the relative strengths of nonlinear refraction and two-photon absorption in a bulk nonlinear optical material is the figure of merit:

$$\text{FOM}_{\text{bulk}} = n_2/\lambda\beta_{\text{TPA}}. \tag{13.16}$$

At 1.55 μm, silicon has a poor $\text{FOM}_{\text{bulk}} \approx 0.4$. The presence of TPA places a significant constraint on the performance of nonlinear optical devices in silicon, since it limits the maximum intensity within waveguides. Moreover, the quasi-free carriers generated by TPA contribute to an additional nonlinear loss by free carrier absorption (FCA) and additionally slow device response times due to the finite recombination lifetimes of quasi-free carriers.

Several approaches to minimize or eliminate the disadvantages of TPA in silicon photonics have been explored in the literature:

- *Sweep-out generated quasi-free carriers.* FCA resulting from TPA can be largely mitigated through reversed bias carrier removal in *p-i-n* junction diodes. The diode is generally formed from a rib structure with the intrinsic region of the diode

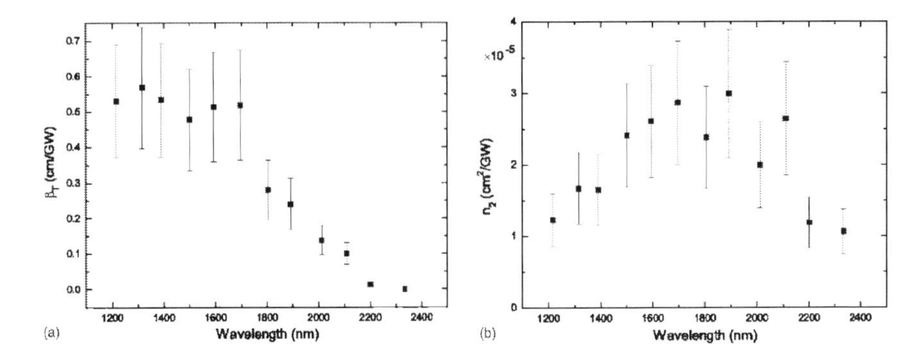

Figure 13.1: Measured dispersion of silicon two-photon absorption coefficient β_{TPA} and nonlinear refractive index n_2 for near-infrared wavelengths. Reprinted with permission from Ref. [46]. © 2007 American Institute of Physics.

acting as the silicon waveguide. The first report of a continuous-wave silicon Raman laser relied on reverse biasing to reduce FCA and achieve high enough gain to demonstrate lasing was given in Ref. [47]. Other groups have shown that the reduced lifetime of quasi-free carriers significantly improves four-wave mixing efficiency [48, 49].

- *Use longer wavelengths to eliminate TPA.* The two-photon absorption coefficient in silicon vanishes at photon energies below half of the band gap energy of 1.11 eV or equivalently at wavelengths greater than 2200 nm. This means that the nonlinear figure of merit becomes much more favorable at these longer wavelengths. Several research directions advocate the use of wavelengths closer to 2 μm for integrated nonlinear silicon photonics [50–52]. Other considerations like choice of cladding [53] and higher-order multi-photon absorption losses will come into play at longer wavelengths approaching 4 μm [54].
- *Adopt hybrid material platforms.* Several other highly nonlinear materials that are compatible with the silicon photonics platform have low TPA at 1.55 μm. Some examples include hydrogenated amorphous-Si [55], silicon nitride [27], silicon-rich nitride [56], and highly nonlinear optical polymers [57]. A hybrid material platform can allow nonlinear functionalities to be performed within these low TPA materials, while transitioning back to silicon for linear or low-power operations.

Raman scattering is an important electron mediated photon-phonon coupling interaction in crystalline silicon. The first demonstration of lasing in silicon used stimulated Raman scattering as a means to produce optical gain [58]. Lasing had not been shown prior to that due to the indirect band gap nature of crystalline silicon. When the so-called co-propagating pump and Stokes waves are separated by the frequency of atomic vibrations in silicon (phonon-polaritons), which is 15.6 THz, they are able to exchange energy through stimulated Raman scattering [59]. The linewidth of this interaction is 105 GHz. Following the crystal direction convention $\hat{x} \equiv [100], \hat{y} \equiv [010], \hat{z} \equiv [001]$, the Raman tensors at the center of the Brillouin zone are:

$$R_x = \begin{pmatrix} 0 & 0 & 0 \\ 0 & 0 & 1 \\ 0 & 1 & 0 \end{pmatrix}, \quad R_y = \begin{pmatrix} 0 & 0 & 1 \\ 0 & 0 & 0 \\ 1 & 0 & 0 \end{pmatrix}, \quad R_z = \begin{pmatrix} 0 & 1 & 0 \\ 1 & 0 & 0 \\ 0 & 0 & 0 \end{pmatrix}. \tag{13.17}$$

The third-order susceptibility describing stimulated Raman scattering is $\chi^{(3)}_{ijkl} \sim (R_m)_{ij}(R_m)_{kl}$, which has 12 non-zero components [60]. The nonlinear polarization is of the form

$$P_i(\omega_s) \approx \varepsilon_0 \chi^{(3)}_{ijkl} E_j(\omega_p) E_k(-\omega_p) E_l(\omega_s), \tag{13.18}$$

where ω_p and ω_s are the pump and Stokes frequencies, respectively. Using this set of equations, we can derive as before a set of coupled amplitude equations describing stimulated Raman scattering in silicon waveguides [61].

Brillouin scattering is another photon-phonon interaction in silicon related to the lower frequency phonons. Two types of stimulated Brillouin scattering (SBS) have been observed in waveguides: backward SBS (on phonons), where the pump and Stokes waves are counter-propagating, and forward SBS (on phonon-polariton), where the pump and Stokes waves are co-propagating. There has been a surge of interest in stimulated Brillouin scattering in silicon waveguides in recent years due to the discovery of new physical mechanisms contributing to SBS in nanoscale waveguides [62]. Beyond the bulk material effect of electro-stiction producing density waves that alter permittivity, it was found that radiation pressure also produces a boundary nonlinearity that was previously neglected in larger waveguides. SBS in silicon waveguides is currently an active field of theoretical and experimental research [63, 64], having the potential for applications in lasers, microwave photonics, and nonreciprocal devices [65].

13.2.2 Application of nonlinearities

The study of the nonlinear optical properties of integrated silicon waveguides is not only an academically interesting pursuit; it also has many potential applications. Here we try to broadly outline and categorize these applications and provide a brief discussion and references for further reading.

- *Optical gain and lasing.* Due to the tight eigenmode confinement in micrometer and sub-micrometer cross-section silicon-on-insulator waveguides, the field intensity inside the waveguide core is greatly enhanced. It was proposed that such silicon waveguides could display highly nonlinear characteristics [66]. A source of optical gain in silicon was considered elusive because of its indirect band gap. However, several groups were able to harness the large Raman gain of silicon to demonstrate optically pumped Raman lasers [47, 58, 60]. Later, four-wave mixing (FWM) was also used to demonstrate optical gain in silicon waveguides [48, 67, 68]. However, the FWM gain was largely offset by waveguide losses. In an effort to enhance FWM gain by eliminating TPA, several groups advocated the use of mid-infrared wavelengths nearer to 2 μm [50, 69]. Optical parametric oscillators have been demonstrated using silicon as the gain medium in mid-infrared wavelengths [70]. More recently, hydrogenated amorphous silicon (a-Si) waveguides have emerged as an alternative to standard crystalline silicon for nonlinear optics applications due to the larger bandgap of $\sim$1.6 eV [55, 71]. Such a-Si waveguides have potential for very large optical gain at mid-infrared and telecommunications wavelengths [72]. All such sources of optical gain (including Brillouin gain [63]) could become increasingly useful with the introduction of integrated on-chip lasers [73, 74].

- *Wavelength conversion, generation, and frequency combs.* Wavelength conversion in silicon waveguides can be accomplished with a variety of $\chi^{(3)}$ processes. The most common are the Kerr-effect-based processes such as FWM, SPM, and XPM. Signal translation from mid-infrared wavelengths to telecommunications band and vice versa can be useful in combining the respective strengths of each wavelength domain. For example, mid-infrared is particularly useful for environment gas sensing and spectroscopy applications. Signal translation of mid-infrared light to the telecom band would eliminate the need for mid-infrared detectors and allow the use of telecom wavelength detectors (InGaAs), that exhibit better performance and do not require special cooling setups [51]. Wavelength conversion in silicon waveguides and micro-resonators can also be used to generate supercontinuums and frequency combs. Supercontinuum generation is useful for applications requiring ultra-broadband low-coherent sources, for example in spectroscopy and optical coherence tomography. Supercontinuums can also be useful for wavelength division multiplexing (WDM), where the broadband source is spectrally filtered into individual WDM channels [53, 75, 76]. Optical frequency combs have ultra-high precision applications in metrology and spectroscopy. On-chip supercontinuum generation and frequency combs are especially attractive for field use due to the compact and portable form factor [52, 77].
- *Signal processing, modulation, and switching.* All-optical signal processing is one of the key applications of nonlinear silicon photonics, as it enables high speed signal manipulation by utilizing the femtosecond response time of several nonlinear optical processes [78]. This femtosecond response has been used to accomplish time-domain demultiplexing of high-speed optical data using FWM in silicon waveguides [79, 80]. Silicon waveguides have also been used to demonstrate wavelength multicasting via FWM, which is a low-power and efficient way of distributing information simultaneously to multiple WDM channels [81]. Optical signal regeneration is an important function in long-distance or complex optical networks, where signals affected by noise or loss can be returned to the original state. Phase-insensitive optical gain, through fiber-based or on-chip amplifiers, can be used to re-amplify loss-impacted signals. On the other hand, phase-modulated signals (e.g., phase shift keying) will be degraded by optical phase noise and cannot be simply re-amplified by phase-insensitive means. To this end, phase-sensitive amplification and phase regeneration have been demonstrated through FWM in silicon waveguides [82, 83]. Nonlinear optical absorption mechanisms in silicon waveguides are also useful for all-optical signal processing purposes. For example, Raman scattering has been used as a loss mechanism in high-Q

silicon micro-resonators to demonstrate single resonance switching [84]. TPA in silicon waveguides has been used for mode locking and cross-absorption modulation [85, 86].

13.2.3 *Engineering waveguide structures for nonlinear photonics*

In waveguide design for nonlinear photonics, the goal is usually to maximize the optical interaction so as to enhance inherently weak nonlinear optical effects. Broadly speaking, this entails maximizing intensity (i.e., polariton confinement), minimizing phase mismatch between eigenwaves and maximizing interaction length (i.e., minimizing waveguide losses). We will use the example of a degenerate pump FWM process in silicon waveguides to illustrate in detail the techniques adopted to maximize the efficiency of such nonlinear optical phenomena. The strength of $\chi^{(3)}$ processes in silicon waveguides is determined by the waveguide nonlinear parameter γ_{wg} which is given as [87]

$$\gamma_{\mathrm{wg}} = \frac{3\omega}{4cZ_0} \frac{\iint \chi^{(3)} |\mathbf{E}|^4 \, \mathrm{d}x \, \mathrm{d}y}{\left|\iint \mathrm{Re}\{\mathbf{E} \times \mathbf{H}^*\} \, \mathrm{d}x \, \mathrm{d}y\right|^2}. \tag{13.19}$$

Note that $\chi^{(3)}$ is generally a function of the coordinates x, y due to the different nonlinearities of the silicon core and the cladding layer. The real and imaginary parts of γ_{wg} separately contribute to the nonlinear refractive index and TPA [see Eq. (13.15)]. The silicon waveguide TPA figure of merit, which is different from that of bulk silicon due to the influence of the cladding layer, is defined as

$$\mathrm{FOM}_{\mathrm{TPA}} = -\frac{1}{4\pi} \frac{\mathrm{Re}(\gamma_{\mathrm{wg}})}{\mathrm{Im}(\gamma_{\mathrm{wg}})}. \tag{13.20}$$

It can be considered as the ratio of the nonlinear phase shift to the TPA loss in the waveguide. The coupled amplitude equations describing a continuous wave (CW) or quasi-CW degenerate pump FWM process with wavelength conversion from signal wave to idler wave are as follows:

$$\frac{\partial A_p}{\partial z} = -\frac{\alpha}{2} A_p - i\gamma |A_p|^2 A_p - \gamma' |A_p|^2 A_p - N_c \left(\frac{\sigma}{2} + i \frac{2\pi}{\lambda_p} k_c \right) A_p, \tag{13.21}$$

$$\frac{\partial A_s}{\partial z} = -\frac{\alpha}{2} A_s - 2i\gamma |A_p|^2 A_s - 2\gamma' |A_p|^2 A_s - N_c \left(\frac{\sigma}{2} + i \frac{2\pi}{\lambda_s} k_c \right) A_s - i\gamma A_p^2 A_i^* e^{i\Delta k}, \tag{13.22}$$

$$\frac{\partial A_i}{\partial z} = -\frac{\alpha}{2} A_i - 2i\gamma |A_p|^2 A_i - 2\gamma' |A_p|^2 A_i - N_c \left(\frac{\sigma}{2} + i \frac{2\pi}{\lambda_i} k_c \right) A_i - i\gamma A_p^2 A_s^* e^{i\Delta k}. \tag{13.23}$$

We have assumed a strong undepleted pump, which accounts for most practical cases. Here, N_c is the averaged concentration of quasi-free carriers, σ and k_c are the FCA and FCD coefficients, and γ and γ' are the real and imaginary parts of γ_{wg}; $\sigma =$

$1.45 \times 10^{-21}\text{m}^2$ and $k_c = 1.35 \times 10^{-27}\text{m}^3$ [88]. In order to calculate N_c, we first import the TPA photogeneration rate profile,

$$G(x,y) = \frac{\beta_{\text{TPA}} I(x,y)^2}{2h\nu}, \tag{13.24}$$

into a semiconductor technology computer aided design (TCAD) software. The TCAD can then solve the carrier transport equations to give a carrier (hole or electron) concentration distribution $n(x,y)$. N_c is then calculated as a weighted spatial average,

$$N_c = \frac{\iint n(x,y) I(x,y)\, \mathrm{d}x\, \mathrm{d}y}{\iint I(x,y)\, \mathrm{d}x\, \mathrm{d}y}, \tag{13.25}$$

using the mode intensity profile $I(x,y)$ as a normalised distribution. The terms on the right side of Eq. (13.23) going from left to right are: linear loss, XPM, TPA, FCA and FCD, and finally FWM. More sophisticated equations will include non-degenerate pump terms as well as cross-TPA and XPM terms from the signal and idler waves [13, 89, 90]. To maximize FWM efficiency, we have to reduce loss terms (α, γ', N_c), minimize phase mismatch Δk and maximize the gain coefficient γ. The engineering challenge is to find a balanced trade-off between these optimization parameters that may be in conflict with each other.

Methods of reducing TPA and FCA have been discussed in Section 13.2.1. The main contribution to linear loss α in sub-micrometer silicon waveguides is optical scattering due to etched sidewall roughness. Hence, linear scattering loss is very sensitive to the waveguide geometry and the mode overlap with etched waveguide sidewalls [91]. A straightforward technique to reduce roughness-induced scattering is to reduce the field intensity at the sidewalls by using wider waveguides or decreasing etch depth in ridge waveguides [92]. Advanced CMOS fabrication technology has allowed demonstrations of fully etched rib waveguides with propagation loss of 0.5 dB/cm [93, 94]. The propagation loss also determines the optimum waveguide length that maximizes FWM conversion efficiency. To see this, we can analytically solve a simplified version of Eq. (13.23) that neglects TPA and FCA. The FWM conversion efficiency η in that case is [8, 95, 96]

$$\eta = (\gamma P L')^2 \left[\frac{\sinh(gL)}{gL}\right]^2 \exp(-\alpha L), \tag{13.26}$$

with g and L' defined as

$$g = \sqrt{\left(\gamma P \frac{L'}{L}\right)^2 - \left(\frac{\Delta k L + 2\gamma P L'}{2L}\right)^2}, \qquad L' = \frac{1 - e^{-\alpha L}}{\alpha},$$

where L' is the *effective interaction length*. By assuming zero phase mismatch in Eq. (13.26), the optimum length is found as

$$L_{\text{opt}} = \frac{\log(3)}{\alpha} \approx \frac{1}{\alpha}. \tag{13.27}$$

For a degenerate pump FWM process, the total phase mismatch is

$$\Delta k_{\text{tot}} = \Delta k + \Delta k_{NL} = -2k_p + k_s + k_i + 2\gamma P \approx 2\gamma P + k_2(\Delta\omega)^2 + \frac{1}{12}k_4(\Delta\omega)^4, \quad (13.28)$$

which includes the effects of self- and cross-phase modulation. In the power series expansion of the phase mismatch, k_2 is the group velocity dispersion (GVD), k_4 is the fourth-order dispersion (FOD), and $\Delta\omega = \omega_p - \omega$ is the detuning from the pump frequency. Equation (13.28) allows several ways to minimize phase mismatch. For waveguides with anomalous dispersion ($k_2 < 0$), phase matching can be observed near the pump such that $\Delta\omega^2 \gg \Delta\omega^4$. For waveguides with anomalous GVD and positive FOD, phase matching can be observed far away from the pump, which is known as discrete band phase matching. For normal GVD waveguides ($k_2 > 0$), phase matching can also be observed under the condition of negative FOD [97]. Waveguide dispersion engineering, the process of choosing the waveguide cross-section dimensions to obtain certain desired dispersion characteristics, has enabled various groups to observe ultra-broadband frequency conversion and frequency translation from the telecom band to mid-infrared wavelengths [98, 99]. Some groups have demonstrated quasi-phase matched FWM in silicon waveguides through periodic waveguide width modulation by choosing the grating period $\Lambda = 2\pi/\Delta k$. In this way, the phase matching condition becomes [100]

$$2\gamma P - 2k_p + k_s + k_i - \frac{2\pi}{\Lambda} = 0. \quad (13.29)$$

Two complementary approaches can be adopted to maximize the silicon waveguide gain coefficient $\gamma = \text{Re}(\gamma_{\text{wg}})$. The first is to engineer the waveguide nonlinear parameter γ_{wg} by optimizing the cross-section dimensions or by using other highly nonlinear materials such as a-Si [55] and highly nonlinear polymers [57] in combination with crystalline silicon. The second approach involves using resonant structures to increase the local field intensity, thereby enhancing the effective waveguide gain coefficient.

The quintessential micro-resonator structure in silicon photonics is the micro-ring resonator. The micro-ring is a circular loop of waveguide of length $L = 2\pi R$ that forms a ring with a radius R on the order of micrometers. The resonance condition of a micro-ring is

$$\phi = \frac{2\pi}{\lambda_m} n_{\text{eff}} L = 2\pi m, \qquad m = 1, 2, 3, \ldots, \quad (13.30)$$

where the round-trip phase accumulated is an integer multiple of 2π. Excitation of the micro-ring resonances is achieved by placing a bus waveguide in close proximity to the ring so that waves can evanescently couple in and out. The FWM conversion efficiency in a micro-ring is given by [101]

$$\eta = (\gamma P L')^2 (\text{FE}_p)^4 (\text{FE}_s)^2 (\text{FE}_i)^2 \exp(-\alpha L). \quad (13.31)$$

For the sake of simplicity, we ignored TPA effects and assumed zero phase mismatch and a strong undepleted pump. In comparison with Eq. (13.26), there is extra field-enhancement factor,

$$\mathrm{FE} = \left| \frac{\kappa}{1 - t\exp(-\alpha L/2 + i\phi)} \right|, \tag{13.32}$$

that leads to an *effective* enhancement of the gain coefficient $\gamma_{\mathrm{eff}} = \gamma(\mathrm{FE})^4$. Here, κ is the cross-coupling coefficient and t is the through-coupling coefficient. The field-enhancement factor is a function of wavelength due to the round-trip phase factor ϕ. To maximize the enhancement factor, the (constant) frequency difference between pump, signal, and idler waves, $\Delta\omega = \omega_p - \omega_s = \omega_i - \omega_p$, has to coincide with the ring free spectral range (FSR). However, the FSR is given by $\Delta\omega_{\mathrm{FSR}} = 2\pi c/(n_g L)$, which is non-constant due to the group velocity dispersion. This results in some impairments on the maximum achievable γ_{eff}. Several groups used coupled resonators to induce resonance splitting in order to compensate for this dispersion effect [102–104].

Micro-resonators and other resonant devices such as photonic crystal cavities and waveguides fall in a class of dielectric photonic structures called structural slow-light devices. Such slow-light phenomena result from the patterning of the dielectric material, quite unlike slow light observed in atomic media. The interest in slow-light devices arises from the possibility of enhancing various linear and nonlinear optical phenomena, as evidenced by the field-enhancement factors observed in micro-ring resonators. Such slow-light devices are usually characterized by a slow-down factor S defined as the ratio of the group velocities of the unpatterned waveguide to the slow-light waveguide [105]. As a general rule, linear phenomena are enhanced by a factor $\sim S$, while nonlinear phenomena are enhanced by $\sim S^2$. Heuristically, the two S factors can be understood as originating separately from the increased interaction time and the increased intensity in the waveguide [106]. FWM conversion efficiency is ideally enhanced by a factor $\eta \propto S^4\gamma^2$ [107]. Unfortunately, the concurrent enhancement of detrimental linear and nonlinear loss effects in such slow-light waveguides ultimately limits the experimentally observed optical interaction length [108]. The primary benefit of slow-light waveguides is thus to greatly reduce required device length by confining the optical interaction to compact resonant structures and hence limiting the impairment of losses.

13.3 Integrated nonlinear quantum photonics

Quantum information technologies that explicitly tap into the unique properties of quantum systems are predicted to greatly outperform their classical counterparts. Quantum communications promises unconditionally secure key distribution guaranteed by the laws of quantum physics [109]. Quantum computation, although far from reaching technological maturity, promises an exponential speedup on certain computational tasks as compared to classical computers [110].

In many implementations of such quantum technologies, quantum states of light will need to be generated, manipulated, and detected. Integrated photonics is an ideal technological platform to perform such operations in a low-loss and highly compact optical device. Reconfigurable integrated photonic circuits have been shown to be versatile enough to implement all possible linear optical operations or equivalently all unitary operations on a specified number of optical modes [111]. Highly efficient single photon detectors are also being developed [112, 113]. On the other hand, the generation of photonic quantum states remains a key challenge for the successful implementation of quantum optical technology. An ideal source of quantum light should fulfill the following criteria:

- *Bright and efficient* source that can provide a high flux of photons at low pump powers
- *Clean* source with low noise or unwanted photons
- *Deterministic* source that provides photons at desired time intervals
- *Scalable* source architecture

Harnessing the nonlinear properties of waveguides enables the probabilistic generation of heralded single photons through nonlinear optical processes such as spontaneous parametric down-conversion (SPDC) and spontaneous four-wave mixing (SFWM). Generating the photon pairs within the waveguide is advantageous since there is no need for collection optics to further guide the photons. In addition, the generated states are pure in the waveguide modal degree-of-freedom. However, due to the probabilistic nature of SPDC and SFWM, photons arrive at non-deterministic intervals. This severely limits schemes for quantum computation and communication. We will discuss SFWM in silicon photonic waveguides and current efforts in the development of an SFWM-based silicon photonic quantum light source in more detail below.

13.3.1 Photon pair generation via nonlinear silicon photonics

SFWM is a spontaneous and probabilistic photon pair generation process resulting from the nonlinear Kerr interaction in silicon. This is distinct from deterministic or "on-demand" sources of photons such as vacancy centers in diamond [114] and quantum dots [115]. At the quantum level, two pump photons (ω_{p1} and ω_{p2}) are absorbed, and a pair of photons called the signal (ω_s) and idler (ω_i) photons are created. The total energy of the photons is conserved such that $\omega_{p1}+\omega_{p2}=\omega_s+\omega_i$. The number of photon pairs generated per second (pair flux) via SFWM in the low pump-power regime is given by [116]:

$$F=\Delta\nu(\gamma PL')\mathrm{sinc}^2(k_2\Delta\omega^2L/2+\gamma PL')\exp(-\alpha L). \tag{13.33}$$

To minimize the probability of multi-photon pair generation [117], which contributes noise photons, pump powers for SFWM are usually much reduced compared to a "classical" FWM experiment. In this case, nonlinear loss mechanisms such as TPA and FCA, which are highly detrimental to FWM in silicon waveguides, will not have high impact on photon pair generation [118]. Moreover, spontaneous Raman scattering photons, a broadband source of noise for SFWM in silica-based optical fiber, are easily filtered out in silicon waveguides due to the narrowband nature of the Raman gain.

Early demonstrations of SFWM photon pair generation in silicon utilized conventional rectangular channel waveguides [119, 120]. Subsequently, micro-resonators were used to further reduce the required pump power down to microwatt levels [118, 121]. Silicon micro-resonators with size on the order of several micrometers have a number of inherent advantages as photon pair sources. Generally, a small polariton volume enhances nonlinear optical processes [122]. Moreover, small micro-rings with a free spectral range of several nanometers facilitates demultiplexing of specific signal-idler pairs using readily available telecommunications optical filters [123]. Integrated silicon photonics, with its chip-scale optical circuits, allows combining an array of such micrometer-sized photon pair sources onto a compact footprint. Actively multiplexing a large number of such probabilistic sources using fast optical switching and high-efficiency heralding detectors introduces the possibility of forming high-probability quasi-deterministic heralded single-photon sources with minimal noise photon contamination [124, 125]. Much work remains to be done on understanding the fundamental limitations of such a multiplexing scheme and what needs to be optimized to obtain a nearly deterministic source [126, 127].

13.3.2 Experiments using entangled photon pairs

Entanglement is a fundamental quantum property that underlies many implementations of quantum communications and computing. Photon pairs are entangled when their quantum states cannot be described independently, but must be taken as a whole. Photon pairs generated by SFWM are inherently energy-time entangled as a result of the generation process. Due to energy conservation, we can know the sum of the energies of the two generated photons but not the individual energies unless one of them is measured. In the time domain, this means we do not know the exact time of generation of the two photons. Generally, proof of entanglement hinges on violation of the Bell-type inequalities that imply non-classical correlations. Several groups have shown entanglement of SFWM photon pairs in the energy-time degree of freedom by using Franson-type interferometry [128]. These include photon pairs generated from micro-rings, photonic-crystal cavities, and coupled micro-rings [129–132].

Photon pairs generated by SFWM may be manipulated to become entangled in other degrees of freedom. For example, polarization entanglement can be generated by integrating a polarization rotator in the middle of a photon pair source [133]. Moreover, such polarization entangled photon pairs are inherently energy-time entangled and thus shown to be hyper-entangled [134]. Hyper-entangled photon pairs are entangled in multiple degrees of freedom and are a promising resource of quantum information.

13.3.3 Quantum frequency conversion

Individual sub-systems in a future quantum technology may operate optimally at different frequencies. Quantum frequency conversion allows each constituent subsystem to retain its performance benefits while providing functionality to a separate component that may be operating at different frequencies. For example, frequency translation enables future quantum communications networks that may consist of quantum photon sources and quantum memories connected over large distances by low-loss optical fibers [135]. The low-loss window of optical fiber is in the telecom wavelength regime, whereas quantum memories operate via atomic transitions in the visible wavelengths. Furthermore, quantum repeater schemes, which are required for long-distance distribution of quantum information, will likely rely on quantum interference between indistinguishable photons of the same frequency [136]. Thus, each source, memory, and inter-connect sub-system will separately operate at different frequencies with the quantum frequency converter acting as a mediating interface. Another example is the frequency up-conversion of single photons at telecommunications wavelengths to visible wavelengths [137, 138]. This allows the use of silicon-based single-photon avalanche photodetectors that provide better detection efficiencies and better noise performance than InGaAs-based detectors.

Transferring the quantum state of a single photon from one frequency to another can be accomplished through quantum frequency conversion using $\chi^{(2)}$ and $\chi^{(3)}$ optical materials. It is vital that the frequency conversion process preserve quantum information and produces minimal noise amplification. The second-order processes such as sum and difference frequency generation and the third-order ones such as Bragg-scattering four-wave mixing are inherently *background free* [139, 140]. This means that vacuum fluctuations are not amplified during the frequency conversion process. This property is not present in degenerate pump four-wave mixing [141]. The remaining noise comes from frequency conversion of background Raman scattered photons and other competing nonlinear optical processes. Experimental results have shown that quantum frequency conversion preserves the quantum properties of the converted light, such as entanglement, quantum correlations, and photon statistics [142–144]. Several groups have demonstrated that frequency converted light from separate sources can be made indistinguishable [145–147].

Current experimental demonstrations have generally used non-integrated optics solutions for frequency conversion. A chip-scale quantum frequency conversion interface is desirable, since it can be integrated with other electronic or photonic circuits in a compact quantum communications device. Several promising results in integrated silicon photonic waveguides have shown frequency conversion of light at the single-photon level, but have yet to explicity demonstrate the translation of quantum states [148, 149]. The outstanding challenges include reducing contaminating noise processes and improving waveguide propagation loss [150, 151].

13.4 Future outlook

Integrated nonlinear photonics exhibit a large variety of potential applications that have already shown promising results. Several outstanding challenges, however, remain, such as reduction of limiting nonlinear loss effects and minimizing the required pump power. In addition, quantum photonics has a need in integrated nonlinear optical devices for future quantum communications interfaces. Further developments require eliminating spurious noise processes. Only after these obstacles are resolved can nonlinear integrated photonics be considered a mainstream technology and not just a fringe technique.

Bibliography

[1] R. W. Boyd, *Nonlinear Optics*. Academic Press, 2008.

[2] P. Franken, A. Hill, C. Peters, and G. Weinreich, "Generation of optical harmonics," *Phys. Rev. Lett.*, vol. 7, no. 4, p. 118, 1961.

[3] R. Terhune, P. Maker, and C. Savage, "Optical harmonic generation in calcite," *Phys. Rev. Lett.*, vol. 8, no. 10, p. 404, 1962.

[4] J. Armstrong, N. Bloembergen, J. Ducuing, and P. Pershan, "Interactions between light waves in a nonlinear dielectric," *Phys. Rev.*, vol. 127, no. 6, p. 1918, 1962.

[5] E. Ippen and R. Stolen, "Stimulated Brillouin scattering in optical fibers," *Appl. Phys. Lett.*, vol. 21, no. 11, pp. 539–541, 1972.

[6] R. H. Stolen, "Phase-matched-stimulated four-photon mixing in silica-fiber waveguides," *IEEE J. Quantum Electron.*, vol. 11, no. 3, pp. 100–103, 1975.

[7] R. Stolen and C. Lin, "Self-phase-modulation in silica optical fibers," *Phys. Rev. A*, vol. 17, no. 4, p. 1448, 1978.

[8] G. Agrawal, *Nonlinear Fiber Optics*. Academic Press, 2012.

[9] P. Tien, "Integrated optics and new wave phenomena in optical waveguides," *Rev. Mod. Phys.*, vol. 49, no. 2, p. 361, 1977.

[10] G. I. Stegeman and C. T. Seaton, "Nonlinear integrated optics," *J. Appl. Phys.*, vol. 58, no. 12, pp. R57–R78, 1985.

[11] M. Danckwerts and L. Novotny, "Optical frequency mixing at coupled gold nanoparticles," *Phys. Rev. Lett.*, vol. 98, no. 2, p. 026104, 2007.

[12] P. Apiratikul, J. J. Wathen, G. A. Porkolab, B. Wang, L. He, T. E. Murphy, and C. J. Richardson, "Enhanced continuous-wave four-wave mixing efficiency in nonlinear algaas waveguides," *Opt. Express*, vol. 22, no. 22, pp. 26814–26824, 2014.

[13] R. Osgood, N. Panoiu, J. Dadap, X. Liu, X. Chen, I.-W. Hsieh, E. Dulkeith, W. Green, and Y. Vlasov, "Engineering nonlinearities in nanoscale optical systems: physics and applications in dispersion-engineered silicon nanophotonic wires," *Adv. Opt. Photonics*, vol. 1, no. 1, pp. 162–235, 2009.

[14] J. Leuthold, C. Koos, and W. Freude, "Nonlinear silicon photonics," *Nat. Photonics*, vol. 4, no. 8, pp. 535–544, 2010.

[15] A. Politi, J. C. Matthews, M. G. Thompson, and J. L. O'Brien, "Integrated quantum photonics," *IEEE J. Sel. Topics Quantum Electron.*, vol. 15, no. 6, pp. 1673–1684, 2009.

[16] P. G. Kwiat, K. Mattle, H. Weinfurter, A. Zeilinger, A. V. Sergienko, and Y. Shih, "New high-intensity source of polarization-entangled photon pairs," *Phys. Rev. Lett.*, vol. 75, no. 24, p. 4337, 1995.

[17] N. Bloembergen, R. K. Chang, S. Jha, and C. Lee, "Optical second-harmonic generation in reflection from media with inversion symmetry," *Phys. Rev.*,

vol. 174, no. 3, p. 813, 1968.

[18] M. Scalora, M. Vincenti, D. De Ceglia, V. Roppo, M. Centini, N. Akozbek, and M. Bloemer, "Second-and third-harmonic generation in metal-based structures," *Phys. Rev. A*, vol. 82, no. 4, p. 043828, 2010.

[19] K. Okamoto, *Fundamentals of Optical Waveguides*. Academic Press, 2010.

[20] A. Yariv and P. Yeh, *Photonics: Optical Electronics in Modern Communications*. Oxford University Press, 2006.

[21] F. De Leonardis, B. Troia, and V. M. Passaro, "Mid-IR optical and nonlinear properties of germanium on silicon optical waveguides," *J. Lightw. Technol.*, vol. 32, no. 22, pp. 3747–3757, 2014.

[22] K. Hammani, M. A. Ettabib, A. Bogris, A. Kapsalis, D. Syvridis, M. Brun, P. Labeye, S. Nicoletti, D. J. Richardson, and P. Petropoulos, "Optical properties of silicon germanium waveguides at telecommunication wavelengths," *Opt. Express*, vol. 21, no. 14, pp. 16690–16701, 2013.

[23] C. Lacava, V. Pusino, P. Minzioni, M. Sorel, and I. Cristiani, "Nonlinear properties of AlGaAs waveguides in continuous wave operation regime," *Opt. Express*, vol. 22, no. 5, pp. 5291–5298, 2014.

[24] U. D. Dave, B. Kuyken, F. Leo, S.-P. Gorza, S. Combrie, A. De Rossi, F. Raineri, and G. Roelkens, "Nonlinear properties of dispersion engineered ingap photonic wire waveguides in the telecommunication wavelength range," *Opt. Express*, vol. 23, no. 4, pp. 4650–4657, 2015.

[25] M. R. Lamont, B. Luther-Davies, D.-Y. Choi, S. Madden, and B. J. Eggleton, "Supercontinuum generation in dispersion engineered highly nonlinear (γ = 10/w/m) As2S3 chalcogenide planar waveguide," *Opt. Express*, vol. 16, no. 19, pp. 14938–14944, 2008.

[26] X. Lu, J. Y. Lee, S. Rogers, and Q. Lin, "Optical Kerr nonlinearity in a high-q silicon carbide microresonator," *Opt. Express*, vol. 22, no. 25, pp. 30826–30832, 2014.

[27] D. J. Moss, R. Morandotti, A. L. Gaeta, and M. Lipson, "New CMOS-compatible platforms based on silicon nitride and hydex for nonlinear optics," *Nat. Photonics*, vol. 7, no. 8, pp. 597–607, 2013.

[28] C. Xiong, W. Pernice, K. K. Ryu, C. Schuck, K. Y. Fong, T. Palacios, and H. X. Tang, "Integrated GaN photonic circuits on silicon (100) for second harmonic generation," *Opt. Express*, vol. 19, no. 11, pp. 10462–10470, 2011.

[29] T. Ohara, H. Takara, I. Shake, K. Mori, S. Kawanishi, S. Mino, T. Yamada, M. Ishii, T. Kitoh, T. Kitagawa, *et al.*, "160-gb/s optical-time-division multiplexing with PPLN hybrid integrated planar lightwave circuit," *IEEE Photon. Technol. Lett.*, vol. 15, no. 2, pp. 302–304, 2003.

[30] B. Hausmann, I. Bulu, V. Venkataraman, P. Deotare, and M. Lončar, "Diamond nonlinear photonics," *Nat. Photonics*, vol. 8, no. 5, pp. 369–374, 2014.

[31] H. Jung, R. Stoll, X. Guo, D. Fischer, and H. X. Tang, "Green, red, and IR

frequency comb line generation from single IR pump in AlN microring resonator," *Optica*, vol. 1, no. 6, pp. 396–399, 2014.

[32] D. Duchesne, M. Ferrera, L. Razzari, R. Morandotti, B. E. Little, S. T. Chu, and D. J. Moss, "Efficient self-phase modulation in low loss, high index doped silica glass integrated waveguides," *Opt. Express*, vol. 17, no. 3, pp. 1865–1870, 2009.

[33] S. Zhu, G. Lo, and D. Kwong, "Theoretical investigation of silicon MOS-type plasmonic slot waveguide based MZI modulators," *Opt. Express*, vol. 18, no. 26, pp. 27802–27819, 2010.

[34] A. Melikyan, L. Alloatti, A. Muslija, D. Hillerkuss, P. Schindler, J. Li, R. Palmer, D. Korn, S. Muehlbrandt, D. Van Thourhout, *et al.*, "High-speed plasmonic phase modulators," *Nat. Photonics*, vol. 8, no. 3, pp. 229–233, 2014.

[35] C. Haffner, W. Heni, Y. Fedoryshyn, J. Niegemann, A. Melikyan, D. Elder, B. Baeuerle, Y. Salamin, A. Josten, U. Koch, *et al.*, "All-plasmonic mach–zehnder modulator enabling optical high-speed communication at the microscale," *Nat. Photonics*, vol. 9, no. 8, pp. 525–528, 2015.

[36] M. Kauranen and A. V. Zayats, "Nonlinear plasmonics," *Nat. Photonics*, vol. 6, no. 11, pp. 737–748, 2012.

[37] J. Butet, P.-F. Brevet, and O. J. Martin, "Optical second harmonic generation in plasmonic nanostructures: From fundamental principles to advanced applications," *ACS Nano*, vol. 9, no. 11, pp. 10545–10562, 2015.

[38] L.-J. Black, P. R. Wiecha, Y. Wang, C. de Groot, V. Paillard, C. Girard, O. L. Muskens, and A. Arbouet, "Tailoring second-harmonic generation in single l-shaped plasmonic nanoantennas from the capacitive to conductive coupling regime," *ACS Photonics*, vol. 2, no. 11, pp. 1592–1601, 2015.

[39] Z. Dong, M. Asbahi, J. Lin, D. Zhu, Y. M. Wang, K. Hippalgaonkar, H.-S. Chu, W. P. Goh, F. Wang, Z. Huang, *et al.*, "Second-harmonic generation from sub-5 nm gaps by directed self-assembly of nanoparticles onto template-stripped gold substrates," *Nano Lett.*, vol. 15, no. 9, pp. 5976–5981, 2015.

[40] M. Celebrano, X. Wu, M. Baselli, S. Großmann, P. Biagioni, A. Locatelli, C. De Angelis, G. Cerullo, R. Osellame, B. Hecht, *et al.*, "Mode matching in multiresonant plasmonic nanoantennas for enhanced second harmonic generation," *Nat. Nanotechnol.*, vol. 10, no. 5, pp. 412–417, 2015.

[41] B. Metzger, L. Gui, J. Fuchs, D. Floess, M. Hentschel, and H. Giessen, "Strong enhancement of second harmonic emission by plasmonic resonances at the second harmonic wavelength," *Nano Lett.*, vol. 15, no. 6, pp. 3917–3922, 2015.

[42] N. K. Hon, K. K. Tsia, D. R. Solli, and B. Jalali, "Periodically poled silicon," *Appl. Phys. Lett.*, vol. 94, no. 9, p. 091116, 2009.

[43] B. Chmielak, M. Waldow, C. Matheisen, C. Ripperda, J. Bolten, T. Wahlbrink, M. Nagel, F. Merget, and H. Kurz, "Pockels effect based fully integrated,

strained silicon electro-optic modulator," *Opt. Express*, vol. 19, no. 18, pp. 17212–17219, 2011.

[44] M. Cazzanelli, F. Bianco, E. Borga, G. Pucker, M. Ghulinyan, E. Degoli, E. Luppi, V. Véniard, S. Ossicini, D. Modotto, *et al.*, "Second-harmonic generation in silicon waveguides strained by silicon nitride," *Nat. Mater.*, vol. 11, no. 2, pp. 148–154, 2012.

[45] A. D. Bristow, N. Rotenberg, and H. M. Van Driel, "Two-photon absorption and Kerr coefficients of silicon for 850–2200 nm," *Appl. Phys. Lett.*, vol. 90, no. 19, p. 191104, 2007.

[46] Q. Lin, J. Zhang, G. Piredda, R. W. Boyd, P. M. Fauchet, and G. P. Agrawal, "Dispersion of silicon nonlinearities in the near infrared region," *Appl. Phys. Lett.*, vol. 91, no. 2, pp. 21111–21111, 2007.

[47] H. Rong, R. Jones, A. Liu, O. Cohen, D. Hak, A. Fang, and M. Paniccia, "A continuous-wave Raman silicon laser," *Nature*, vol. 433, no. 7027, pp. 725–728, 2005.

[48] H. Rong, Y.-H. Kuo, A. Liu, M. Paniccia, and O. Cohen, "High efficiency wavelength conversion of 10 gb/s data in silicon waveguides," *Opt. Express*, vol. 14, no. 3, pp. 1182–1188, 2006.

[49] J. R. Ong, R. Kumar, R. Aguinaldo, and S. Mookherjea, "Efficient CW four-wave mixing in silicon-on-insulator micro-rings with active carrier removal," *IEEE Photon. Technol. Lett.*, vol. 25, no. 17, pp. 1699–1702, 2013.

[50] S. Zlatanovic, J. S. Park, S. Moro, J. M. C. Boggio, I. B. Divliansky, N. Alic, S. Mookherjea, and S. Radic, "Mid-infrared wavelength conversion in silicon waveguides using ultracompact telecom-band-derived pump source," *Nat. Photonics*, vol. 4, no. 8, pp. 561–564, 2010.

[51] X. Liu, B. Kuyken, G. Roelkens, R. Baets, R. M. Osgood Jr, and W. M. Green, "Bridging the mid-infrared-to-telecom gap with silicon nanophotonic spectral translation," *Nat. Photonics*, vol. 6, no. 10, pp. 667–671, 2012.

[52] A. G. Griffith, R. K. Lau, J. Cardenas, Y. Okawachi, A. Mohanty, R. Fain, Y. H. D. Lee, M. Yu, C. T. Phare, C. B. Poitras, *et al.*, "Silicon-chip mid-infrared frequency comb generation," *Nat. Commun.*, vol. 6, 2015.

[53] N. Singh, D. D. Hudson, Y. Yu, C. Grillet, S. D. Jackson, A. Casas-Bedoya, A. Read, P. Atanackovic, S. G. Duvall, S. Palomba, *et al.*, "Midinfrared supercontinuum generation from 2 to 6 μm in a silicon nanowire," *Optica*, vol. 2, no. 9, pp. 797–802, 2015.

[54] T. Wang, N. Venkatram, J. Gosciniak, Y. Cui, G. Qian, W. Ji, and D. T. Tan, "Multi-photon absorption and third-order nonlinearity in silicon at mid-infrared wavelengths," *Opt. Express*, vol. 21, no. 26, pp. 32192–32198, 2013.

[55] C. Grillet, L. Carletti, C. Monat, P. Grosse, B. B. Bakir, S. Menezo, J. Fedeli, and D. Moss, "Amorphous silicon nanowires combining high nonlinearity, FOM and optical stability," *Opt. Express*, vol. 20, no. 20, pp. 22609–22615, 2012.

[56] T. Wang, D. K. Ng, S.-K. Ng, Y.-T. Toh, A. Chee, G. F. Chen, Q. Wang, and D. T. Tan, "Supercontinuum generation in bandgap engineered, back-end CMOS compatible silicon rich nitride waveguides," *Laser Photonics Rev.*, vol. 9, no. 5, pp. 498–506, 2015.

[57] C. Koos, P. Vorreau, T. Vallaitis, P. Dumon, W. Bogaerts, R. Baets, B. Esembeson, I. Biaggio, T. Michinobu, F. Diederich, *et al.*, "All-optical high-speed signal processing with silicon–organic hybrid slot waveguides," *Nat. Photonics*, vol. 3, no. 4, pp. 216–219, 2009.

[58] O. Boyraz and B. Jalali, "Demonstration of a silicon Raman laser," *Opt. Express*, vol. 12, no. 21, pp. 5269–5273, 2004.

[59] B. Jalali, V. Raghunathan, D. Dimitropoulos, and Ö. Boyraz, "Raman-based silicon photonics," *IEEE J. Sel. Topics Quantum Electron.*, vol. 12, no. 3, pp. 412–421, 2006.

[60] Y. Takahashi, Y. Inui, M. Chihara, T. Asano, R. Terawaki, and S. Noda, "A micrometre-scale Raman silicon laser with a microwatt threshold," *Nature*, vol. 498, no. 7455, pp. 470–474, 2013.

[61] D. Dimitropoulos, B. Houshmand, R. Claps, and B. Jalali, "Coupled-mode theory of the Raman effect in silicon-on-insulator waveguides," *Opt. Lett.*, vol. 28, no. 20, pp. 1954–1956, 2003.

[62] P. T. Rakich, C. Reinke, R. Camacho, P. Davids, and Z. Wang, "Giant enhancement of stimulated Brillouin scattering in the subwavelength limit," *Phys. Rev. X*, vol. 2, no. 1, p. 011008, 2012.

[63] R. Van Laer, B. Kuyken, D. Van Thourhout, and R. Baets, "Interaction between light and highly confined hypersound in a silicon photonic nanowire," *Nat. Photonics*, vol. 9, no. 3, pp. 199–203, 2015.

[64] C. Wolff, M. J. Steel, B. J. Eggleton, and C. G. Poulton, "Stimulated Brillouin scattering in integrated photonic waveguides: Forces, scattering mechanisms, and coupled-mode analysis," *Phys. Rev. A*, vol. 92, no. 1, p. 013836, 2015.

[65] B. J. Eggleton, C. G. Poulton, and R. Pant, "Inducing and harnessing stimulated Brillouin scattering in photonic integrated circuits," *Adv. Opt. Photonics*, vol. 5, no. 4, pp. 536–587, 2013.

[66] R. Claps, D. Dimitropoulos, Y. Han, and B. Jalali, "Observation of Raman emission in silicon waveguides at 1.54 μm," *Opt. Express*, vol. 10, no. 22, pp. 1305–1313, 2002.

[67] M. A. Foster, A. C. Turner, J. E. Sharping, B. S. Schmidt, M. Lipson, and A. L. Gaeta, "Broad-band optical parametric gain on a silicon photonic chip," *Nature*, vol. 441, no. 7096, pp. 960–963, 2006.

[68] H. Fukuda, K. Yamada, T. Shoji, M. Takahashi, T. Tsuchizawa, T. Watanabe, J.-i. Takahashi, and S.-i. Itabashi, "Four-wave mixing in silicon wire waveguides," *Opt. Express*, vol. 13, no. 12, pp. 4629–4637, 2005.

[69] X. Liu, R. M. Osgood, Y. A. Vlasov, and W. M. Green, "Mid-infrared optical parametric amplifier using silicon nanophotonic waveguides," *Nat. Photonics*,

vol. 4, no. 8, pp. 557–560, 2010.

[70] B. Kuyken, X. Liu, R. M. Osgood, R. Baets, G. Roelkens, and W. M. Green, "A silicon-based widely tunable short-wave infrared optical parametric oscillator," *Opt. Express*, vol. 21, no. 5, pp. 5931–5940, 2013.

[71] B. Kuyken, H. Ji, S. Clemmen, S. Selvaraja, H. Hu, M. Pu, M. Galili, P. Jeppesen, G. Morthier, S. Massar, *et al.*, "Nonlinear properties of and nonlinear processing in hydrogenated amorphous silicon waveguides," *Opt. Express*, vol. 19, no. 26, pp. B146–B153, 2011.

[72] K.-Y. Wang, M. A. Foster, and A. C. Foster, "Wavelength-agile near-IR optical parametric oscillator using a deposited silicon waveguide," *Opt. Express*, vol. 23, no. 12, pp. 15431–15439, 2015.

[73] S. Keyvaninia, G. Roelkens, D. Van Thourhout, C. Jany, M. Lamponi, A. Le Liepvre, F. Lelarge, D. Make, G.-H. Duan, D. Bordel, *et al.*, "Demonstration of a heterogeneously integrated III-V/SOI single wavelength tunable laser," *Opt. Express*, vol. 21, no. 3, pp. 3784–3792, 2013.

[74] S. Wirths, R. Geiger, N. Von Den Driesch, G. Mussler, T. Stoica, S. Mantl, Z. Ikonic, M. Luysberg, S. Chiussi, J. Hartmann, *et al.*, "Lasing in direct-bandgap GeSn alloy grown on Si," *Nat. Photonics*, vol. 9, no. 2, pp. 88–92, 2015.

[75] I.-W. Hsieh, X. Chen, X. Liu, J. I. Dadap, N. C. Panoiu, C.-Y. Chou, F. Xia, W. M. Green, Y. A. Vlasov, and R. M. Osgood, "Supercontinuum generation in silicon photonic wires," *Opt. Express*, vol. 15, no. 23, pp. 15242–15249, 2007.

[76] B. Kuyken, X. Liu, R. M. Osgood, R. Baets, G. Roelkens, and W. M. Green, "Mid-infrared to telecom-band supercontinuum generation in highly nonlinear silicon-on-insulator wire waveguides," *Opt. Express*, vol. 19, no. 21, pp. 20172–20181, 2011.

[77] B. Kuyken, T. Ideguchi, S. Holzner, M. Yan, T. W. Hänsch, J. Van Campenhout, P. Verheyen, S. Coen, F. Leo, R. Baets, *et al.*, "An octave-spanning mid-infrared frequency comb generated in a silicon nanophotonic wire waveguide," *Nat. Commun.*, vol. 6, 2015.

[78] A. E. Willner, S. Khaleghi, M. R. Chitgarha, and O. F. Yilmaz, "All-optical signal processing," *J. Lightw. Technol.*, vol. 32, no. 4, pp. 660–680, 2014.

[79] F. Li, M. Pelusi, D. Xu, A. Densmore, R. Ma, S. Janz, and D. Moss, "Error-free all-optical demultiplexing at 160gb/s via FWM in a silicon nanowire," *Opt. Express*, vol. 18, no. 4, pp. 3905–3910, 2010.

[80] H. Ji, M. Pu, H. Hu, M. Galili, L. K. Oxenløwe, K. Yvind, J. M. Hvam, and P. Jeppesen, "Optical waveform sampling and error-free demultiplexing of 1.28 tb/s serial data in a nanoengineered silicon waveguide," *J. Lightw. Technol.*, vol. 29, no. 4, pp. 426–431, 2011.

[81] A. Biberman, B. G. Lee, A. C. Turner-Foster, M. A. Foster, M. Lipson, A. L. Gaeta, and K. Bergman, "Wavelength multicasting in silicon photonic

nanowires," *Opt. Express*, vol. 18, no. 17, pp. 18047–18055, 2010.

[82] Y. Zhang, C. Husko, J. Schröder, S. Lefrancois, I. H. Rey, T. F. Krauss, and B. J. Eggleton, "Phase-sensitive amplification in silicon photonic crystal waveguides," *Opt. Lett.*, vol. 39, no. 2, pp. 363–366, 2014.

[83] F. Da Ros, D. Vukovic, A. Gajda, K. Dalgaard, L. Zimmermann, B. Tillack, M. Galili, K. Petermann, and C. Peucheret, "Phase regeneration of DPSK signals in a silicon waveguide with reverse-biased pin junction," *Opt. Express*, vol. 22, no. 5, pp. 5029–5036, 2014.

[84] Y. H. Wen, O. Kuzucu, T. Hou, M. Lipson, and A. L. Gaeta, "All-optical switching of a single resonance in silicon ring resonators," *Opt. Lett.*, vol. 36, no. 8, pp. 1413–1415, 2011.

[85] T. Liang, L. Nunes, T. Sakamoto, K. Sasagawa, T. Kawanishi, M. Tsuchiya, G. Priem, D. Van Thourhout, P. Dumon, R. Baets, *et al.*, "Ultrafast all-optical switching by cross-absorption modulation in silicon wire waveguides," *Opt. Express*, vol. 13, no. 19, pp. 7298–7303, 2005.

[86] E.-K. Tien, N. S. Yuksek, F. Qian, and O. Boyraz, "Pulse compression and modelocking by using TPA in silicon waveguides," *Opt. Express*, vol. 15, no. 10, pp. 6500–6506, 2007.

[87] C. Koos, L. Jacome, C. Poulton, J. Leuthold, and W. Freude, "Nonlinear silicon-on-insulator waveguides for all-optical signal processing," *Opt. Express*, vol. 15, no. 10, pp. 5976–5990, 2007.

[88] L. Yin and G. P. Agrawal, "Impact of two-photon absorption on self-phase modulation in silicon waveguides," *Opt. Lett.*, vol. 32, no. 14, pp. 2031–2033, 2007.

[89] Q. Lin, O. J. Painter, and G. P. Agrawal, "Nonlinear optical phenomena in silicon waveguides: modeling and applications," *Opt. Express*, vol. 15, no. 25, pp. 16604–16644, 2007.

[90] W. Mathlouthi, H. Rong, and M. Paniccia, "Characterization of efficient wavelength conversion by four-wave mixing in sub-micron silicon waveguides," *Opt. Express*, vol. 16, no. 21, pp. 16735–16745, 2008.

[91] J. R. Ong and V. H. Chen, "Optimal geometry of nonlinear silicon slot waveguides accounting for the effect of waveguide losses," *Opt. Express*, vol. 23, no. 26, pp. 33622–33633, 2015.

[92] P. Dong, W. Qian, S. Liao, H. Liang, C.-C. Kung, N.-N. Feng, R. Shafiiha, J. Fong, D. Feng, A. V. Krishnamoorthy, *et al.*, "Low loss shallow-ridge silicon waveguides," *Opt. Express*, vol. 18, no. 14, pp. 14474–14479, 2010.

[93] S. K. Selvaraja, P. D. Heyn, G. Winroth, P. Ong, G. Lepage, C. Cailler, A. Rigny, K. Bourdelle, W. Bogaerts, D. VanThourhout, J. V. Campenhout, and P. Absil, "Highly uniform and low-loss passive silicon photonics devices using a 300mm CMOS platform," *in 2014 Optical Fiber Communication Conf.*, p. Th2A.33.

[94] D. Shimura, T. Horikawa, H. Okayama, S.-H. Jeong, M. Tokushima,

H. Sasaki, and T. Mogami, "High precision Si waveguide devices designed for 1.31 μm and 1.55 μm wavelengths on 300mm-SOI," *in 2014 IEEE 11th International Conf. on Group IV Photonics*, pp. 31–32.

[95] M. Ebnali-Heidari, C. Monat, C. Grillet, and M. Moravvej-Farshi, "A proposal for enhancing four-wave mixing in slow light engineered photonic crystal waveguides and its application to optical regeneration," *Opt. Express*, vol. 17, no. 20, pp. 18340–18353, 2009.

[96] N. Matsuda, T. Kato, K.-i. Harada, H. Takesue, E. Kuramochi, H. Taniyama, and M. Notomi, "Slow light enhanced optical nonlinearity in a silicon photonic crystal coupled-resonator optical waveguide," *Opt. Express*, vol. 19, no. 21, pp. 19861–19874, 2011.

[97] B. Kuyken, P. Verheyen, P. Tannouri, X. Liu, J. Van Campenhout, R. Baets, W. Green, and G. Roelkens, "Generation of 3.6 μm radiation and telecom-band amplification by four-wave mixing in a silicon waveguide with normal group velocity dispersion," *Opt. Lett.*, vol. 39, no. 6, pp. 1349–1352, 2014.

[98] M. A. Foster, A. C. Turner, R. Salem, M. Lipson, and A. L. Gaeta, "Broad-band continuous-wave parametric wavelength conversion in silicon nanowaveguides," *Opt. Express*, vol. 15, no. 20, pp. 12949–12958, 2007.

[99] A. C. Turner-Foster, M. A. Foster, R. Salem, A. L. Gaeta, and M. Lipson, "Frequency conversion over two-thirds of an octave in silicon nanowaveguides," *Opt. Express*, vol. 18, no. 3, pp. 1904–1908, 2010.

[100] J. B. Driscoll, N. Ophir, R. R. Grote, J. I. Dadap, N. C. Panoiu, K. Bergman, and R. M. Osgood, "Width-modulation of Si photonic wires for quasi-phase-matching of four-wave-mixing: experimental and theoretical demonstration," *Opt. Express*, vol. 20, no. 8, pp. 9227–9242, 2012.

[101] P. Absil, J. Hryniewicz, B. Little, P. Cho, R. Wilson, L. Joneckis, and P.-T. Ho, "Wavelength conversion in GaAs micro-ring resonators," *Opt. Lett.*, vol. 25, no. 8, pp. 554–556, 2000.

[102] C. M. Gentry, X. Zeng, and M. A. Popović, "Tunable coupled-mode dispersion compensation and its application to on-chip resonant four-wave mixing," *Opt. Lett.*, vol. 39, no. 19, pp. 5689–5692, 2014.

[103] J. Ong, R. Kumar, and S. Mookherjea, "Triply resonant four-wave mixing in silicon-coupled resonator microring waveguides," *Opt. Lett.*, vol. 39, no. 19, pp. 5653–5656, 2014.

[104] S. A. Miller, Y. Okawachi, S. Ramelow, K. Luke, A. Dutt, A. Farsi, A. L. Gaeta, and M. Lipson, "Tunable frequency combs based on dual microring resonators," *Opt. Express*, vol. 23, no. 16, pp. 21527–21540, 2015.

[105] F. Morichetti, A. Canciamilla, C. Ferrari, A. Samarelli, M. Sorel, and A. Melloni, "Travelling-wave resonant four-wave mixing breaks the limits of cavity-enhanced all-optical wavelength conversion," *Nat. Commun.*, vol. 2, p. 296, 2011.

[106] T. F. Krauss, "Slow light in photonic crystal waveguides," *J. Phys. D: Appl.*

Phys., vol. 40, no. 9, p. 2666, 2007.

[107] C. Monat, M. Ebnali-Heidari, C. Grillet, B. Corcoran, B. Eggleton, T. White, L. OFaolain, J. Li, and T. Krauss, "Four-wave mixing in slow light engineered silicon photonic crystal waveguides," *Opt. Express*, vol. 18, no. 22, pp. 22915–22927, 2010.

[108] J. Li, L. OFaolain, I. H. Rey, and T. F. Krauss, "Four-wave mixing in photonic crystal waveguides: slow light enhancement and limitations," *Opt. Express*, vol. 19, no. 5, pp. 4458–4463, 2011.

[109] C. H. Bennett and G. Brassard, "Quantum cryptography: Public key distribution and coin tossing," *Theoret. Comp. Sci.*, vol. 560, pp. 7–11, 2014.

[110] M. A. Nielsen and I. L. Chuang, *Quantum Computation and Quantum Information*. Cambridge University Press, 2010.

[111] J. Carolan, C. Harrold, C. Sparrow, E. Mart´in-López, N. J. Russell, J. W. Silverstone, P. J. Shadbolt, N. Matsuda, M. Oguma, M. Itoh, *et al.*, "Universal linear optics," *Science*, vol. 349, no. 6249, pp. 711–716, 2015.

[112] W. H. Pernice, C. Schuck, O. Minaeva, M. Li, G. Goltsman, A. Sergienko, and H. Tang, "High-speed and high-efficiency travelling wave single-photon detectors embedded in nanophotonic circuits," *Nat. Commun.*, vol. 3, p. 1325, 2012.

[113] F. Marsili, V. Verma, J. Stern, S. Harrington, A. Lita, T. Gerrits, I. Vayshenker, B. Baek, M. Shaw, R. Mirin, *et al.*, "Detecting single infrared photons with 93% system efficiency," *Nat. Photonics*, vol. 7, no. 3, pp. 210–214, 2013.

[114] C. Kurtsiefer, S. Mayer, P. Zarda, and H. Weinfurter, "Stable solid-state source of single photons," *Phys. Rev. Lett.*, vol. 85, no. 2, p. 290, 2000.

[115] A. J. Shields, "Semiconductor quantum light sources," *Nat. Photonics*, vol. 1, no. 4, pp. 215–223, 2007.

[116] C. Xiong, G. D. Marshall, A. Peruzzo, M. Lobino, A. S. Clark, D.-Y. Choi, S. J. Madden, C. M. Natarajan, M. G. Tanner, R. H. Hadfield, *et al.*, "Generation of correlated photon pairs in a chalcogenide As2S3 waveguide," *Appl. Phys. Lett.*, vol. 98, no. 5, p. 051101, 2011.

[117] Q. Lin and G. P. Agrawal, "Silicon waveguides for creating quantum-correlated photon pairs," *Opt. Lett.*, vol. 31, no. 21, pp. 3140–3142, 2006.

[118] M. Savanier, R. Kumar, and S. Mookherjea, "Photon pair generation from compact silicon microring resonators using microwatt-level pump powers," *Opt. Express*, vol. 24, pp. 3313–3328, Feb 2016.

[119] J. E. Sharping, K. F. Lee, M. A. Foster, A. C. Turner, B. S. Schmidt, M. Lipson, A. L. Gaeta, and P. Kumar, "Generation of correlated photons in nanoscale silicon waveguides," *Opt. Express*, vol. 14, no. 25, pp. 12388–12393, 2006.

[120] K.-i. Harada, H. Takesue, H. Fukuda, T. Tsuchizawa, T. Watanabe, K. Yamada, Y. Tokura, and S.-i. Itabashi, "Generation of high-purity entangled photon pairs using silicon wire waveguide," *Opt. Express*, vol. 16, no. 25, pp. 20368–

20373, 2008.

[121] S. Clemmen, K. P. Huy, W. Bogaerts, R. G. Baets, P. Emplit, and S. Massar, "Continuous wave photon pair generation in silicon-on-insulator waveguides and ring resonators," *Opt. Express*, vol. 17, no. 19, pp. 16558–16570, 2009.

[122] S. Azzini, D. Grassani, M. Galli, L. C. Andreani, M. Sorel, M. J. Strain, L. Helt, J. Sipe, M. Liscidini, and D. Bajoni, "From classical four-wave mixing to parametric fluorescence in silicon microring resonators," *Opt. Lett.*, vol. 37, no. 18, pp. 3807–3809, 2012.

[123] R. Kumar, J. R. Ong, J. Recchio, K. Srinivasan, and S. Mookherjea, "Spectrally multiplexed and tunable-wavelength photon pairs at 1.55 μm from a silicon coupled-resonator optical waveguide," *Opt. Lett.*, vol. 38, no. 16, pp. 2969–2971, 2013.

[124] M. J. Collins, C. Xiong, I. H. Rey, T. D. Vo, J. He, S. Shahnia, C. Reardon, T. F. Krauss, M. Steel, A. S. Clark, *et al.*, "Integrated spatial multiplexing of heralded single-photon sources," *Nat. Commun.*, vol. 4, 2013.

[125] G. J. Mendoza, R. Santagati, J. Munns, E. Hemsley, M. Piekarek, E. Mart´in-López, G. D. Marshall, D. Bonneau, M. G. Thompson, and J. L. OBrien, "Active temporal and spatial multiplexing of photons," *Optica*, vol. 3, no. 2, pp. 127–132, 2016.

[126] D. Bonneau, G. J. Mendoza, J. L. OBrien, and M. G. Thompson, "Effect of loss on multiplexed single-photon sources," *New J. Phys.*, vol. 17, no. 4, p. 043057, 2015.

[127] Z. Vernon, M. Liscidini, and J. E. Sipe, "No free lunch: the trade-off between heralding rate and efficiency in microresonator-based heralded single photon sources," *Opt. Lett.*, vol. 41, pp. 788–791, Feb 2016.

[128] J. D. Franson, "Bell inequality for position and time," *Phys. Rev. Lett.*, vol. 62, no. 19, p. 2205, 1989.

[129] D. Grassani, S. Azzini, M. Liscidini, M. Galli, M. J. Strain, M. Sorel, J. Sipe, and D. Bajoni, "Micrometer-scale integrated silicon source of time-energy entangled photons," *Optica*, vol. 2, no. 2, pp. 88–94, 2015.

[130] R. Wakabayashi, M. Fujiwara, K.-I. Yoshino, Y. Nambu, M. Sasaki, and T. Aoki, "Time-bin entangled photon pair generation from Si micro-ring resonator," *Opt. Express*, vol. 23, no. 2, pp. 1103–1113, 2015.

[131] H. Takesue, N. Matsuda, E. Kuramochi, and M. Notomi, "Entangled photons from on-chip slow light," *Sci. Rep.*, vol. 4, 2014.

[132] R. Kumar, M. Savanier, J. R. Ong, and S. Mookherjea, "Entanglement measurement of a coupled silicon microring photon pair source," *Opt. Express*, vol. 23, no. 15, pp. 19318–19327, 2015.

[133] N. Matsuda, H. Le Jeannic, H. Fukuda, T. Tsuchizawa, W. J. Munro, K. Shimizu, K. Yamada, Y. Tokura, and H. Takesue, "A monolithically integrated polarization entangled photon pair source on a silicon chip," *Sci. Rep.*, vol. 2, 2012.

[134] J. Suo, S. Dong, W. Zhang, Y. Huang, and J. Peng, "Generation of hyper-entanglement on polarization and energy-time based on a silicon micro-ring cavity," *Opt. Express*, vol. 23, no. 4, pp. 3985–3995, 2015.

[135] H. J. Kimble, "The quantum internet," *Nature*, vol. 453, no. 7198, pp. 1023–1030, 2008.

[136] N. Sangouard, C. Simon, H. De Riedmatten, and N. Gisin, "Quantum repeaters based on atomic ensembles and linear optics," *Rev. Mod. Phys.*, vol. 83, no. 1, p. 33, 2011.

[137] H. Xu, L. Ma, A. Mink, B. Hershman, and X. Tang, "1310-nm quantum key distribution system with up-conversion pump wavelength at 1550 nm," *Opt. Express*, vol. 15, no. 12, pp. 7247–7260, 2007.

[138] J. Pelc, P. S. Kuo, O. Slattery, L. Ma, X. Tang, and M. Fejer, "Dual-channel, single-photon upconversion detector at 1.3 μm," *Opt. Express*, vol. 20, no. 17, pp. 19075–19087, 2012.

[139] P. Kumar, "Quantum frequency conversion," *Opt. Lett.*, vol. 15, no. 24, pp. 1476–1478, 1990.

[140] C. McKinstrie, J. Harvey, S. Radic, and M. Raymer, "Translation of quantum states by four-wave mixing in fibers," *Opt. Express*, vol. 13, no. 22, pp. 9131–9142, 2005.

[141] C. McKinstrie, M. Yu, M. Raymer, and S. Radic, "Quantum noise properties of parametric processes," *Opt. Express*, vol. 13, no. 13, pp. 4986–5012, 2005.

[142] S. Tanzilli, W. Tittel, M. Halder, O. Alibart, P. Baldi, N. Gisin, and H. Zbinden, "A photonic quantum information interface," *Nature*, vol. 437, no. 7055, pp. 116–120, 2005.

[143] L. Ma, M. T. Rakher, M. J. Stevens, O. Slattery, K. Srinivasan, and X. Tang, "Temporal correlation of photons following frequency up-conversion," *Opt. Express*, vol. 19, no. 11, pp. 10501–10510, 2011.

[144] H. McGuinness, M. Raymer, C. McKinstrie, and S. Radic, "Quantum frequency translation of single-photon states in a photonic crystal fiber," *Phys. Rev. Lett.*, vol. 105, no. 9, p. 093604, 2010.

[145] H. Takesue, "Erasing distinguishability using quantum frequency up-conversion," *Phys. Rev. Lett.*, vol. 101, no. 17, p. 173901, 2008.

[146] M. Raymer, S. Van Enk, C. McKinstrie, and H. McGuinness, "Interference of two photons of different color," *Opt. Commun.*, vol. 283, no. 5, pp. 747–752, 2010.

[147] R. Ikuta, T. Kobayashi, H. Kato, S. Miki, T. Yamashita, H. Terai, M. Fujiwara, T. Yamamoto, M. Koashi, M. Sasaki, *et al.*, "Nonclassical two-photon interference between independent telecommunication light pulses converted by difference-frequency generation," *Phys. Rev. A*, vol. 88, no. 4, p. 042317, 2013.

[148] I. Agha, S. Ates, M. Davanço, and K. Srinivasan, "A chip-scale,

telecommunications-band frequency conversion interface for quantum emitters," *Opt. Express*, vol. 21, no. 18, pp. 21628–21638, 2013.

[149] Q. Li, M. Davanço, and K. Srinivasan, "Efficient and low-noise single-photon-level frequency conversion interfaces using silicon nanophotonics," *Nat. Photonics*, 2016.

[150] P. S. Kuo, J. S. Pelc, O. Slattery, Y.-S. Kim, M. Fejer, and X. Tang, "Reducing noise in single-photon-level frequency conversion," *Opt. Lett.*, vol. 38, no. 8, pp. 1310–1312, 2013.

[151] B. A. Bell, J. He, C. Xiong, and B. J. Eggleton, "Frequency conversion in silicon in the single photon regime," *Opt. Express*, vol. 24, no. 5, pp. 5235–5242, 2016.

Chapter 14

Integrated nanophotonics for multi-user quantum key distribution networks

Han Chuen Lim
DSO National Laboratories, Singapore;
Nanyang Technological University, Singapore

Mao Tong Liu
Nanyang Technological University, Singapore

This chapter will overview developments in the field of integrated nanophotonics for multi-user quantum key distribution (QKD) networks. Section 14.1 begins with the significance of QKD and presents some of the challenges faced by researchers and system developers, especially regarding the key rate, transmission distance, and implementation cost. Section 14.2 will consider various options to implement a multi-user QKD network compatible with the existing fiber-optic network infrastructure. Section 14.3 will describe entanglement-based QKD and its wavelength-multiplexed implementation enabling efficient use of the available optical fiber bandwidth for QKD. We will show how a centralized broadband entangled-photon source can be shared by many users in an access network and explain the benefits of wavelength-multiplexed entanglement distribution. Section 14.4 will look at recent advances in integrated broadband entangled-photon sources, waveguide-integrated photon detectors, and low-loss integrated passive devices. Section 14.5 will provide an outlook for future research and development in this direction.

14.1 Introduction and background

The security of QKD has been proven theoretically. After more than two decades of research and development, QKD technology is now mature and ready for practical use. There are, however, challenges ahead with regard to its adoption in metropolitan area networks. For widespread adoption, QKD implementation must be compatible with the existing fiber-optic network infrastructure built for the Internet. Such networks typically consist of long-range backbone optical fiber links connecting a number of core nodes, and shorter-range optical fiber access networks that connect many remote users to these core nodes. Ad hoc installation of point-to-point QKD systems is unlikely to scale well in a metropolitan area network. In this review, we consider the implementation of multi-user QKD networks based on an approach called *wavelength-multiplexed entanglement distribution*, leveraging on integrated nanophotonics technology for scalability and cost reduction.

Integrated nanophotonics has progressed tremendously in recent years and integrated quantum photonics has become an active research field. There have been experimental demonstrations of on-chip entangled-photon generation, on-chip multiphoton interference, on-chip single-photon detection, and so on. The benefits of nanophotonic integration are smaller form factor, better stability, and lower cost, as compared with conventional bulk-optic or fiber-optic approaches. These benefits become apparent when we consider implementing large-scale multi-user QKD networks, where system scalability, reliability and cost are important factors. This review highlights emerging integrated nanophotonic technologies that have potential to drive the development of practical multi-user QKD networks, in which any pair of users connected to the network can benefit from QKD-secured communications.

14.1.1 Significance of QKD

The invention of QKD by Bennett and Brassard in 1984 is a significant milestone in the development of modern cryptography [1]. Deriving its security from physics principles, QKD enables two remote parties that share an initial authentication key to expand their key securely without computational assumptions. Using the expanded key as one-time-pad to encrypt and decrypt secret messages, the overall cipher scheme would be information-theoretically secure. This means that security is not compromised, even if eavesdroppers have unlimited computing power. The automated and secure delivery of fresh keys to remote parties without computational assumptions is a benefit that mathematical ciphers have not been able to offer [2–4].

Today, many different types of QKD schemes exist [5–7]. They can be broadly divided into *prepare-and-measure* and *entanglement-based* schemes. In a prepare-and-measure QKD scheme, one of the two communicating parties prepares quantum states randomly based on the outputs of a random number generator, while the other party measures the transmitted states with measurement bases randomly chosen from a predetermined set. In an entanglement-based QKD scheme, the quantum states used are entangled and they need not be prepared by any of the communicating parties. All these schemes require transmission of quantum states between remote parties and classical communication over an authenticated channel for extracting secure keys. Both free-space and fiber-optic QKD schemes have been widely studied. One of the earliest QKD schemes implemented over a point-to-point (P2P) optical fiber link is a *plug-and-play* version of the Bennett-Brassard 1984 (BB84) scheme [8, 9]. The plug-and-play implementation applies phase coding to photons from an attenuated laser, and single-photon detectors are used to detect the photons. This scheme features phase and polarization stability and has been commercialized by market leader, ID Quantique SA. While the key rates are not high enough to support one-time-pad encryption of Gbps data, commercial QKD systems enable frequent renewal of symmetric keys over a dedicated dark fiber linking two remote parties. The shared symmetric keys can be used as Advanced Encryption Standard (AES) keys for securing bank transactions, data transfers across data centers, and other tasks.

14.1.2 Challenges

QKD offers a high level of security. The price to pay is low key rate, limited transmission distance, and high cost. Commercial QKD systems offer key rates on the order of kbps for reasonable transmission distances of a few tens of kilometers. Next generation systems may reach a higher key rate of 1 Mbps. This is still very low compared to the multi-Gbps data communication rate that we enjoy today. The main reason for a low QKD key rate is photon loss during transmission due to absorption and scattering in the optical fiber material. Optical attenuation increases exponentially with respect to the distance travelled, so a significant proportion of the photons is lost during transmission over long lengths of optical fiber. In the telecommunication bands, the attenuation coefficient of silica fiber is 0.2 dB/km for C-band (1530 to 1565 nm) and 0.35 dB/km for O-band (1260 to 1360 nm). Single-photon detectors based on InGaAs/InP avalanche photodiodes (APDs) are commonly used in commercial QKD systems. They have detection quantum efficiencies $<30\%$ in the telecommunication bands, and this contributes to additional photon loss. Other contributions to a low key rate include insertion loss of optical components at the receiver and the discarding of raw key bits during the post-processing step of the QKD protocol.

If photon detectors are noiseless, a user might just wait long enough to obtain a key. However, in practice, photon detectors have thermally excited dark counts. When photon arrival rate drops below a certain level, noise arising from detector dark counts dominates and it becomes impossible to extract a secure key. Typical dark count rates for well-designed thermo-electrically-cooled InGaAs/InP APDs are on the order of a few hundred Hz (with 10 μs deadtime). It is helpful to use low-noise superconducting single-photon detectors with dark count rates of a few Hz to increase the transmission distance [10], but the disadvantage is that these detectors require cryogenic temperatures to operate. Optical amplifiers cannot be used at intermediate locations to compensate for photon loss because optical amplification attempts to make copies of the quantum states and the *no-cloning theorem* guarantees that noise is added in the process [11, 12]. If no noise was added, we would be able to gain full information of the photon state by simultaneously measuring complementary properties on perfect copies, and this violates Heisenberg's uncertainty principle.

There are two ways to overcome the distance limitation. One way is to use quantum repeaters based on the concept of entanglement swapping. This involves quantum teleportation of entangled photons, and quantum memories are needed for storing quantum states [13, 14]. These technologies are not practical yet and under active research today [15, 16]. Another way is to use a *trusted node architecture*. In this architecture, all the intermediate nodes within a QKD network are secure and trusted. QKD keys are used to relay a secret key to remote nodes, such that two nodes that do not have a direct optically transparent connection between them can still share the same key. Typically, a key management layer obtains keys from the physical layer and manages the keys. QKD networks based on the trusted-node architecture have been demonstrated in United States [17], Europe [18, 19], Japan [20], and China [21]. China has further announced plans to install a QKD backbone link between Beijing and Shanghai, which are 2,000 kilometers apart, with 32 trusted nodes [22].

The cost of installing and operating QKD is an important factor to consider when implementing a multi-user QKD network. As QKD systems are highly sensitive to noise, most operate only on dark fibers that do not carry any other signal. Moreover, QKD requires optical transparency between sender and receiver because quantum information is encoded onto optical signals. Any optical-to-electronic conversion along the transmission line destroys the quantum information. Hence, QKD is not compatible with network switches and routers, and it would be necessary for the QKD signals to avoid or bypass them if they are present in the network. Optical access networks have many fiber connections with each fiber leading to one remote user. Operating QKD on these fiber lines to reach all users may be highly costly if each fiber line must be a dark fiber leased solely for the purpose of QKD. It is also not cost-effective to have one dedicated set of QKD modules for each pair of users. A set of QKD modules consisting of a QKD transmitter module and a QKD receiver module now costs more than US$100,000. This includes the material costs of single-photon detectors, optical modulators, high-speed electronics, and various optical components.

14.2 Multi-user QKD network

Figure 14.1 illustrates a typical metropolitan area optical network that consists of a backbone network linking multiple core nodes and access networks that connect many remote users to the backbone network. A multi-user QKD network should be designed to be compatible with this network architecture to leverage on existing infrastructure. It is therefore likely that a QKD network is also organized into a backbone QKD network with QKD access networks, as depicted in the figure. A satellite QKD link may also be useful for key expansion with remote parties beyond the reach of fiber-based QKD. In this case, the satellite ground station must have a QKD link with the main network.

14.2.1 Backbone QKD links

For backbone QKD links, it is advantageous to operate QKD in the C-band to take advantage of the lowest loss telecommunication window, as link lengths may be as long as 100 km. O-band is not suited for backbone QKD links because of higher transmission loss. To the best of our knowledge, all the QKD network demonstrations performed so far have used P2P QKD systems and adopted the trusted-node architecture in which keys are relayed securely across trusted nodes via a key management layer [17–21]. These demonstrations provide strong evidence that QKD is compatible with the existing fiber-optic backbone network, although some modifications to the network may still be necessary for long-term stability and network re-configurability [23–27].

Most field demonstrations used dark fibers for QKD. Leasing a dark fiber solely for QKD over a single-wavelength channel is clearly not optimal because most of the

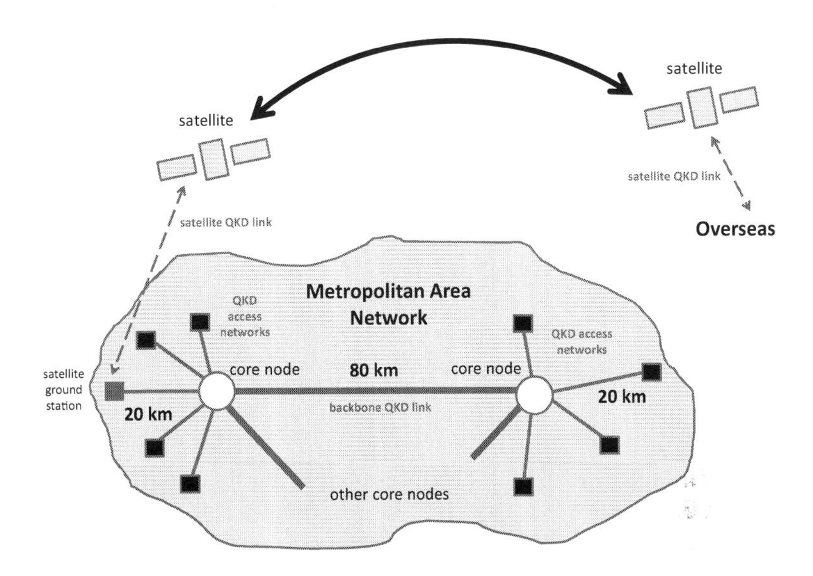

Figure 14.1: QKD implementation in a metropolitan area network. The QKD network consists of backbone QKD links connecting the core nodes, QKD access networks that connect remote users to the core nodes, and satellite link for QKD with remote parties that are not reachable via optical fiber.

available optical bandwidth is wasted. There are two possible approaches to improve the bandwidth utilization. The first approach is to wavelength-multiplex the single-wavelength QKD channel with other Internet data channels all within the C-band. In this way, the available transmission bandwidth in the C-band is shared by QKD and Internet data transmission. The feasibility of combining QKD and classical data channels together on the same optical fiber has been studied and demonstrated in a laboratory environment [28–33]. Peters et al. showed that it is possible to avoid *four-wave mixing* (FWM) crosstalk from the classical wavelength-division-multiplexed (WDM) channels [28]. FWM occurs, when two data channels of different angular frequencies ω_1 and ω_2 interact nonlinearly in optical fiber to generate new frequencies, $2\omega_1 - \omega_2$ and $2\omega_2 - \omega_1$. Higher order FWM processes generating other frequencies may also occur. Avoidance of FWM crosstalk is possible by proper choice of QKD channel wavelength and reduction of launched power for the data channels. FWM is also less efficient in non-dispersion-shifted fibers. A more detrimental noise source is *spontaneous Raman scattering*, which is caused by photon-phonon interaction during light propagation in optical fibers. Raman noise generally covers a broad spectrum at room temperature [34]. With tight temporal and spectral filtering of the QKD channels and restriction on the launched power of the classical data channels, it is possible to suppress Raman noise photons to acceptable levels and operate QKD in a dense WDM network spanning tens of kilometers [30–32].

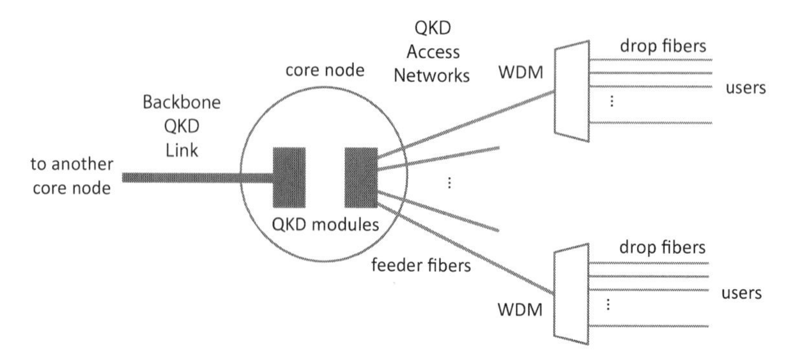

Figure 14.2: Backbone QKD link and QKD access network. While backbone QKD link connects two core nodes, the QKD access network connects many remote users to a core node through the use of feeder fibers and drop fibers.

An alternative approach is to use the entire C-band of a dedicated fiber link for QKD. This can be done by wavelength multiplexing a large number of QKD channels, each operating on a slightly different wavelength [35]. The higher key rate obtainable with this approach is desirable for backbone link because they need to support many remote users. The NEC group has demonstrated a WDM QKD system with impressive real-time electronic processing capability [36–38]. The system is an eight-channel WDM system designed to achieve 1 Mbps key rate at a transmission distance of 50 km. Using only three wavelength channels and clocking at 1.25 GHz, the group demonstrated 200 kbps real-time key generation for a total transmission loss of 14.5 dB over 45 km in a field trial. It should be noted that at full speed, NEC's eight-channel QKD system has to handle 50 Gbits of truly random bits per second [37]. The random bits should be generated in real time using hardware random bit generators, stored in electronic memory and retrieved during the post-processing step necessary to extract the secure key from photon detection results. This is rather challenging to achieve in practice. In Section 14.3, we will discuss a wavelength-multiplexed entanglement-based QKD scheme that is less demanding in this aspect.

14.2.2 QKD access networks

For access networks, the transmission distance is usually shorter. Figure 14.2 shows the concept of a QKD access network. A feeder fiber connects the core node to a power splitter or wavelength demultiplexer, where multiple drop fibers branch out to individual users. The total length of a feeder fiber and a drop fiber could be on the order of 20 km. The large number of remote users makes it very costly to operate P2P QKD in an access network. The situation is further complicated by the optical power splitters, electronic routers, and switches that may be present between user terminals and the core node. It would be very costly to lease a dark fiber for each remote user to implement P2P QKD. There are two main approaches to save cost. The first approach is for expensive QKD hardware to be shared among the users. The second approach is to combine QKD and classical data channels via wavelength multiplexing on the same fiber, so that QKD does not operate on a different fiber. It is possible to combine both approaches.

As early as in 1994, Townsend described a quantum network architecture where the QKD transmitter is shared by multiple QKD receivers [39–42]. More recently, Frohlich et al. proposed an upstream QKD architecture, where multiple quantum transmitters send photons to a centralized quantum receiver [43]. The advantage of this approach is that an expensive and complicated QKD receiver is shared by many users, while each user can use slower-speed quantum transmitters to save cost. They showed that it is possible to support 64 users with secure keys at a distance of 20 km. For access networks that use power splitters, high optical loss is inevitable. To avoid incurring high optical loss, the power splitter should be replaced by an arrayed waveguide grating (AWG) to combine QKD channels, each operating on a different wavelength.

Another cost-saving strategy is to wavelength multiplex QKD and classical data channels on the same fiber such that the cost of leasing the fiber is shared. The main challenge of this approach is the noise from spontaneous Raman scattering of the classical data channels. Spontaneous Raman scattering occurs in both forward and backward propagation directions. Choi et al. proposed to send QKD photons at a wavelength close to that of the upstream data channel and in the same propagation direction so that the photons can exploit the gaps in the Raman noise [44]. The gaps arise because classical data channels are intensity modulated, and spontaneous Raman scattering occurs only where the optical intensity is high. As a result, the Raman pulse train in the forward direction has the same pattern as the data, which contains positions where the optical intensity is low. Another approach to circumvent the effect of Raman noise in quantum access network is to use a second feeder fiber between the splitter and core node for transporting QKD channels [45]. In this way, the QKD channels are immune to noise induced by spontaneous Raman scattering of classical data channels in the first feeder fiber [46].

Implementing prepare-and-measure QKD in an access network may still be complicated and costly, as decoy states are needed to prevent photon-number splitting attacks, and high-speed true random number generators (RNGs) are required for cryptographic assurance [47]. A simpler and potentially less costly approach we will consider in Sec. 14.3, is wavelength-multiplexed entanglement distribution.

For access networks in which fiber lengths are shorter than those of backbone links, we have the option to implement QKD in the C-band or in the O-band [25] because the loss difference is small (only about 3 dB for 20 km of fiber). There are two considerations that may favor operating QKD in the O-band. First, spontaneous Raman scattering is more efficient at the longer wavelength side (Stokes) compared to the shorter wavelength side (anti-Stokes) of the spectrum. Therefore, it is advantageous to place classical channels at wavelengths longer than the QKD channels [48–53]. Second, for existing access networks that already use the C-band for data transmission, adding a QKD service in the O-band involves less disruption to the existing network. Ciurana et al. have shown how to construct an uninterrupted optical path between any two remote users in a network [25].

14.3 Wavelength-multiplexed entanglement-based QKD

14.3.1 Prepare-and-measure QKD

The original BB84 protocol is a prepare-and-measure scheme in which a sender, named Alice prepares the quantum states and sends them to the receiver, named Bob who performs a measurement on the received quantum states [1]. The interested reader is referred to review papers for more detailed descriptions on how QKD works [5, 6]. Prepare-and measure QKD has a few disadvantages in practice. First, it requires large amounts of random bits during execution of the protocol and consequently very high-speed random number generators must be used. For cryptographic assurance, it is not acceptable to use pseudo-random bits generated by mathematical algorithms. The randomness employed in QKD should have a physical origin. ID Quantique SA uses quantum random number generators (QRNGs) that operate at few MHz for their QKD systems. These QRNGs are based on quantum randomness associated with the unpredictability of photon detection. To operate QKD at GHz repetition rate, a user would need to generate and handle true random bits at a rate of multiple Gbits/s in real-time. It is non-trivial to run QRNGs at such high output bit rates [54, 55]. The generation and storage of large numbers of true random bits would place substantial burden on the electronic hardware.

The original BB84 protocol assumes an ideal photon source. However, the most commonly used photon source for QKD is attenuated laser. When an attenuated laser instead of ideal photon source is used for QKD, there are multi-photon components that may potentially leak information to eavesdroppers. The photon number distribution of an attenuated laser is well described by a Poissonian distribution, and so there is non-zero probability of obtaining more than one photon in a time slot. One method to remove the undesired effect of multi-photon components is to implement decoy states [47, 56, 57]. This method imposes additional requirements such as random insertion of decoy pulses and randomization of the phase of the laser. These requirements increase the complexity and cost of system implementation.

14.3.2 Entanglement-based QKD

Ekert proposed to use the unique correlation properties of quantum entanglement for QKD [58]. Quantum correlation is stronger than classical correlation; this is demonstrated in a violation of Bell's inequality [59, 60]. One may use violation of Bell's inequality to verify the presence of entanglement; it is useful for QKD because any eavesdropping attempt on the entangled photons would have destroyed or degraded the entanglement, and a non-entangled photon pair would not be able to violate Bell's inequality. Therefore, the loss of entanglement can be used to indicate eavesdropping. In some implementations, Alice creates pairs of entangled photons, immediately detects one photon of each pair at her location, and sends the other photon to Bob. This scheme is then equivalent to the BB84 protocol [61]. The two parties measure the transmission error rate and extract a secure key via post-processing [62–64]. A fully automated entanglement-based QKD system has been demonstrated by a group in Vienna [65].

Entanglement-based QKD has several advantages with regards to implementation. First, there is no need to use random bits. Generating large numbers of truly random bits and storing them is non-trivial. In an entanglement-based QKD scheme, randomness is intrinsic in the detection of entangled photons. It is not possible for anyone to predict the measurement outcomes before the actual measurement has been carried out. Second, multiple entangled pairs produced within the detection time window but outside the photon coherence time are distinguishable and therefore not related to one another. Although the presence of more than one pair of entangled photons within a detection window degrades entanglement quality, it does not leak information in the same way as the multi-photon components of a weak coherent state do. Therefore, there is no requirement to introduce additional measures such as decoy states for entanglement-based QKD. A third advantage of entanglement-based QKD is that it is relatively simple to wavelength-multiplex multiple quantum channels via the use of a broadband entangled photon source. This approach would enable efficient use of available transmission bandwidth without incurring high cost.

14.3.3 Entanglement distribution in a network

The key rate of entanglement-based QKD has been studied in detail by Ma et al. [66]. They showed that entanglement-based QKD systems can tolerate high losses as compared to those that use weak coherent sources. When the source is located symmetrically between two users, the transmission distance is maximized. The possibility of long-distance QKD over 300 km has been studied by experimental groups [67, 68].

Lim et al. considered a generic model of entanglement distribution over a star-type optical fiber network [69]. An entangled photon source is placed at a core node of a network and multiple optical fiber lines branch out from the node to many remote users. The users are equipped with photon detectors for detecting the photons. The advantage of this type of network is that the core node provides QKD service with entangled photon sources shared among many users. The number of users can be more than the number of entangled photon sources at the core node, as one entangled photon source can be time-shared by many users. Herbauts et al. proposed an active routing scheme using optical switches. It allows a single centralized entangled photon source to dynamically allocate entangled photons to any pair of users in a multi-user network setting [70]. More sophisticated designs of entanglement distribution optical network have been discussed in detail by Ciurana et al. [71]

In QKD, the aim is to increase the key rate by sending out more entangled photon pairs in unit time. However, entanglement quality is better when mean photon-pair number is kept low. This is because the photon-pair generation process is a probabilistic process. As the mean photon-pair number is increased, there is also increased probability that multiple pairs will be produced at the same time. Since the different pairs are unrelated to one another, this leads to errors [72, 73]. The quality of en-

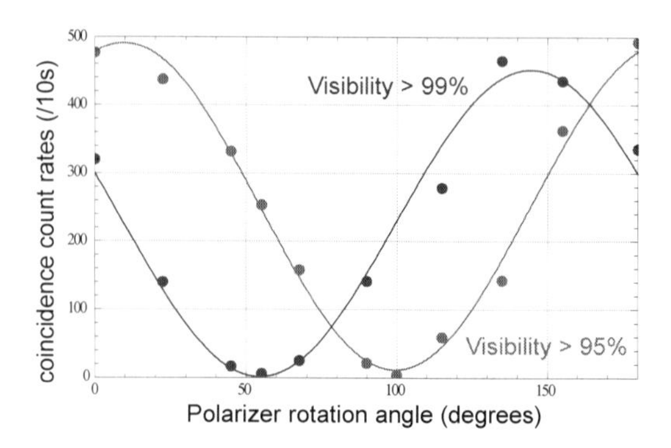

Figure 14.3: Two-photon interference results obtained from measuring polarization-entangled photons.

tangled photon pairs is further degraded after transmission due to optical attenuation and noise.

The two-photon interference fringe visibility is a good indication of entanglement quality. To observe this interference for polarization-entanglement, we typically place two polarizers, one before each photon detector. The coincidence count rate measurement is taken for one polarizer's rotation angle fixed at 0 degree, while the other polarizer is rotated from 0 to 180 degrees. This process is then repeated with rotation angle of the first polarizer fixed at 45 degrees.

Figure 14.3 shows a typical two-photon interference result that we obtained from a PPLN waveguide source of polarization-entangled photons. The interference visibility V can be calculated from

$$V = \frac{C_{\max} - C_{\min}}{C_{\max} + C_{\min}}, \tag{14.1}$$

where $C_{\max}$ and $C_{\min}$ are the maximum and minimum coincidence count rates. To violate Bell's inequality, V must be >0.71. It has been shown that V is a function of the mean photon-pair number, optical transmittance of the channel between the users and the source, and dark count probabilities of the photon detectors. The optical transmittance takes into account the absorption coefficient of the optical fiber line connecting the source to user receiver, optical loss of the receiver, and detector quantum efficiency. Different users have different optical transmittance values because their distances to the source are different, and they may use different types of photon detectors with different quantum efficiencies and dark count rates. With knowledge of channel transmittance and the photon detector dark count rate for each user, the service provider is able to optimize the pump laser power to maximize the two-photon-interference visibility or maximize the photon rate for a desired visibility [69].

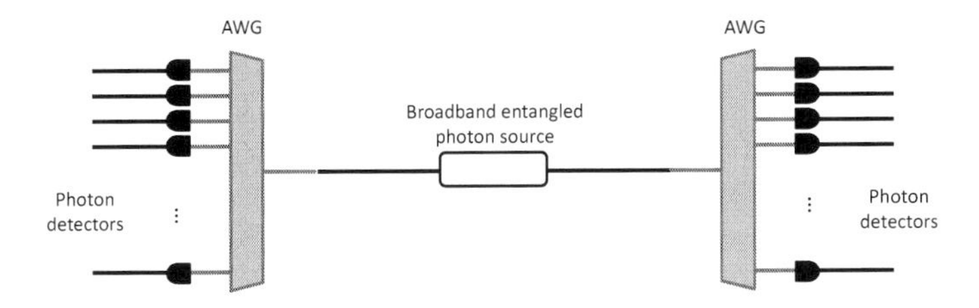

Figure 14.4: Concept of wavelength-multiplexed entanglement distribution. AWG: arrayed waveguide grating.

14.3.4 Wavelength-multiplexed entanglement distribution

Wavelength-multiplexing many prepare-and-measure QKD channels for increasing key rate is possible, but in this case, the system needs to handle large amounts of random bits in real-time. Entanglement-based QKD circumvents this problem, as it does not require the use of random bits. When two users choose the same basis for measurement of an entangled photon pair, the outcome would be perfectly correlated even though the bit values are unpredictable. We can wavelength-multiplex many entangled photon sources to increase the key rate, but the complexity and cost of such a system will be high. One promising approach is to make use of a broadband entangled photon source that produces wavelength-multiplexed entangled photons. Wavelength-demultiplexing is carried out only at the receiver before photon detection to define the wavelength channels [74]. This concept is illustrated in Fig. 14.4.

Entangled photons are often produced via spontaneous parametric down-conversion (SPDC) in second-order nonlinear materials such as lithium niobate or spontaneous four-wave mixing (SFWM) in third-order nonlinear materials such as silica or silicon. As energy is transferred from pump photons to the photon pairs, energy conservation requires $\omega_p = \omega_s + \omega_i$ for SPDC and $2\omega_p = \omega_s + \omega_i$ for SFWM, to be satisfied. Here, ω_p, ω_s, and ω_i denote the angular frequencies of the pump, signal, and idler, respectively. The spectral bandwidth, over which SPDC or SFWM occurs efficiently, is determined by the phase-matching condition.

A good material for generating broadband entangled photons is type-0 quasi-phase-matched (QPM) periodically-poled lithium niobate (PPLN) waveguide. The phase-matching condition for type-0 PPLN waveguide can be expressed as $\Delta k = 0$, where

$$\Delta k \equiv k(\omega_p) - k(\omega_s) - k(\omega_i) - 2\pi/\Lambda \tag{14.2}$$

is called the phase-mismatch with $k(\omega_j)$ denoting the wavenumber at angular frequency ω_j and Λ being the poling period of the PPLN. The efficiency of energy transfer from the pump mode to the signal and idler modes depends on the phase relation between the propagating modes. The product of phase-mismatch and interaction length determines the bandwidth of the SPDC or SFWM process.

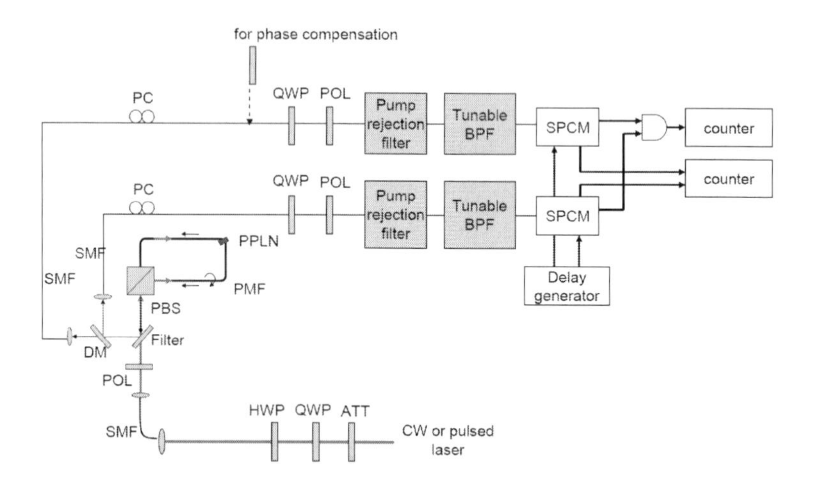

Figure 14.5: Broadband source of polarization-entangled photons using short PPLN waveguide. Reprinted with permission from Reference [75].© 2008 Optical Society of America.

By using a short PPLN waveguide that is 1-mm-long, Lim et al. obtained polarization-entangled photon pairs covering a very broad bandwidth in the C-band [75]. The experimental schematic is shown in Fig. 14.5. The laser pumps the waveguide bidirectionally in a Sagnac loop configuration and the photon pairs produced are coherently combined at the polarization beam splitter (PBS) to obtain polarization entanglement. The entangled photon bandwidth is very broad because the short waveguide length allows the phase-matching condition to be satisfied over a wide range of wavelengths. Using this source, Lim et al. demonstrated wavelength-multiplexed entanglement distribution over 10 km of standard single-mode optical fiber [76].

Subsequently, Arahira and Murai used cascaded optical nonlinearities in a PPLN ridge waveguide to obtain broadband polarization-entangled photons [77]. Kang et al. used an AlGaAs Bragg reflection waveguide to obtain broadband entangled photon pairs [78]. Zhu et al., on the other hand, proposed the use of a poled optical fiber, which also provided a broad entangled photon bandwidth [79]. Telecommunication-grade arrayed waveguide gratings (AWGs) have proven suitable for demultiplexing the photons into different wavelength channels before photon detection [77, 80, 81]. In a recent report, wavelength-multiplexed time-energy entangled photons generated from a PPLN waveguide were transmitted over 150 km of optical fiber [82]. In the next section we consider the potential of realizing wavelength-multiplexed entanglement distribution with integrated nanophotonic chips.

14.4 Integrated nanophotonics for QKD applications

14.4.1 Ideal single-photon source and weak coherent source

The original BB84 protocol assumes the use of an ideal single-photon source that emits only one photon at a time into a well-defined spectral and spatial mode. Isolated quantum dots and diamond color centers are good approximations to an ideal single-photon source; they have been used in QKD experiments [83–89]. Research efforts to integrate such photon sources on chips continues. The biggest challenge today is obtaining a high photon emission rate and a high photon collection efficiency since these are important factors that affect the final achievable key rate. While there have been significant advances in recent years [90, 91], quantum dots and diamond color centers are still unable to meet the demands of practical QKD systems. The trend is toward GHz system clock rates.

The most commonly used photon source in prepare-and-measure QKD is the weak coherent source, which is simply a strongly attenuated laser source. Its output photon stream is described by Poissonian statistics with the probability $P(n)$ of finding n photons within a given time slot expressed by

$$P(n) = \frac{\mu^n \exp(-\mu)}{n!}, \tag{14.3}$$

where μ is the mean photon number [92]. There is non-zero probability of finding more than one photon in a time slot, and it increases with μ. In QKD, we wish to maximize $P(1)$, while at the same time suppressing $P(>1)$. If more than one photon is available, an eavesdropper may use them to learn about the encoded bit value. To effectively suppress the probability of multiple photons, we usually use a low mean photon number, sacrificing the photon rate. It would be feasible to integrate high-speed phase and intensity modulators and optical attenuators on chips for prepare-and-measure QKD using weakly coherent photons. However, for reasons discussed in the previous section, prepare-and-measure QKD schemes are more difficult to scale compared to entanglement-based QKD systems.

14.4.2 On-chip entangled photon generation

Spontaneous four-wave mixing (SFWM) in silicon waveguides is an attractive source of entangled photons because silicon waveguides can be fabricated by processing techniques that are mature and compatible with the CMOS industry. This approach could potentially lead to very compact and low-cost devices. Silicon waveguides have single crystal silicon cores surrounded by lower-index silicon dioxide cladding. The cross-sectional dimensions of the silicon core are typically a few hundreds of nanometers. The core-cladding index difference can be as large as 40%; this leads to tight confinement of the fields propagating in the core. The main advantage of using silicon waveguides for entangled photon generation is their large effective nonlinearity due to a large material nonlinearity and tight optical confinement made possible by a large index contrast. The nonlinear parameter of a silicon waveguide may be as large as 300,000 $W^{-1}km^{-1}$. This should be compared to the 20 $W^{-1}km^{-1}$ of

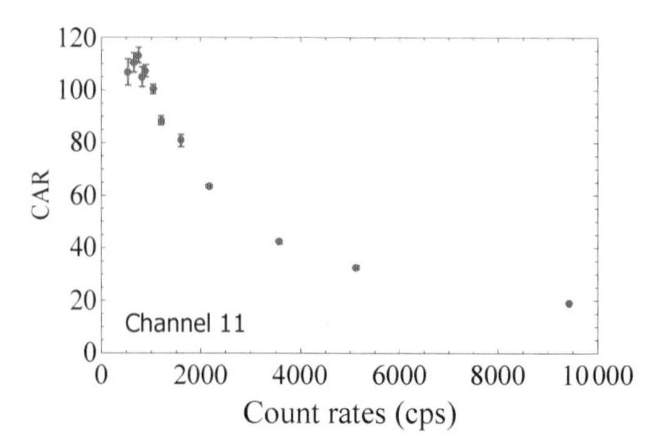

Figure 14.6: CAR obtained from a 2.6-mm-long straight silicon waveguide.

highly nonlinear optical fibers. The waveguide dispersion can be tailored to fulfill the phase-matching condition needed for SFWM to occur efficiently [93–95]. The Raman line is distinct and about 15.6 THz away from the pump frequency. The full-width-half-maximum (FWHM) of the Raman peak is about 105 GHz at room temperature [96–98]. For SFWM, the pump and photon wavelengths are very close; therefore walk-off between pump and photons is negligible. Challenges associated with silicon waveguides include a relatively high propagation loss of a few dB/cm and fiber-coupling loss of 2 to 3 dB per facet, even with use of inverse tapering method [99, 100].

The first experimental demonstration of quantum-correlated photon-pair generation in a silicon waveguide was reported in 2006 [101], following a theoretical study by Lin et al. [102, 103]. After that, several groups went ahead to generate quantum-correlated photon pairs and entangled photon pairs using silicon waveguides. In particular, Takesue and co-workers at NTT Corporation in Japan performed a series of experiments, including the generation of time-bin entanglement and polarization entanglement using silicon waveguides [104–107], and the interference of indistinguishable photons from two different silicon waveguides [108].

Figure 14.6 shows the result of a coincidence-to-accidental ratio (CAR) measurement that we performed on an integrated photon-pair source that used a 440 nm $\times$ 220 nm $\times$ 2.6 mm straight silicon waveguide. CAR is a measure that is commonly used to compare the qualities of photon-pair sources. The highest CAR we observed was 114. It is likely to be limited by optical loss in the system. The NTT group obtained higher maximum CAR value of 325 in a similar experiment with lower loss in their optical system [107]. A review of the experimental results obtained at NTT is available [109]. We outline the ways to generate time-bin- and polarization-entangled photons using straight silicon waveguides below.

Time-bin entanglement is generated by using a pair of pump pulses. Figure 14.7 shows the creation of the pump pulses by a delayed interferometer. Takesue et al. used intensity modulation to create pairs of pump pulses that are separated in time

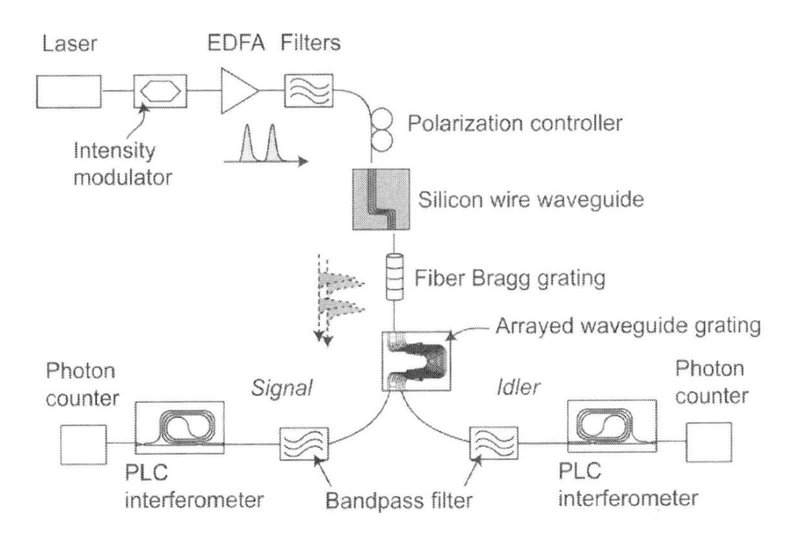

Figure 14.7: Method to produce time-bin entanglement. DI=delayed interferometer. WG=waveguide. WDM=wavelength division multiplexing coupler. Reprinted with permission from Reference [104]. © 2007 AIP Publishing.

by T [104]. The coherence time of the laser is much longer than T, and therefore the two pulses have a fixed phase relation. The pump pulses are injected into a silicon waveguide to produce photon pairs via SFWM. At the output port of the silicon waveguide, the pump pulses are filtered, and a WDM coupler separates the signal and idler photons. The time-bin-entangled state is expressed as

$$|\psi\rangle = \frac{1}{\sqrt{2}}\left(|1\rangle_s|1\rangle_i + \mathrm{e}^{\mathrm{i}\phi_p}|2\rangle_s|2\rangle_i\right), \tag{14.4}$$

where $|k\rangle$ denotes the quantum state whereby a photon is at the k-th time-slot. The subscripts s and i refer to signal and idler, respectively, while ϕ_p is the phase difference between the two pump pulses. To observe the two-photon interference, two delay interferometers are used in a Franson-type setup [110], where the time differences between the long and short arms of the interferometers are set to T as well. The resultant state after passing the photons through the delay interferometers is

$$|1\rangle_s|1\rangle_i + \left(\mathrm{e}^{\mathrm{i}(\theta_s+\theta_i)} + \mathrm{e}^{2\mathrm{i}\phi_p}\right)|2\rangle_s|2\rangle_i + \mathrm{e}^{\mathrm{i}\left(\theta_s+\theta_i+2\phi_p\right)}|3\rangle_s|3\rangle_i, \tag{14.5}$$

where θ_s and θ_i are the phase differences between the two paths in the interferometer for the signal and idler photon, respectively. Note that in the above expression, non-coincident terms and normalization coefficients are omitted for clarity. Two-photon

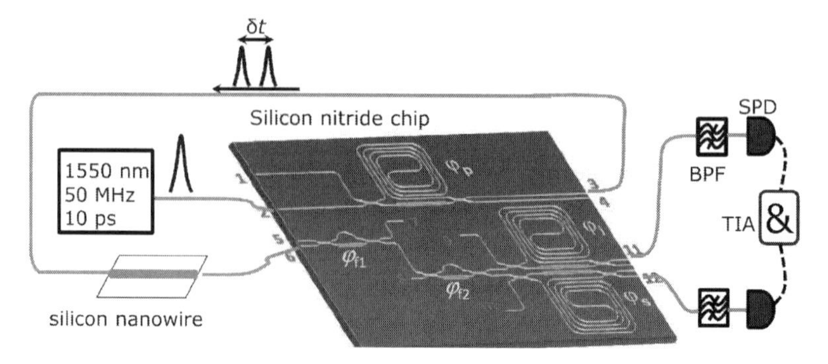

Figure 14.8: Producing and measuring time-bin-entangled photons on chip. Reprinted with permission from Reference [111]. © 2015 Optical Society of America.

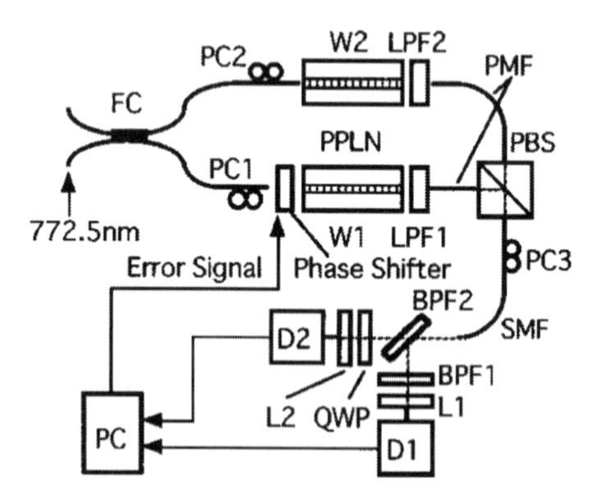

Figure 14.9: Mach-Zehnder-interferometer configuration with real-time phase stabilization for producing polarization-entangled photons. Reprinted with permission from Reference [112]. © 2004 AIP Publishing.

interference can be observed by taking the coincidence count rate of the middle time-slot and changing the temperature of one of the interferometers.

Xiong et al. demonstrated on-chip time-bin entanglement generation and measurement [111]. Figure 14.8 shows the experimental schematic. They used a chip containing a silicon nitride asymmetric Mach-Zehnder interferometer to create a double pulse from a single input pulse. The double pulse was then coupled into a second silicon waveguide chip for producing photon pairs via SFWM. The output photon pairs were coupled back to the first chip that also contained two silicon nitride delayed interferometers for observing two-photon interference.

To obtain polarization-entangled photons, we must create a superposition of two polarized two-photon states. This can be achieved using SPDC in two nonlinear optical waveguides. Yoshizawa et al. demonstrated this concept with a pair of PPLN

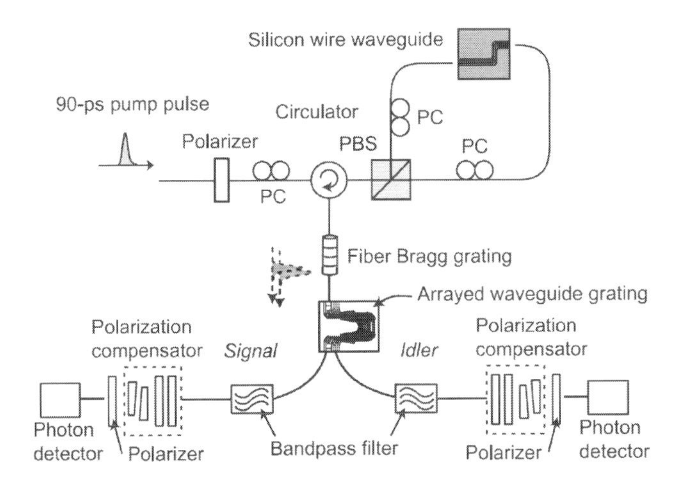

Figure 14.10: Sagnac loop configuration for producing polarization-entangled photons. Reprinted with permission from Reference [105]. © 2008 Optical Society of America.

waveguides, as shown in Fig. 14.9 [112]. A pump pulse was divided into two paths with each path pumping a PPLN waveguide to generate co-polarized photon pairs via SPDC. Horizontally-polarized and vertically-polarized photon pairs produced in the two PPLN waveguides were combined coherently at a PBS to create a polarization-entangled photon-pair state. The two arms of the Mach-Zehnder interferometer were stabilized via a feedback circuit to maintain the relative phase.

Figure 14.10 shows another method that makes use of a Sagnac loop configuration and is considerably more stable than the Mach-Zehnder interferometer configuration without electronic feedback. By bidirectional pumping one silicon waveguide and with the help of in-loop polarization controllers, this scheme produces polarization-entangled photon pairs at the output port of the PBS [105]. A slight disadvantage is that an optical circulator is needed to extract the photon pairs. The input pump polarization must be stabilized to 45 degrees with respect to the principal axes of the PBS.

A fully monolithic entangled-photon source on silicon chip has been reported by Matsuda et al. [113]. As shown in Fig. 14.11, this scheme utilizes a polarization rotator sandwiched between two silicon waveguides. Monolithic integration is possible because the polarization rotator can be realized on a silicon chip [114]. A large difference in the SFWM bandwidth for TE- and TM-polarized pumps is essential to generate photon pairs in only one polarization. At the output of the device, TE- and TM-polarized photon-pair states are in a superposition giving the desired entangled state.

A high pump power increases photon-pair generation rate, but it also induces two-photon absorption and free-carrier absorption that increases optical loss detrimen-

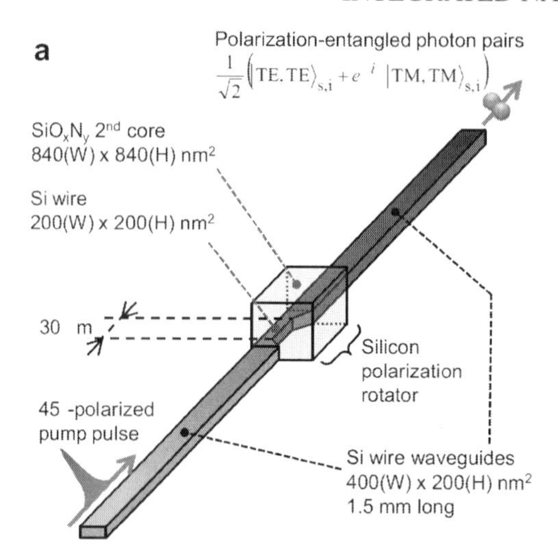

Figure 14.11: Sandwiched polarization rotator method for producing polarization-entangled photon pairs. Reprinted from with permission from Reference [113].

tal to photon-pair generation in silicon waveguides in the telecommunication-band wavelengths. Based on such considerations, pump pulses that are much shorter than the free carrier lifetime are often used in experiments. Clemmen et al. were the first to investigate a micro-ring resonator approach for photon-pair generation [115]. The nonlinear interaction between pump and the SFWM photons was greatly enhanced because the interacting light fields traveled many rounds within the ring resonator. This allows continuous-wave pumping at very low power levels. The rings usually have diameters of $< 10\ \mu$m and are evanescently coupled to a straight bus waveguide. The frequencies of the pump and the photon pairs must satisfy the resonance condition $M\lambda = n_{\text{eff}}L$, where L is the circumference of the ring and n_{eff} is the effective polariton index at vacuum wavelength λ. The free spectral range is therefore determined by the properties of the ring, and quantum correlated photon pairs are emitted with narrow linewidths. The ring can take the shape of a circle or a racetrack. The dimensions of the waveguide can be dispersion engineered for phase-matching. The circumference of the ring and the gap between the ring and the bus waveguide determine the field enhancement. CAR values as high as 600 have been demonstrated with removal of free carriers [116].

By exploiting the field enhancement in a resonator, several experiments have shown that it is possible to produce photon pairs with sub-mW or even μWs of continuous-wave pumping [117, 118]. A good comparison of the experiments performed by different groups in the past few years is available [119]. One drawback of the ring resonator scheme is sensitivity to temperature changes. A slight change in the refractive index is sufficient to shift the resonance wavelengths. It is therefore often necessary to monitor such changes and adjust the pump laser wavelength accordingly to maintain photon-pair generation efficiency.

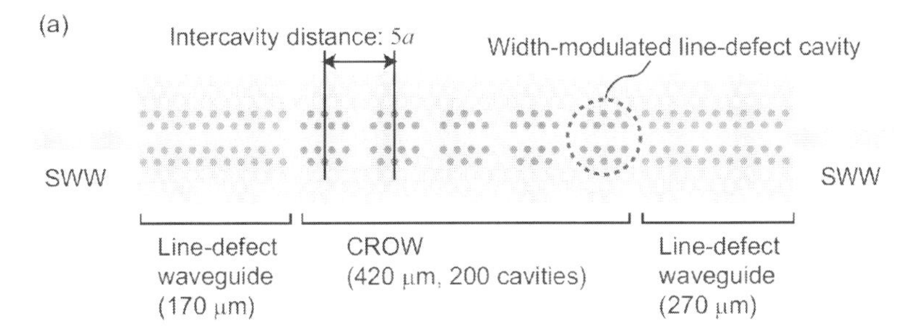

Figure 14.12: Coupled resonator optical waveguide (CROW) for slow-light-enhanced time-bin-entangled photon generation. Reprinted with permission from Reference [122].

It is possible to create time-bin entanglement using micro-ring resonators. Wakabayashi et al. generated time-bin-entangled photons using a 7 μm silicon micro-ring resonator. The intrinsic photon-pair generation rate was estimated to be 21 MHz with 0.41 mW of pump power [120]. In another work, Grassani et al. produced energy-time entanglement with micro-ring resonators [121]. Energy-time entanglement is produced by continuous-wave pumping. Differing from the pulse-pumped case, the photon emission time is indeterminate to within the coherence time of the pump laser. The time-difference between the long and short arms of the interferometers denoted by T should be set to be greater than the coherence time of the SWFM photons to avoid single-photon interference and shorter than the photon-pair coherence time to observe two-photon interference.

Another approach to enhance nonlinear interaction with reduced device size is to make use of slow-light that prolongs light-matter interaction on chip. A few groups have demonstrated slow-light enhanced photon-pair generation [113, 122–128]. A popular choice is a coupled resonator optical waveguide (CROW) based on a sequence of coupled resonators with high quality factors. The resonators are weakly coupled to form a collectively resonant supermode. The SFWM efficiency is enhanced by a factor of N^2, where N is the number of rings. The NTT group generated time-bin-entangled photon pairs via slow-light-enhancement with 200 silicon photonic crystal nanocavities [113, 122]. Figure 14.12 shows the design of the device used in their experiment. The length of the CROW device was only 420 μm. The CAR values obtained from CROW photon-pair sources are usually lower than those from straight waveguides because the slow-light bandwidth is rather narrow, and it is challenging to filter off the pump photons. Kumar et al. obtained spectrally multiplexed photon pairs from a silicon CROW with free-spectral range of 7 nm [127]. They obtained higher CAR values than cited in previous works by reducing the propagation loss and using fewer cascaded micro-rings.

14.4.3 On-chip photon detection

A commonly used single-photon detector for near-infrared telecommunication-band photons is the InGaAs/InP APD operated in Geiger mode. In this mode, the APD is reverse-biased above its breakdown voltage. Avalanche breakdown as a result of the absorption of a photon, produces a measurable voltage pulse. Typical quantum efficiencies of commercial products are $<30\%$ with dark count rates of more than a few hundred Hz. Both gated and free-running operations are possible, although a deadtime on the order of 10 μs is usually recommended to reduce after-pulsing probability down to the level of a few percent. Timing resolution is typically hundreds of picoseconds. A review of the state of the art is available [129]. There are few reports on large-scale integration of InGaAs/InP single-photon detector arrays; current interest seems to be aimed at imaging applications rather than QKD. Princeton Lightwave has developed near-infrared photon-counting cameras primarily for lidar applications.

Superconducting nanowire single-photon detector (SNSPD) has gained popularity in recent years [130, 131]. These detectors are usually made of niobium nitride (NbN) nanowires that are sub-100-nm wide and a few nanometers thick. They are operated under cryogenic temperatures such that the nanowire material becomes superconducting. The absorption of a photon disrupts the Cooper pairs in the superconductor and creates a hot-spot where the local superconductivity is destroyed. This leads to a change in the detector resistance detectable as a voltage change across the nanowire [132]. The commercial product from Single Quantum has a timing resolution of <20 ps, a quantum efficiency of $>75\%$, and a dark count rate of only a few Hz, with negligible afterpulsing. The deadtime can be as short as 20 ns. The sensitivity of SNSPD typically covers a broad wavelength range from UV to mid-IR.

The use of low-noise SNSPDs for QKD has enabled greater tolerance to optical loss and hence allows longer transmission distances [10, 133, 134]. In practice, it is important to maintain low-loss optical coupling between the SNSPD and optical fiber. Mechanical positioning of optical fiber generally suffers from poor alignment stability. Bachar et al. fabricated an SNSPD directly on the tip of a single-mode optical fiber [135]. This method allows precise alignment of the detector to the fiber core. Unfortunately, it cannot be extended to integration with other on-chip photonic devices.

To move toward large-scale integration, it is desirable to integrate multiple SNSPDs with photonic waveguides on a chip. Hu et al. first proposed a waveguide-integrated SNSPD [136]. On top of a straight NbN nanowire is a Si_3N_4 waveguide integrated with a SiON cladding layer. Light is coupled evanescently from the Si_3N_4 waveguide to the NbN nanowire. Sprengers et al. fabricated NbN nanowires on top of a GaAs ridge waveguide [137]. Kahl et al. demonstrated SNSPDs made of NbN thin films deposited on top of silicon nitride waveguides. The detection efficiencies at telecommunication wavelengths are $>80\%$ [138]. Such waveguide-integrated SNSPDs may also be used to provide photon number resolution [139]. It was reported that other superconducting materials such as NbTiN and WSi are able to achieve very high system quantum efficiencies at telecommunication wavelengths with very low dark count rates [140–142]. By using amorphous NbTiN nanowires deposited directly on top of Si_3N_4 waveguides, Schuck et al. demonstrated higher quantum efficiencies for telecommunication-band photons with sub-Hz-level dark count rate [143, 144]. Waki et al. achieved internal quantum efficiency of 90% with NbTiN fabricated on a silicon waveguide [145].

The difficulty with the conventional approach to waveguide-integrated SNSPD has been poor yield due to defects. This presents an obstacle to the scaling up of the number of SNSPDs on a single chip. Najafi et al. reported a micrometer-scale flip-chip process that overcomes the yield problem by separating the photonic waveguide and SNSPD fabrication processes [146]. Multiple SNSPDs are fabricated on a silicon nitride layer over a silicon substrate. The substrate layer is then etched to leave membranes carrying the SNSPDs. After a characterization process that identifies low-jitter and high-efficiency SNSPDs, only selected membranes carrying good SNSPD are transferred to the target waveguides under an optical microscope, using tungsten microprobes coated with polydimethylsiloxane (PDMS) adhesive. Via this method, Najafi et al. successfully assembled and operated ten good quality SNSPDs on a single chip.

One potential application of on-chip integrated photon detectors that is worth mentioning is a multiplexed heralded photon source. A *heralded photon source* is a type of photon source whereby the presence of an output photon is known via the detection of a correlated photon. Typically, it is a photon-pair source, in which one photon of each pair is detected immediately to herald the presence of the other photon of the same pair. Migdall et al. proposed multiplexing an array of heralded photon sources with an active optical switching network to increase the output single-photon probability [147]. This concept is illustrated in Fig. 14.13 [148]. An array of photon detectors is used to detect the heralding photons. By keeping the mean photon-pair number of each heralded photon source low, multi-pair probability is suppressed. At each time slot, the optical switch selects the input port where a heralded photon is present, such that at the output port of the switch is a train of single photons. The optical switch should have a low insertion loss, and the quantum efficiency of the heralding photon detectors should be high for this scheme to work well [149]. Meany et al. combined high efficiency PPLN waveguides, low-loss laser inscribed circuits, and fast ($>$1 MHz) fiber coupled electro-optic switches to demonstrate multiplexing of four identical sources of single photons to one output [150]. Another potential benefit of using heralded single photons for QKD is improved noise tolerance [151, 152].

14.4.4 On-chip photon wavelength demultiplexing

An important device for realizing wavelength-multiplexed entanglement distribution is the arrayed waveguide grating (AWG) for demultiplexing photons into individual channels. Matsuda demonstrated broadband photon-pair generation and wavelength-demultiplexing on the same chip [153]. The wavelength-demultiplexer he used is an AWG fabricated with silica waveguide. Low nonlinearity of silica prevents the generation of photons by the pump laser. The maximum CAR value obtained was 30

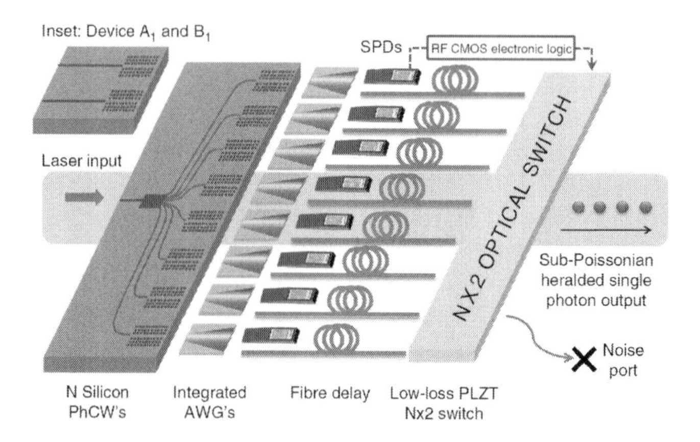

Figure 14.13: Concept of an integrated multiplexed heralded photon source. Reprinted with permission from Reference [148]. © 2013 Macmillan Publishers Ltd.

and this was limited by device loss. Reducing the insertion loss to the AWG would be one way to improve the CAR.

Another material for fabricating AWG is silicon nitride, which has a refractive index closer to that of silicon, and therefore a lower insertion loss can be expected. Silicon nitride is CMOS-compatible and does not have the nonlinear absorption problem of silicon [154]. Low-loss AWGs have been fabricated on silicon nitride platforms [155–158]. Dai et al. proposed a structure with an ultra-thin core layer that helps reduce scattering due to sidewall roughness. They demonstrated very low loss of 0.4 to 0.8 dB/m. The crosstalks from adjacent and non-adjacent channels were about 30 dB and 40 dB, respectively, at the wavelength range around 1310 nm [156].

14.4.5 Toward fully monolithic integration

On-chip generation of broadband entangled photons from a silicon straight waveguide has been demonstrated in several experiments [107, 113]. Narrow linewidth time-bin-entangled photons in the form of frequency combs have also been demonstrated with micro-ring resonators [159]. These sources are well suited for wavelength-multiplexed entanglement distribution because the photons are emitted over a broad spectral width, and therefore are intrinsically wavelength-multiplexed. However, to the best of our knowledge, none of these experimental demonstrations integrated the peripheral devices that are often needed for extracting useful entangled photons onto the same chip. Examples of peripheral devices include optical filters for rejecting the pump laser after photon-pair generation, wavelength-division-multiplexing couplers (WDMCs) for separating the signal and idler bands, and op-

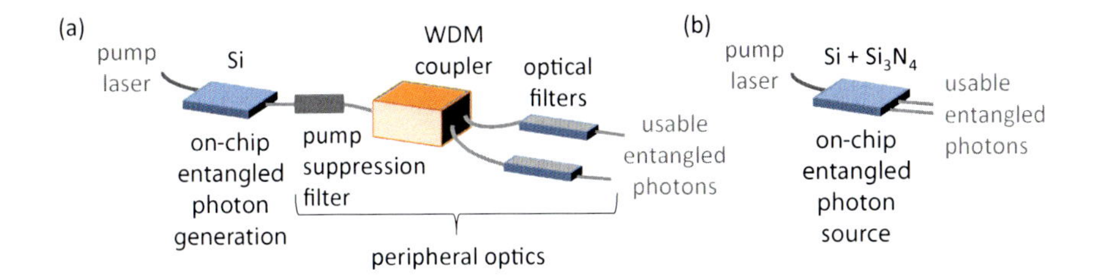

Figure 14.14: (a) On-chip entangled photon generation with peripheral optics connected externally. (b) Fully-integrated on-chip entangled photon source.

tical filters for suppressing out-of-band noise photons. These devices are usually attached externally, as illustrated in Fig. 14.14(a). To achieve true scalability of entangled photon sources, it would be important to integrate all the peripheral optics on the same chip to achieve true monolithic integration, as shown in Fig. 14.14(b). The peripheral optical components to be integrated on the same chip should have low optical loss and not produce noise photons that degrade entanglement quality.

Peripheral devices are passive, and therefore it is advantageous to employ silicon nitride platforms for them due to much lower loss than silicon [160–163]. A hybrid Si/Si_3N_4 platform in which low-loss devices are fabricated in the Si_3N_4 layer while the photon-pair sources are fabricated in the Si layer would enable the full monolithic integration of an entangled photon source. The transfer of optical power between layers can be through evanescent field or propagation field coupling. The coupling loss between the Si and Si_3N_4 layers is typically <0.5 dB per transition [161].

Figures 14.15 and 14.16 depict two designs for producing broadband entangled photons. The first design is for polarization entanglement, while the second design is for time-bin entanglement. In the first design in Fig. 14.15, pump pulses are split into four pulses using cascaded beam splitters (BS), and these pulses pump two independent entangled photon sources on the same chip. Each straight waveguide generates photon pairs that are co-polarized with the pump. Immediately after the straight waveguides, pump suppression filters (PSFs) are placed to prevent the strong pump pulses from producing addition photon pairs in the polarization rotator (PR), PBS, and wavelength-division-multiplexing coupler (WDMC), especially where silicon nitride demonstrates high nonlinearity. These additional photon pairs are undesirable as they contribute to background noise and degrade the quality of entanglement. Such on-chip PSFs have been demonstrated by Pierarek et al. [164]. They generated quantum correlated photon pairs in a micro-ring resonator and rejected the co-propagating laser pump with on-chip, fully passive PSFs based on multi-stage lattice filters. The estimated extinction was more than 100 dB.

The PR rotates photon polarization by 90 degrees and the PBS combines the orthogonally-polarized photon pairs. The WDMC separates the signal and idler bands. An additional fiber WDMC is used to combine signal and idler bands from

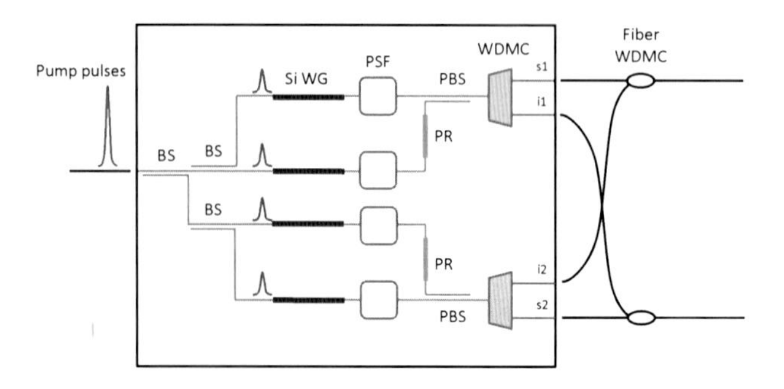

Figure 14.15: Integrated broadband source of polarization-entangled photons. BS=beam splitter. WG=waveguide. PSF=pump suppression filter. PR=polarization rotator. PBS=polarization beam splitter. WDMC=wavelength-division-multiplexing coupler.

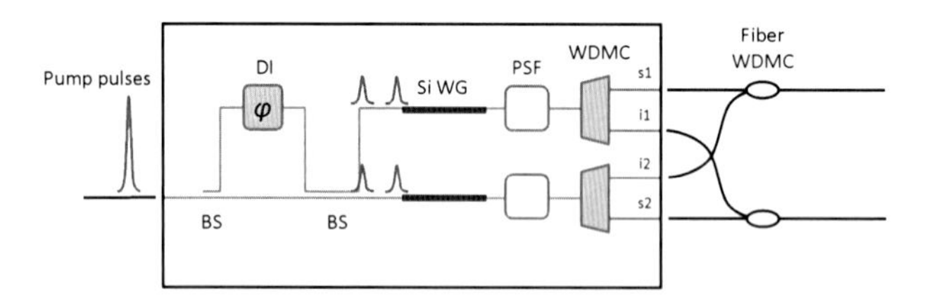

Figure 14.16: Integrated broadband source of time-bin-entangled photons. DI=delayed interferometer.

the two independent entangled photon sources to better utilize the transmission bandwidth of the fiber.

The design shown in Fig. 14.16 is for generating time-bin-entangled photons. A delayed interferometer creates dual pulses that pump two straight waveguides to produce time-bin-entangled photon pairs. The PSFs, WDMCs and fiber WDMCs play the same roles as described above. We note that multilayer crossing is realizable on-chip using silicon nitride [165], and therefore the fiber WDMC can also be integrated on the same chip with the rest of the components.

At the receiver, it is very useful to integrate AWG and SNSPD array on the same chip to realize wavelength-demultiplexing and photon detection together in a compact form-factor. Figures 14.17 and 14.18 show two basic receiver designs for the wavelength-demultiplexed detection of polarization-entangled and time-bin-entangled photons, respectively.

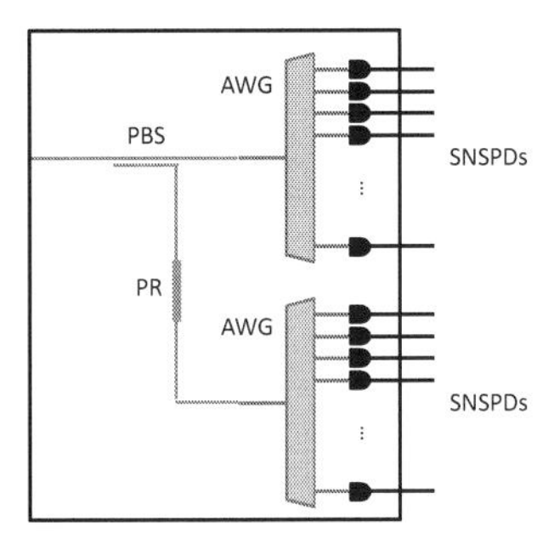

Figure 14.17: Integrated wavelength-demultiplexing receiver for polarization-entangled photons.

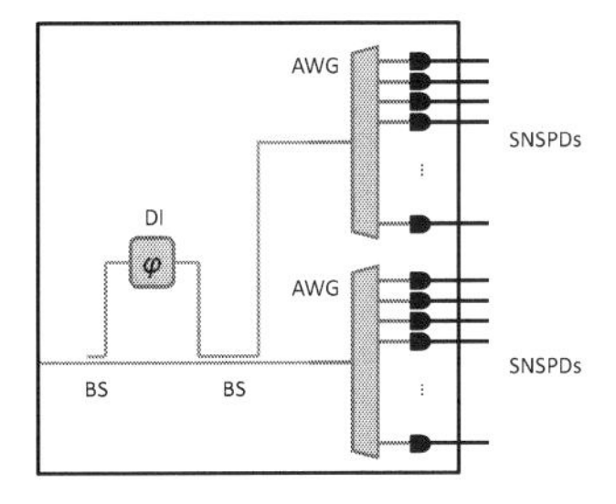

Figure 14.18: Integrated wavelength-demultiplexing receiver for time-bin-entangled photons.

14.5 Future outlook

There have been spectacular advances in the applications of integrated nanophotonics to quantum photonics since the first demonstration of quantum-correlated photon-pair generation in straight silicon waveguide a decade ago. New additions to our toolbox include on-chip broadband entangled-photon sources, on-chip waveguide-integrated SNSPD arrays, low-loss integrated passive devices such as polarization beam splitters, polarization rotators, wavelength-demultiplexers and delay lines. Fully monolithic integration of a combination of these devices into complex photonics circuits on SOI and silicon nitride hybrid platform is already within reach.

Wavelength-multiplexed entanglement distribution has many advantages that make it a strong candidate for realizing a scalable and low-cost multi-user QKD network within metropolitan areas. The main advantage of this scheme is that it can be implemented with broadband entangled-photon sources and there is no need to generate and store large numbers of random bits. Another significant advantage is compatibility with nanophotonic integration technology. Along with integrated broadband entangled-photon source, at the receiver side, low-loss silicon nitride AWGs and waveguide-integrated SNSPD arrays can be integrated into ultra-compact QKD receiver modules. Once these technologies are established, scaling up of wavelength-multiplexed entanglement distribution for a multi-user QKD network will be in sight.

There are other approaches to integrated QKD that deserve to be mentioned. There is a recent report on the incorporation of integrated nanophotonics technology into distributed phase reference QKD schemes aiming to achieve low-cost QKD with high performance [166]. Distributed-phase-reference schemes such as the coherent-one-way (COW) and the differential phase-shift-keying (DPSK) have potential for attaining high key rates [167–170]. Sibson et al. implemented a transmitter that monolithically integrated a tunable laser, optical interferometers, electro-optic modulators, and photodiodes on an InP platform [166]. They implemented a receiver consisting of a photonic circuit with thermo-optic phase-shifters and reconfigurable delay lines on a SiON platform. Using off-chip photon detectors, they operated the devices at a rate of 1.7 GHz and measured QBER of $<1.4\%$ for BB84, COW, and DPS protocols. Another approach is to realize continuous-variable (CV) QKD with integrated photonics. In CV-QKD, the quantum information is carried by the optical field quadratures instead of single photons [171, 172]. There is also ongoing effort to achieve integrated transmitters and receivers for CV-QKD.

With different approaches converging toward an integrated solution, ultimately it will be important to compare the cost per secure key bit and the key rates that can be offered to users. An approach compatible with the existing fiber network infrastructure and at the same time offering sufficiently high key rate at low cost to remote users will gain the upper hand. It is clear that integrated nanophotonics will play an increasingly important role in cost-effective implementation and operation of QKD systems and it is likely that integrated wavelength-multiplexed QKD, given its unique advantages, will emerge as a preferred solution for widespread QKD adoption in metropolitan area networks.

Bibliography

[1] C. H. Bennett and G. Brassard, "Quantum cryptography: Public key distribution and coin tossing," in *Proc. IEEE Intl. Conf. Comp., Syst. and Signal Processing*, vol. 1, pp. 175–179, 1984.

[2] R. Alleaume, C. Branciard, J. Bouda, T. Debuisschert, M. Dianati, N. Gisin, M. Godfrey, P. Grangier, T. Langer, N. Lutkenhaus, C. Monyk, P. Painchault, M. Peev, A. Poppe, T. Pornin, J. Rarity, R. Renner, G. Ribordy, M. Riguidel, L. Salvail, A. Shields, H. Weinfurter, and A. Zeilinger, "Using quantum key distribution for cryptographic purposes: a survey," *Theo. Comp. Sci.*, vol. 560, pp. 62–81, 2014.

[3] D. Stebila, M. Mosca, and N. Lutkenhaus, "The case for quantum key distribution," *Lecture Notes Inst. Comput. Sci. Soc. Informat. Telecommun. Eng.*, vol. 36, pp. 283–296, 2010.

[4] G. van Assche, *Quantum Cryptography and Secret-Key Distillation (Reprint Edition)*. Cambridge University Press, 2012.

[5] N. Gisin, G. Ribordy, W. Tittel, and H. Zbinden, "Quantum cryptography," *Rev. Mod. Phys.*, vol. 74, pp. 145–195, 2002.

[6] V. Scarani, H. Bechmann-Pasquinucci, N. J. Cerf, M. Dusek, N. Lutkenhaus, and M. Peev, "The security of practical quantum key distribution," *Rev. Mod. Phys.*, vol. 81, pp. 1301–1350, 2009.

[7] H.-K. Lo, M. Curty, and K. Tamaki, "Secure quantum key distribution," *Nat. Photon.*, vol. 8, pp. 595–604, 2014.

[8] G. Ribordy, J.-D. Gautier, N. Gisin, O. Guinnard, and H. Zbinden, "Automated "plug & play" quantum key distribution," *Electron. Lett.*, vol. 34, pp. 2116–2117, 1998.

[9] D. Stucki, N. Gisin, O. Guinnard, G. Ribordy, and H. Zbinden, "Quantum key distribution over 67 km with a plug & play system," *New J. Phys.*, vol. 4, pp. 41.1–41.8, 2002.

[10] H. Takesue, S. W. Nam, Q. Zhang, R. H. Hadfield, T. Honjo, K. Tamaki, and Y. Yamamoto, "Quantum key distribution over a 40-db channel loss using superconducting single-photon detectors," *Nat. Photon.*, vol. 1, pp. 343–348, 2007.

[11] W. K. Wootters and W. H. Zurek, "A single quantum cannot be cloned," *Nature*, vol. 299, pp. 802–803, 1982.

[12] D. Dieks, "Communication by EPR devices," *Phys. Lett. A*, vol. 92, pp. 271–272, 1982.

[13] H. J. Briegel, W. Dur, J. I. Cirac, and P. Zoller, "Quantum repeaters: The role of imperfect local operations in quantum communications," *Phys. Rev. Lett.*, vol. 81, pp. 5932–5935, 1998.

[14] L.-M. Duan, M. D. Lukin, J. I. Cirac, and P. Zoller, "Long-distance quantum communication with atomic ensembles and linear optics," *Nature*, vol. 414, pp. 413–418, 2001.

[15] N. Gisin and R. Thew, "Quantum communication," *Nat. Photon.*, vol. 1, pp. 165–171, 2007.

[16] N. Sangouard, C. Simon, H. de Riedmatten, and N. Gisin, "Quantum repeaters based on atomic ensembles and linear optics," *Rev. Mod. Phys.*, vol. 83, pp. 33–80, 2011.

[17] C. Elliot, "Building the quantum network," *New J. Phys.*, vol. 4, pp. 46.1–46.12, 2002.

[18] M. Peev, C. Pacher, R. Allaume, C. Barreiro, J. Bouda, W. Boxleitner, T. Debuisschert, E. Diamanti, M. Dianati, J. F. Dynes, S. Fasel, S. Fossier, M. Frst, J. D. Gautier, O. Gay, N. Gisin, P. Grangier, A. Happe, Y. Hasani, M. Hentschel, H. Hbel, G. Humer, T. Lnger, M. Legr, R. Lieger, J. Lodewyck, T. Lornser, N. Ltkenhaus, A. Marhold, T. Matyus, O. Maurhart, L. Monat,

S. Nauerth, J. B. Page, A. Poppe, E. Querasser, G. Ribordy, S. Robyr, L. Salvail, A. W. Sharpe, A. J. Shields, D. Stucki, M. Suda, C. Tamas, T. Themel, R. T. Thew, Y. Thoma, A. Treiber, P. Trinkler, R. Tualle-Brouri, F. Vannel, N. Walenta, H. Weier, H. Weinfurter, I. Wimberger, Z. L. Yuan, H. Zbinden, and A. Zeilinger, "The SECOQC quantum key distribution network in Vienna," *New J. Phys.*, vol. 11, p. 075001, 2009.

[19] D. Stucki, M. Legre, F. Buntschu, B. Clausen, N. Felber, N. Gisin, L. Henzen, P. Junod, G. Litzistorf, P. Monbaron, L. Monat, J.-B. Page, D. Perroud, G. Ribordy, A. Rochas, S. Robyr, J. Tavares, R. Thew, P. Trinkler, S. Ventura, R. Voirol, N. Walenta, and H. Zbinden, "Long-term performance of the SwissQuantum quantum key distribution network in a field environment," *New J. Phys.*, vol. 13, p. 123001, 2011.

[20] M. Sasaki, M. Fujiwara, H. Ishizuka, W. Klaus, K. Wakui, M. Takeoka, S. Miki, T. Yamashita, Z. Wang, A. Tanaka, K. Yoshino, Y. Nambu, S. Takahashi, A. Tajima, A. Tomita, T. Domeki, T. Hasegawa, Y. Sakai, H. Kobayashi, T. Asai, K. Shimizu, T. Tokura, T. Tsurumaru, M. Matsui, T. Honjo, K. Tamaki, H. Takesue, Y. Tokura, J. F. Dynes, A. R. Dixon, A. W. Sharpe, Z. L. Yuan, A. J. Shields, S. Uchikoga, M. Legr, S. Robyr, P. Trinkler, L. Monat, J.-B. Page, G. Ribordy, A. Poppe, A. Allacher, O. Maurhart, T. Lnger, M. Peev, and A. Zeilinger, "Field test of quantum key distribution in the Tokyo QKD network," *Opt. Express*, vol. 19, pp. 10387–10409, 2011.

[21] S. Wang, W. Chen, Z.-Q. Yin, H.-W. Li, D.-Y. He, Y.-H. Li, Z. Zhou, X.-T. Song, F.-Y. Li, D. Wang, H. Chen, Y.-G. Han, J.-Z. Huang, J.-F. Guo, P.-L. Hao, M. Li, C.-M. Zhang, D. Liu, W.-Y. Liang, C.-H. Miao, P. Wu, G.-C. Guo, and Z.-F. Han, "Field and long-term demonstration of a wide area quantum key distribution network," *Opt. Express*, vol. 22, pp. 21739–21756, 2014.

[22] J. Qiu, "Quantum communications leap out of the lab," *Nature*, vol. 508, pp. 441–442, 2014.

[23] W. Maeda, A. Tanaka, S. Takahashi, A. Tajima, and A. Tomita, "Technologies for quantum key distribution networks integrated with optical communication networks," *IEEE J. Sel. Top. Quantum Electron.*, vol. 15, pp. 1591–1601, 2009.

[24] D. Lancho, J. Martinez, D. Elkouss, M. Soto, and V. Martin, *QKD in Standard Optical Telecommunications Networks*, vol. 36. Springer, 2010.

[25] A. Ciurana, J. J. Martinez-Mateo, M. Peev, A. Poppe, N. Walenta, H. Zbinden, and V. Martin, "Quantum metropolitan optical network based on wavelength division multiplexing," *Opt. Express*, vol. 22, pp. 1576–1593, 2014.

[26] S. Aleksic, F. Hipp, D. Winkler, A. Poppe, B. Schrenk, and G. Franzl, "Perspectives and limitations of QKD integration in metropolitan area networks," *Opt. Express*, vol. 23, pp. 10359–10373, 2015.

[27] M. Sasaki, M. Fujiwara, R.-B. Jin, M. Takeoka, T. S. Han, H. Endo, K. Yoshino, T. Ochi, S. Asami, and A. Tajima, "Quantum photonic network: Concept, basic tools, and future issues," *IEEE J. Sel. Top. Quantum Electron.*,

vol. 21, p. 6400313, 2015.

[28] N. A. Peters, P. Toliver, T. E. Chapuran, R. J. Runser, S. R. McNown, C. G. Peterson, D. Rosenberg, N. Dallmann, R. J. Hughes, K. P. McCabe, J. E. Nordholt, and K. T. Tyagi, "Dense wavelength multiplexing of 1550 nm QKD with strong classical channels in reconfigurable networking environments," *New J. Phys.*, vol. 11, p. 045012, 2009.

[29] P. Eraerds, N. Walenta, M. Legre, N. Gisin, and H. Zbinden, "Quantum key distribution and 1 gbps data encryption over a single fibre," *New J. Phys.*, vol. 12, p. 063027, 2010.

[30] K. A. Patel, J. F. Dynes, I. Choi, A. W. Sharpe, A. R. Dixon, Z. L. Yuan, R. V. Penty, and A. J. Shields, "Coexistence of high-bit-rate quantum key distribution and data on optical fiber," *Phys. Rev. X*, vol. 2, p. 041010, 2012.

[31] K. A. Patel, J. F. Dynes, M. Lucamarini, I. Choi, A. W. Sharpe, Z. L. Yuan, R. V. Penty, and A. J. Shields, "Quantum key distribution for 10 Gb/s dense wavelength division multiplexing networks," *Appl. Phys. Lett.*, vol. 104, p. 051123, 2014.

[32] I. Choi, Y. R. Zhou, J. F. Dynes, Z. Yuan, A. Klar, A. Sharpe, A. Plews, M. Lucamarini, C. Radig, J. Neubert, H. Griesser, M. Eiselt, C. Chunnilall, G. Lepert, A. Sinclair, J.-P. Elbers, A. Lord, and A. Shields, "Field trial of a quantum secured 10 Gb/s DWDM transmission system over a single installed fiber," *Opt. Express*, vol. 22, pp. 23121–23128, 2014.

[33] L.-J. Wang, L.-K. Chen, L. Ju, M.-L. Xu, Y. Zhao, K. Chen, Z.-B. Chen, T.-Y. Chen, and J.-W. Pan, "Experimental multiplexing of quantum key distribution with classical optical communication," *Appl. Phys. Lett.*, vol. 106, p. 081108, 2015.

[34] H. Kawahara, A. Medhipour, and K. Inoue, "Effect of spontaneous Raman scattering on quantum channel wavelength-multiplexed with classical channel," *Opt. Commun.*, vol. 284, pp. 691–696, 2011.

[35] G. Brassard, F. Bussieres, N. Godbout, and S. Lacroix, "Multi-user quantum key distribution using wavelength division multiplexing," in *Proc. SPIE*, vol. 5260, pp. 149–153, 2003.

[36] K. Yoshino, M. Fujiwara, A. Tanaka, S. Takahashi, Y. Nambu, A. Tomita, S. Miki, T. Yamashita, Z. Wang, M. Sasaki, and A. Tajima, "High-speed wavelength-division-multiplexing quantum key distribution system," *Opt. Lett.*, vol. 37, pp. 223–225, 2012.

[37] A. Tanaka, M. Fujiwara, K. Yoshino, S. Takahashi, Y. Nambu, A. Tomita, S. Miki, T. Yamashita, Z. Wang, M. Sasaki, and A. Tajima, "High-speed quantum key distribution system for 1-Mbps real-time key generation," *IEEE J. Quantum Electron.*, vol. 48, pp. 542–550, 2012.

[38] K. Yoshino, T. Ochi, M. Fujiwara, M. Sasaki, and A. Tajima, "Maintenance-free operation of WDM quantum key distribution system through a field fiber over 30 days," *Opt. Express*, vol. 21, pp. 31395–31401, 2013.

[39] P. D. Townsend, S. J. D. Phoenix, K. J. Blow, and S. M. Barnett, "Design of quantum cryptography systems for passive optical networks," *Electron. Lett.*, vol. 30, pp. 1875–1877, 1994.

[40] P. D. Townsend, "Quantum cryptography on multiuser optical fibre networks," *Nature*, vol. 385, pp. 47–49, 1997.

[41] V. Fernandez, R. J. Collins, K. J. Gordon, P. D. Townsend, and G. S. Buller, "Passive optical network approach to gigahertz-clocked multiuser quantum key distribution," *IEEE J. Quantum Electron.*, vol. 43, pp. 130–138, 2007.

[42] J. Bogdanski, N. Rafiei, and M. Bourennane, "Multiuser quantum key distribution over telecom fiber networks," *Opt. Commun.*, vol. 282, pp. 258–262, 2009.

[43] B. Frohlich, J. F. Dynes, M. Lucamarini, A. W. Sharpe, Z. Yuan, and A. J. Shields, "A quantum access network," *Nature*, vol. 501, pp. 69–72, 2013.

[44] I. Choi, R. J. Young, and P. D. Townsend, "Quantum information to the home," *New J. Phys.*, vol. 13, p. 063039, 2011.

[45] I. Choi, R. J. Young, and P. D. Townsend, "Quantum key distribution on a 10 Gb/s WDM-PON," *Opt. Express*, vol. 18, pp. 9600–9612, 2010.

[46] B. Frohlich, J. F. Dynes, M. Lucamarini, A. W. Sharpe, S. W.-B. Tam, Z. Yuan, and A. J. Shields, "Quantum secured gigabit optical access networks," *Sci. Rep.*, vol. 5, p. 18121, 2015.

[47] M. Lucamarini, K. A. Patel, J. F. Dynes, B. Frohlich, A. W. Sharpe, A. R. Dixon, Z. L. Yuan, R. V. Penty, and A. J. Shields, "Efficient decoy-state quantum key distribution with quantified security," *Opt. Express*, vol. 21, pp. 24550–24565, 2013.

[48] N. I. Nweke, P. Toliver, R. J. Runser, S. R. McNown, J. B. Khurgin, T. E. Chapuran, M. S. Goodman, R. J. Hughes, C. G. Peterson, K. McCabe, J. E. Nordholt, K. Tyagi, P. Hiskett, and N. Dallmann, "Experimental characterization of the separation between wavelength-multiplexed quantum and classical communication channels," *Appl. Phys. Lett.*, vol. 87, p. 174103, 2005.

[49] H. Xu, L. Ma, A. Mink, B. Hershman, and X. Tang, "1310-nm quantum key distribution system with up-conversion pump wavelength at 1550 nm," *Opt. Express*, vol. 15, pp. 7247–7260, 2007.

[50] T. E. Chapuran, P. Toliver, N. A. Peters, J. Jackel, M. S. Goodman, R. J. Runser, S. R. McNown, N. Dallmann, R. J. Hughes, K. P. McCabe, J. E. Nordholt, C. G. Peterson, K. T. Tyagi, L. Mercer, and H. Dardy, "Optical networking for quantum key distribution and quantum communications," *New J. Phys.*, vol. 13, p. 105001, 2009.

[51] M. A. Hall, J. B. Altepeter, and P. Kumar, "Drop-in compatible entanglement for optical-fiber networks," *Opt. Express*, vol. 17, pp. 14558–14566, 2009.

[52] L. Ma, S. Nam, H. Xu, B. Baek, T. Chang, O. Slattery, A. Mink, and X. Tang, "1310 nm differential-phase-shift QKD system using superconducting single-photon detectors," *New J. Phys.*, vol. 11, p. 045020, 2009.

[53] T. Zhong, X. Hu, F. N. C. Wong, K. K. Berggren, T. D. Roberts, and P. Battle, "High-quality fiber-optic polarization entanglement distribution at 1.3 μm telecom wavelength," *Opt. Lett.*, vol. 35, pp. 1392–1394, 2010.

[54] M. Jofre, M. Curty, F. Steinlechner, G. Anzolin, J. P. Torres, M. W. Mitchell, and V. Pruneri, "True random numbers from amplified quantum vacuum," *Opt. Express*, vol. 19, pp. 20665–20672, 2011.

[55] F. Xu, B. Qi, X. Ma, H. Xu, H. Zheng, and H.-K. Lo, "Ultrafast quantum random number generation based on quantum phase fluctuations," *Opt. Express*, vol. 20, pp. 12366–12377, 2012.

[56] W.-Y. Hwang, "Quantum key distribution with high loss: toward global secure communication," *Phys. Rev. Lett.*, vol. 91, p. 057901, 2003.

[57] H.-K. Lo, X. Ma, and K. Chen, "Decoy state quantum key distribution," *Phys. Rev. Lett.*, vol. 94, p. 230504, 2005.

[58] A. K. Ekert, "Quantum cryptography based on bell's theorem," *Phys. Rev. Lett.*, vol. 67, pp. 661–663, 2005.

[59] M. A. Nielsen and I. L. Chuang, *Quantum Computation and Quantum Information*. Cambridge University Press, 2000.

[60] D. Bouwmeester, A. K. Ekert, and A. Zeilinger, *The Physics of Quantum Information: Quantum Cryptography, Quantum Computation, Quantum Computation*. Springer, 2000.

[61] C. H. Bennett, G. Brassard, and N. D. Mermin, "Quantum cryptography without bell's theorem," *Phys. Rev. Lett.*, vol. 68, pp. 557–559, 1992.

[62] T. Jennewein, C. Simon, G. Weihs, H. Weinfurter, and A. Zeilinger, "Quantum cryptography with entangled photons," *Phys. Rev. Lett.*, vol. 84, pp. 4729–4732, 2000.

[63] D. S. Naik, C. G. Peterson, A. G. White, A. J. Berglund, and P. G. Kwiat, "Entangled state quantum cryptography: eavesdropping on the Ekert protocol," *Phys. Rev. Lett.*, vol. 84, pp. 4733–4736, 2000.

[64] W. Tittel, J. Brendel, H. Zbinden, and N. Gisin, "Quantum cryptography using entangled photons in energy-time Bell states," *Phys. Rev. Lett.*, vol. 84, pp. 4737–4740, 2000.

[65] A. Treiber, A. Poppe, M. Hentschel, D. Ferrini, T. Lorunser, E. Querasser, T. Matyus, H. Hubel, and A. Zeilinger, "A fully automated entanglement-based quantum cryptography system for telecom fiber networks," *New J. Phys.*, vol. 11, p. 045013, 2009.

[66] X. Ma, C.-H. F. Fung, and H.-K. Lo, "Quantum key distribution with entangled photon sources," *Phys. Rev. A*, vol. 76, p. 012307, 2007.

[67] T. Scheidl, R. Ursin, A. Fedrizzi, S. Ramelow, X.-S. Ma, T. Herbst, R. Prevedel, L. Ratschbacher, J. Kofler, T. Jennewein, and A. Zeilinger, "Feasibility of 300 km quantum key distribution with entangled states," *New J. Phys.*, vol. 11, p. 085002, 2009.

[68] T. Inagaki, N. Matsuda, O. Tadanaga, M. Asobe, and H. Takesue, "Entanglement distribution over 300 km of fiber," *Opt. Express*, vol. 21, pp. 23241–23249, 2013.

[69] H. C. Lim, A. Yoshizawa, H. Tsuchida, and K. Kikuchi, "Distribution of polarization-entangled photon-pairs produced via spontaneous parametric down-conversion within a local-area fiber network: theoretical model and experiment," *Opt. Express*, vol. 16, pp. 14512–14523, 2008.

[70] I. Herbauts, B. Blauensteiner, A. Poppe, T. Jennewein, and H. Hubel, "Demonstration of active routing of entanglement in a multi-user network," *Opt. Express*, vol. 21, pp. 29013–29024, 2013.

[71] A. Ciurana, V. Martin, J. Martinez-Mateo, B. Schrenk, M. Peev, and A. Poppe, "Entanglement distribution in optical networks," *IEEE J. Sel. Top. Quantum Electron.*, vol. 21, p. 6400212, 2015.

[72] H. C. Lim, A. Yoshizawa, H. Tsuchida, and K. Kikuchi, "Stable source of high quality telecom-band polarization-entangled photon-pairs based on a single, pulse-pumped, short PPLN waveguide," *Opt. Express*, vol. 16, pp. 12460–12468, 2008.

[73] H. Takesue and K. Shimizu, "Effects of multiple pairs on visibility measurements of entangled photons generated by spontaneous parametric processes," *Opt. Commun.*, vol. 283, pp. 276–287, 2010.

[74] H. C. Lim, A. Yoshizawa, H. Tsuchida, and K. Kikuchi, "Wavelength-multiplexed entanglement distribution," *Opt. Fiber Technol.*, vol. 16, pp. 225–235, 2010.

[75] H. C. Lim, A. Yoshizawa, H. Tsuchida, and K. Kikuchi, "Broadband source of telecom-band polarization-entangled photon-pairs for wavelength-multiplexed entanglement distribution," *Opt. Express*, vol. 16, pp. 16052–16057, 2008.

[76] H. C. Lim, A. Yoshizawa, H. Tsuchida, and K. Kikuchi, "Wavelength-multiplexed distribution of highly entangled photon-pairs over optical fiber," *Opt. Express*, vol. 16, pp. 22099–22104, 2008.

[77] S. Arahira and H. Murai, "Nearly degenerate wavelength-multiplexed polarization entanglement by cascaded optical nonlinearities in a PPLN ridge waveguide device," *Opt. Express*, vol. 21, pp. 7841–7850, 2013.

[78] D. Kang, A. Anirban, and A. S. Helmy, "Monolithic semiconductor chips as a source for broadband wavelength-multiplexed polarization entangled photons," *Preprint archive, arXiv:quant-ph/1511.00903v1*, 2015.

[79] E. Y. Zhu, C. Corbari, A. V. Gladyshev, P. G. Kazansky, H.-K. Lo, and L. Qian, "Multi-party agile quantum key distribution network with a broadband fiber-based entangled source," *Preprint archive, arXiv:quant-ph/1506.03896*, 2015.

[80] J. Ghalbouni, I. Agha, R. Frey, E. Diamanti, and I. Zaquine, "Experimental wavelength-division-multiplexed photon-pair distribution," *Opt. Lett.*, vol. 38, pp. 34–36, 2013.

[81] J. Trapateau, J. Ghalbouni, A. Orieux, E. Diamanti, and I. Zaquine, "Multi-user distribution of polarization entangled photon pairs," *J. Appl. Phys.*, vol. 118, p. 143106, 2015.

[82] D. Aktas, B. Fedrici, F. Kaiser, T. Lunghi, L. Labonte, and S. Tanzilli, "Entanglement distribution over 150 km in wavelength division multiplexed channels for quantum cryptography," *Laser Photon. Rev.*, vol. 10, pp. 451–457, 2016.

[83] E. Waks, K. Inoue, C. Santori, D. Fattal, J. Vuckovic, G. S. Solomon, and Y. Yamamoto, "Quantum cryptography with a photon turnstile," *Nature*, vol. 420, p. 762, 2002.

[84] A. Beveratos, R. Brouri, T. Gacoin, A. Villing, J.-P. Poizat, and P. Grangier, "Single photon quantum cryptography," *Phys. Rev. Lett.*, vol. 89, p. 187901, 2002.

[85] R. Alleaume, F. Treussart, G. Messin, Y. Dumiege, J.-F. Roch, A. Beveratos, R. Brouri-Tualle, J.-P. Poizat, and P. Grangier, "Experimental open-air quantum key distribution with a single-photon source," *New J. Phys.*, vol. 6, p. 92, 2004.

[86] T. Heindel, C. A. Kessler, M. Rau, C. Schneider, M. Furst, F. Hargart, W.-M. Schulz, M. Eichfelder, R. Rossbach, and S. Nauerth, "Quantum key distribution using quantum dot single-photon emitting diodes in the red and near infrared spectral range," *New J. Phys.*, vol. 14, p. 083001, 2014.

[87] M. Rau, T. Heindel, S. Unsleber, T. Braun, J. Fischer, S. Frick, S. Nauerth, C. Schneider, G. Vest, S. Reizenstein, M. Kamp, A. Forchel, S. Hofling, and H. Weinfurter, "Free space quantum key distribution over 500 meters using electrically driven quantum dot single-photon sources - a proof of principle experiment," *New J. Phys.*, vol. 16, p. 043003, 2014.

[88] K. Takemoto, Y. Nambu, T. Miyazawa, K. Wakui, S. Hirose, T. Usuki, M. Takatsu, N. Yokoyama, K. Yoshino, and A. Tomita, "Transmission experiment of quantum keys over 50 km using high-performance quantum-dot single-photon source at 1.5 μm wavelength," *Appl. Phys. Express*, vol. 3, p. 092802, 2010.

[89] K. Takemoto, Y. Nambu, T. Miyazawa, Y. Sakuma, T. Yamamoto, S. Yorozu, and Y. Arakawa, "Quantum key distribution over 120 km using ultrahigh purity single-photon source and superconducting single-photon detectors," *Sci. Rep.*, vol. 5, p. 14383, 2015.

[90] A. Schlehahn, R. Schmidt, C. Hopfmann, J.-H. Schulze, A. Strittmatter, T. Heindel, L. Gantz, E. R. Schmidgall, D. Gershoni, and S. Reitzenstein, "Generating single photons at gigahertz modulation-speed using electrically controlled quantum dot microlenses," *Appl. Phys. Lett.*, vol. 108, p. 021104, 2016.

[91] D. Riedel, D. Rohner, M. Ganzhorn, T. Kaldewey, P. Appel, E. Neu, R. J. Warburton, and P. Maletinsky, "Low-loss broadband antenna for efficient photon collection from a coherent spin in diamond," *Phys. Rev. Appl.*, vol. 2,

p. 064011, 2014.

[92] S. M. Barnett and P. M. Radmore, *Methods in Theoretical Quantum Optics*. Oxford University Press, 1997.

[93] J. I. Dadap, N. C. Panoiu, X. Chen, I.-W. Hsieh, X. Liu, C.-Y. Chou, E. Dulkeith, S. J. McNab, F. Xia, W. M. J. Green, L. Sekaric, Y. A. Vlasov, and R. M. Osgood, "Nonlinear-optical phase modification in dispersion-engineered Si photonic wires," *Opt. Express*, vol. 16, pp. 1280–1299, 2008.

[94] R. M. Osgood, N. C. Panoiu, J. I. Dadap, X. Liu, X. Chen, I.-W. Hsieh, E. Dulkeith, W. M. Green, and Y. A. Vlasov, "Engineering nonlinearities in nanoscale optical systems: physics and applications in dispersion-engineered silicon nanophotonic wires," *Adv. Opt. Photon.*, vol. 1, pp. 162–235, 2009.

[95] J. Leuthold, C. Koos, and W. Freude, "Nonlinear silicon photonics," *Nat. Photon.*, vol. 4, pp. 535–544, 2010.

[96] R. Claps, D. Dimitropoulous, Y. Han, and B. Jalali, "Observation of Raman emission in silicon waveguides at 1.54 μm," *Opt. Express*, vol. 10, pp. 1305–1313, 2002.

[97] J. I. Dadap, R. L. Espinola, R. M. Osgood, S. J. McNab, and Y. A. Vlasov, "Spontaneous Raman scattering in ultrasmall silicon waveguides," *Opt. Lett.*, vol. 29, pp. 2755–2757, 2004.

[98] R. L. Espinola, J. I. Dadap, R. M. Osgood, S. J. McNab, and Y. A. Vlasov, "Raman amplification in ultrasmall silicon-on-insulator wire waveguides," *Opt. Express*, vol. 12, pp. 3713–3718, 2004.

[99] T. Shoji, T. Tsuchizawa, T. Watanabe, K. Yamada, and H. Morita, "Low loss mode size converter from 0.3 μm square Si wire waveguides to singlemode fibres," *Electron. Lett.*, vol. 38, pp. 1669–1670, 2002.

[100] V. R. Almeida, R. R. Panepucci, and M. Lipson, "Nanotaper for compact mode conversion," *Opt. Lett.*, vol. 28, pp. 1302–1304, 2003.

[101] J. E. Sharping, K. F. Lee, M. A. Foster, A. C. Turner, B. S. Schmidt, M. Lipson, A. L. Gaeta, and P. Kumar, "Generation of correlated photons in nanoscale silicon waveguides," *Opt. Express*, vol. 14, pp. 12388–12393, 2006.

[102] Q. Lin and G. P. Agrawal, "Silicon waveguides for creating quantum-correlated photon pairs," *Opt. Lett.*, vol. 31, pp. 3140–3142, 2006.

[103] Q. Lin, O. J. Painter, and G. P. Agrawal, "Nonlinear optical phenomena in silicon waveguides: modeling and applications," *Opt. Express*, vol. 15, pp. 16604–16644, 2007.

[104] H. Takesue, Y. Tokura, H. Fukuda, T. Tsuchizawa, T. Watanabe, K. Yamada, and S. Itabashi, "Entanglement generation using silicon wire waveguide," *Appl. Phys. Lett.*, vol. 91, p. 201108, 2007.

[105] H. Takesue, H. Fukuda, T. Tsuchizawa, T. Watanabe, K. Yamada, Y. Tokura, and S. Itabashi, "Generation of polarization entangled photon pairs using silicon wire waveguide," *Opt. Express*, vol. 16, pp. 5721–5727, 2008.

[106] K. Harada, H. Takesue, H. Fukuda, T. Tsuchizawa, T. Watanabe, K. Yamada, Y. Tokura, and S. Itabashi, "Generation of high-purity entangled photon pairs using silicon wire waveguide," *Opt. Express*, vol. 16, pp. 20368–20373, 2008.

[107] K. Harada, H. Takesue, H. Fukuda, T. Tsuchizawa, T. Watanabe, K. Yamada, Y. Tokura, and S. Itabashi, "Frequency and polarization characteristics of correlated photon-pair generation using a silicon wire waveguide," *IEEE J. Sel. Top. Quantum Electron.*, vol. 16, pp. 325–331, 2010.

[108] K. Harada, H. Takesue, H. Fukuda, T. Tsuchizawa, T. Watanabe, K. Yamada, Y. Tokura, and S. Itabashi, "Indistinguishable photon pair generation using two independent silicon wire waveguides," *New J. Phys.*, vol. 13, p. 065005, 2011.

[109] H. Takesue, "Entangled photon pair generation using silicon wire waveguides," *IEEE J. Sel. Top. Quantum Electron.*, vol. 18, pp. 1722–1732, 2012.

[110] J. D. Franson, "Bell inequality for position and time," *Phys. Rev. Lett.*, vol. 62, pp. 2205–2208, 1989.

[111] C. Xiong, X. Zhang, A. Mahendra, J. He, D.-Y. Choi, C. J. Chae, D. Marpaung, A. Leinse, R. G. Heideman, M. Hoekman, C. G. H. Roeloffzen, R. M. Oldenbeuving, P. W. L. van Dijk, C. Taddei, P. H. W. Leong, and B. J. Eggleton, "Compact and reconfigurable silicon nitride time-bin entanglement circuit," *Optica*, vol. 2, pp. 724–727, 2015.

[112] A. Yoshizawa and H. Tsuchida, "Generation of polarization-entangled photon pairs in 1550 nm band by a fiber-optic two-photon interferometer," *Appl. Phys. Lett.*, vol. 85, pp. 2457–2459, 2004.

[113] N. Matsuda, H. L. Jeannic, H. Fukuda, T. Tsuchizawa, W. J. Munro, K. Shimizu, K. Yamada, Y. Tokura, and H. Takesue, "A monolithically integrated polarization entangled photon pair source on a silicon chip," *Sci. Rep.*, vol. 2, p. 817, 2012.

[114] H. Fukuda, K. Yamada, T. Tsuchizawa, T. Watanabe, H. Shinojima, and S. Itabashi, "Polarization rotator based on silicon wire waveguides," *Opt. Express*, vol. 16, pp. 2628–2635, 2008.

[115] S. Clemmen, K. P. Huy, W. Bogaerts, R. G. Baets, P. Emplit, and S. Massar, "Continuous wave photon pair generation in silicon-on-insulator waveguides and ring resonators," *Opt. Express*, vol. 17, pp. 16558–16570, 2009.

[116] E. Engin, D. Bonneau, C. M. Natarajan, A. S. Clark, M. G. Tanner, R. H. Hadfield, S. N. Dorenbos, V. Zwiller, K. Ohira, N. Suzuki, H. Yoshida, N. Iizuka, M. Ezaki, J. L. OBrien, and M. G. Thompson, "Photon pair generation in a silicon micro-ring resonator with reverse bias enhancement," *Opt. Express*, vol. 21, pp. 27826–27834, 2013.

[117] S. Azzini, D. Grassani, M. J. Strain, M. Sorel, I. G. Helt, J. E. Sipe, M. Liscidini, M. Galli, and D. Bajoni, "Ultra-low power generation of twin photons in a compact silicon ring resonator," *Opt. Express*, vol. 20, pp. 23100–23107, 2012.

[118] J. R. Ong and S. Mookherjea, "Quantum light generation on a silicon chip using waveguides and resonators," *Opt. Express*, vol. 21, pp. 5171–5181, 2013.

[119] M. Savanier, R. Kumar, and S. Mookherjea, "Photon pair generation from compact silicon microring resonators using microwatt-level pump powers," *Opt. Express*, vol. 24, pp. 3313–3328, 2016.

[120] R. Wakabayashi, M. Fujiwara, K. Yoshino, Y. Nambu, M. Sasaki, and T. Aoki, "Time-bin entangled photon pair generation from Si micro-ring resonator," *Opt. Express*, vol. 23, pp. 1103–1113, 2015.

[121] D. Grassani, S. Azzini, M. Liscidini, M. Galli, M. J. Strain, M. Sorel, J. E. Sipe, and D. Bajoni, "Micrometer-scale integrated silicon source of time-energy entangled photons," *Optica*, vol. 2, pp. 88–94, 2015.

[122] H. Takesue, N. Matsuda, E. Kuramochi, and M. Notomi, "Entangled photons from on-chip slow light," *Sci. Rep.*, vol. 4, p. 3913, 2014.

[123] C. Xiong, C. Monat, A. S. Clark, C. Grillet, G. D. Marshall, M. J. Steel, J. Li, L. O'Faolain, T. F. Krauss, J. G. Rarity, and B. J. Eggleton, "Slow-light enhanced correlated photon pair generation in a silicon photonic crystal waveguide," *Opt. Lett.*, vol. 36, pp. 3413–3415, 2011.

[124] N. Matsuda, T. Kato, K. Harada, H. Takesue, E. Kuramochi, H. Taniyama, and M. Notomi, "Slow light enhanced optical nonlinearity in a silicon photonic crystal coupled-resonator optical waveguide," *Opt. Express*, vol. 19, pp. 19861–19874, 2011.

[125] M. Davanco, J. R. Ong, A. B. Shehata, A. Tosi, I. Agha, S. Assefa, F. Xia, W. M. J. Green, S. Mookherjea, and K. Srinivasan, "Telecommunications-band heralded single photons from a silicon nanophotonic chip," *Appl. Phys. Lett.*, vol. 100, p. 261104, 2012.

[126] C. Xiong, C. Monat, M. J. Collins, L. Tranchant, D. Petiteau, A. S. Clark, C. Grillet, G. D. Marshall, M. J. Steel, J. Li, L. O'Faolain, T. F. Krauss, and B. J. Eggleton, "Characteristics of correlated photon pairs generated in ultracompact silicon slow-light photonic crystal waveguides," *IEEE J. Sel. Top. Quantum Electron.*, vol. 18, pp. 1676–1683, 2012.

[127] R. Kumar, J. R. Ong, J. Recchio, K. Srinivasan, and S. Mookherjea, "Spectrally multiplexed and tunable-wavelength photon pairs at 1.55 μm from a silicon coupled-resonator optical waveguide," *Opt. Lett.*, vol. 38, pp. 2969–2971, 2013.

[128] R. Kumar, M. Savanier, J. R. Ong, and S. Mookherjea, "Entanglement measurement of a coupled silicon microring photon pair source," *Opt. Express*, vol. 23, pp. 19318–19327, 2015.

[129] J. Zhang, M. A. Itzler, H. Zbinden, and J.-W. Pan, "Advances in InGaAs/InP single-photon detector systems for quantum communication," *Light: Sci. and Appl.*, vol. 4, p. e286, 2015.

[130] G. Goltsman, A. Korneev, A. Divochiy, O. Minaeva, M. Tarkhov, N. Kaurova, V. Seleznev, B. Voronov, O. Okunev, A. Antipov, K. Smirnov, Y. Vachtomin,

I. Milostnaya, and G. Chulkova, "Ultrafast superconducting single-photon detector," *J. Mod. Opt.*, vol. 56, pp. 1670–1680, 2009.

[131] R. H. Hadfield, "Single-photon detectors for optical quantum information applications," *Nat. Photon.*, vol. 3, pp. 696–705, 2009.

[132] G. N. Goltsman, O. Okunev, G. Chulkova, A. Lipatov, A. Semenov, K. Smirnov, B. Voronov, A. Dzardanov, C. Williams, and R. Sobolewski, "Picosecond superconducting single-photon optical detector," *Appl. Phys. Lett.*, vol. 79, pp. 705–707, 2001.

[133] D. Stucki, N. Walenta, F. Vannel, R. T. Thew, N. Gisin, H. Zbinden, S. Gray, C. R. Towery, and S. Ten, "High rate, long-distance quantum key distribution over 250 km of ultra low loss fibres," *New J. Phys.*, vol. 11, p. 075003, 2009.

[134] S. Wang, W. Chen, J. F. Guo, Z. Q. Yin, H. W. Li, Z. Zhou, G. C. Guo, and Z. F. Han, "2 GHz clock quantum key distribution over 260 km of standard telecom fiber," *Opt. Lett.*, vol. 37, pp. 1008–1010, 2012.

[135] G. Bachar, I. Baskin, O. Shtempluck, and E. Buks, "Superconducting nanowire single photon detectors on-fiber," *Appl. Phys. Lett.*, vol. 101, p. 262601, 2012.

[136] X. Hu, C. W. Holzwarth, D. Masciarelli, E. A. Dauler, and K. K. Berggren, "Efficiently coupling light to superconducting nanowire single-photon detectors," *IEEE Trans. Appl. Supercond.*, vol. 19, pp. 336–340, 2009.

[137] J. P. Sprengers, A. Gaggero, D. Sahin, S. Jahanmirinejad, G. Frucci, F. Mattioli, R. Leoni, J. Beetz, M. Lermer, M. Kamp, S. Hfling, R. Sanjines, and A. Fiore, "Waveguide superconducting single-photon detectors for integrated quantum photonic circuits," *Appl. Phys. Lett.*, vol. 99, p. 181110, 2011.

[138] O. Kahl, S. Ferrari, V. Kovalyuk, G. N. Goltsman, A. Korneev, and W. H. P. Pernice, "Waveguide integrated superconducting single-photon detectors with high internal quantum efficiency at telecom wavelengths," *Sci. Rep.*, vol. 5, p. 10941, 2015.

[139] S. Ferrari, O. Kahl, V. Kovalyuk, G. N. Goltsman, A. Korneev, and W. H. P. Pernice, "Waveguide-integrated single- and multi-photon detection at telecom wavelengths using superconducting nanowires," *Appl. Phys. Lett.*, vol. 106, p. 151101, 2015.

[140] S. N. Dorenbos, E. M. Reiger, U. Perinetti, V. Zwiller, T. Zijstra, and T. M. Klapwijk, "Low noise superconducting single photon detectors on silicon," *Appl. Phys. Lett.*, vol. 93, p. 131101, 2008.

[141] F. Marsili, V. B. Verma, J. A. Stern, S. Harrington, A. E. Lita, T. Gerrits, I. Vayshenker, B. Baek, M. D. Shaw, R. P. Mirin, and S. W. Nam, "Detecting single infrared photons with 93% system efficiency," *Nat. Photon.*, vol. 7, pp. 210–214, 2013.

[142] L. Redaelli, G. Bulgarini, S. Dobrovolskiy, S. N. Dorenbos, V. Zwiller, E. Monroy, and J. M. Gerard, "Design of broadband high-efficiency superconducting-nanowire single photon detectors," *Superconductor Sci. and*

Tech., vol. 29, p. 065016, 2016.

[143] C. Schuck, W. H. P. Pernice, and H. X. Tang, "NbTiN superconducting nanowire detectors for visible and telecom wavelengths single photon counting on Si_3N_4 photonic circuits," *Appl. Phys. Lett.*, vol. 102, p. 051101, 2013.

[144] C. Schuck, W. H. P. Pernice, and H. X. Tang, "Waveguide integrated low noise NbTiN nanowire single-photon detectors with milli-Hz dark count rate," *Sci. Rep.*, vol. 3, p. 1893, 2013.

[145] K. Waki, T. Yamashita, S. Inoue, S. Miki, H. Terai, R. Ikuta, T. Yamamoto, and N. Imoto, "Fabrication and characterization of superconducting nanowire single-photon detectors on Si waveguide," *IEEE Trans. Appl. Supercond.*, vol. 25, p. 2200704, 2015.

[146] F. Najafi, J. Mower, N. C. Harris, F. Bellei, A. Dane, C. Lee, X. Hu, P. Kharel, F. Marsili, S. Assefa, K. K. Berggren, and D. Englund, "On-chip detection of non-classical light by scalable integration of single-photon detectors," *Nat. Commun.*, vol. 6, p. 5873, 2014.

[147] A. L. Migdall, D. Branning, and S. Castelletto, "Tailoring single-photon and multiphoton probabilities of a single-photon on-demand source," *Phys. Rev. A*, vol. 66, p. 053805, 2002.

[148] M. J. Collins, C. Xiong, I. H. Rey, T. D. Vo, J. He, S. Shahnia, C. Reardon, T. F. Krauss, M. J. Steel, A. S. Clark, and B. J. Eggleton, "Integrated spatial multiplexing of heralded single-photon sources," *Nat. Commun.*, vol. 4, p. 2582, 2013.

[149] J. H. Shapiro and F. N. C. Wong, "On-demand single-photon generation using a modular array of parametric downconverters with electro-optic polarization controls," *Opt. Lett.*, vol. 32, pp. 2698–2700, 2007.

[150] T. Meany, L. A. Ngah, M. J. Collins, A. S. Clark, R. J. Williams, B. J. Eggleton, M. J. Steel, M. J. Withford, O. Alibart, and S. Tanzilli, "Hybrid photonic circuit for multiplexed heralded single photons," *Laser Photon. Rev.*, vol. 8, pp. L42–L46, 2014.

[151] M. T. Liu and H. C. Lim, "Transmission of O-band wavelength-division-multiplexed heralded photons over a noise-corrupted optical fiber channel," *Opt. Express*, vol. 21, pp. 30358–30369, 2013.

[152] M. T. Liu and H. C. Lim, "Efficient heralding of O-band passively spatial-multiplexed photons for noise-tolerant quantum key distribution," *Opt. Express*, vol. 22, pp. 23261–23275, 2014.

[153] N. Matsuda, P. Karkus, H. Nishi, T. Tsuchizawa, W. J. Munro, H. Takesue, and K. Yamada, "On-chip generation and demultiplexing of quantum correlated photons using a silicon-silica monolithic photonic integration platform," *Opt. Express*, vol. 22, pp. 22831–22840, 2014.

[154] D. J. Moss, R. Morandotti, A. L. Gaeta, and M. Lipson, "New CMOS-compatible platforms based on silicon nitride and Hydex for nonlinear optics," *Nat. Photon.*, vol. 7, pp. 597–607, 2013.

[155] A. A. Goncharov, S. V. Kuzmin, V. V. Svetikov, K. K. Svidzinskii, V. A. Sychugov, and N. V. Trusov, "Integrated optical demultiplexer based on the SiO_2-SiON waveguide structure," *Quantum Electron.*, vol. 35, pp. 1163–1165, 2005.

[156] D. Dai, Z. Wang, J. F. Bauters, M.-C. Tan, M. J. R. Heck, D. J. Blumenthal, and J. E. Bowers, "Low-loss Si_3N_4 arrayed-waveguide grating (de)multiplexer using nano-core optical waveguides," *Opt. Express*, vol. 19, pp. 14130–14136, 2011.

[157] D. Dai, J. Bauters, and J. E. Bowers, "Passive technologies for future large-scale photonic integrated circuits on silicon: polarization handling, light non-reciprocity and loss reduction," *Light: Sci. and Appl.*, vol. 1, p. e1, 2012.

[158] D. Martens, A. Z. Subramanian, S. Pathak, M. Vanslembrouck, P. Bienstman, W. Bogaerts, and R. G. Baets, "Compact silicon nitride arrayed waveguide gratings for very near-infrared wavelengths," *IEEE Photon. Technol.*, vol. 27, pp. 137–140, 2015.

[159] C. Reimer, M. Kues, P. Roztocki, B. Wetzel, F. Grazioso, B. E. Little, S. T. Chu, T. Johnston, Y. Bromberg, L. Caspani, D. J. Moss, and R. Morandotti, "Generation of multiphoton entangled quantum states by means of integrated frequency combs," *Science*, vol. 351, pp. 1176–1180, 2016.

[160] J. F. Bauters, M. J. R. Heck, D. John, D. Dai, M.-C. Tien, J. S. Barton, A. Leinse, R. G. Heideman, D. J. Blumenthal, and J. E. Bowers, "Ultra-low-loss high-aspect-ratio Si_3N_4 waveguides," *Opt. Express*, vol. 19, pp. 3163–3174, 2011.

[161] J. F. Bauters, M. L. Davenport, M. J. R. Heck, J. K. Doylend, A. Chen, A. W. Fang, and J. E. Bowers, "Silicon on ultra-low-loss waveguide photonic integration platform," *Opt. Express*, vol. 21, pp. 544–555, 2013.

[162] L. Zhuang, D. Marpaung, M. Burla, W. Beeker, A. Leinse, and C. Roeloffzen, "Low-loss, high-index-contrast Si_3N_4/SiO_2 optical waveguides for optical delay lines in microwave photonics signal processing," *Opt. Express*, vol. 19, pp. 23162–23170, 2011.

[163] M. J. R. Heck, J. F. Bauters, M. L. Davenport, D. T. Spencer, and J. E. Bowers, "Ultra-low loss waveguide platform and its integration with silicon photonics," *Laser Photon. Rev.*, vol. 5, pp. 667–686, 2014.

[164] M. Piekarek, D. Bonneau, S. Miki, T. Yamashita, M. Fujiwara, M. Sasaki, H. Terai, M. G. Tanner, C. M. Natarajan, R. H. Hadfield, J. L. O-Brien, and M. G. Thompson, "Passive high-extinction integrated photonic filters for silicon quantum photonics," in *Proc. Conf. on Lasers and Electro-Optics*, p. FM1N.6, Optical Society of America, 2016.

[165] K. Shang, S. Pathak, B. Guan, G. Liu, and S. J. B. Yoo, "Low-loss compact multilayer silicon nitride platform for 3D photonic integrated circuits," *Opt. Express*, vol. 23, pp. 21334–21342, 2015.

[166] P. Sibson, C. Erven, M. Godfrey, S. Miki, T. Yamashita, M. Fujiwara,

M. Sasaki, H. Terai, M. G. Tanner, C. M. Natarajan, R. H. Hadfield, J. L. OBrien, and M. G. Thompson, "Chip-based quantum key distribution," *Preprint archive, arXiv:quant-ph/1509.00768*, 2015.

[167] K. Inoue, E. Waks, and Y. Yamamoto, "Differential phase shift quantum key distribution," *Phys. Rev. Lett.*, vol. 89, p. 037902, 2002.

[168] Q. Zhang, H. Takesue, T. Honjo, K. Wen, T. Hirohata, M. Suyama, Y. Takiguchi, H. Kamada, Y. Tokura, O. Tadanaga, Y. Nishida, M. Asobe, and Y. Yamamoto, "Megabits secure key rate quantum key distribution," *New J. Phys.*, vol. 11, p. 045010, 2009.

[169] D. Stucki, N. Brunner, N. Gisin, V. Scarani, and H. Zbinden, "Fast and simple one-way quantum key distribution," *Appl. Phys. Lett.*, vol. 87, p. 194108, 2005.

[170] N. Walenta, A. Burg, D. Caselunghe, J. Constantin, N. Gisin, O. Guinnard, R. Houlmann, P. Junod, B. Korzh, N. Kulesza, M. Legr, C. W. Lim, T. Lunghi, L. Monat, C. Portmann, M. Soucarros, R. T. Thew, P. Trinkler, G. Trolliet, F. Vannel, and H. Zbinden, "A fast and versatile quantum key distribution system with hardware key distillation and wavelength multiplexing," *New J. Phys.*, vol. 16, p. 013047, 2014.

[171] F. Grosshans and P. Grangier, "Continuous variable quantum cryptography using coherent states," *Phys. Rev. Lett.*, vol. 88, p. 057902, 2002.

[172] P. Jouguet, S. Kunz-Jacques, A. Leverrier, P. Grangier, and E. Diamanti, "Experimental demonstration of long-distance continuous-variable quantum key distribution," *Nat. Photon.*, vol. 7, pp. 378–381, 2013.

Index